Ab initio Quantum Monte Carlo Tutorial

Ryo Maezono

Ab initio Quantum Monte Carlo Tutorial

Going beyond DFT

Springer

Ryo Maezono
Graduate Major in Materials
and Information Sciences
Institute of Science Tokyo
Meguro-ku, Tokyo, Japan

ISBN 978-981-95-4133-1 ISBN 978-981-95-4134-8 (eBook)
https://doi.org/10.1007/978-981-95-4134-8

This Springer imprint is published by the registered company Springer Nature Singapore Pte Ltd.
The registered company address is: 152 Beach Road, #21-01/04 Gateway East, Singapore 189721, Singapore

Preface

This book is a guide to electronic structure calculations using ab initio quantum Monte Carlo, targeted at readers in materials science, even those who do not primarily specialize in theoretical methods. It is uniquely designed with a "learning by doing" approach, making it different from other books in the field. In materials science, **ab initio calculations using density functional theory (DFT)** are widely adopted. However, ab initio quantum Monte Carlo is a leading general-purpose method classified as **"Beyond DFT"** because it overcomes challenges that are difficult for DFT to handle. The problems that the quantum Monte Carlo can address include intermolecular binding forces in biomolecular systems, magnetic descriptions in spintronics, and material properties associated with discontinuous density distributions, such as surfaces, interfaces, and defects. These issues are critical in applied and practical materials science. As a general-purpose ab initio calculation tool in materials science, ab initio quantum Monte Carlo has the potential to gradually expand into industrial and experimental fields, eventually replacing some aspects of DFT calculations.

Looking at the current state of the community in Japan, where the authors reside, ab initio quantum Monte Carlo, or more broadly, **electronic structure calculations using quantum Monte Carlo**, remains a relatively unknown field within the country. Japan has a strong tradition of making internationally recognized and significant contributions to the fundamental methodology of quantum Monte Carlo. There exists a well-established research community with a deep expertise in this field. However, the primary focus of these efforts has been on spin models and lattice models. To extend research to real materials such as atoms, molecules, and solids, it is necessary to develop techniques for constructing ab initio many-body wave functions tailored to real materials, including basis function frameworks. This requires the integration of specialized technologies, such as the construction of pseudopotentials for quantum Monte Carlo applications. Due to the complexity of these tasks, the mainstream trend has been to avoid such complications and instead focus on simplified model systems, aiming to refine the methodology itself.

In fact, when I was first hired as a postdoctoral researcher specializing in quantum Monte Carlo, I mistakenly wondered that my work would involve applying quantum

Monte Carlo to spin models. I held this belief until a few days after arriving in the UK, only to realize that I would instead be working on real materials. As such, very few research groups focus on ab initio quantum Monte Carlo for real materials in a comprehensive manner, integrating various methodological aspects into practical applications in materials science. Engaging in this area requires an established workflow that addresses fundamental issues such as "How do we prepare pseudopotentials and basis function sets?". The trial wave functions for ab initio quantum Monte Carlo are typically prepared using density functional theory or molecular orbital methods. At least in Japan, the community specializing in ab initio calculations and the community focused on model calculations have traditionally been quite distant. As a result, the quantum Monte Carlo community, which has mainly belonged to the latter group, has had few opportunities to develop an interest in ab initio calculations.

Since the 2010s, ab initio calculations have become commonplace in materials science, extending beyond academic research to industrial applications and even being used routinely by experimental researchers. By the 2020s, the rapid integration of data science and AI technologies has driven the emergence of **materials informatics** as a major trend. The problem of exchange-correlation descriptions in many-electron systems is no longer just a topic of interest for fundamental scientists. Instead, it has become a widespread concern among materials scientists in the context of **predictive and descriptive materials science**. With the advancement of parallel computing technologies and the continued development and refinement of general-purpose software packages, it seems only a matter of time before ab initio quantum Monte Carlo gains wider adoption in materials science. Given this background, as a specialist in this field, I decided that this was the right time to compile a tutorial book on the subject.

Several books explain the theoretical principles of ab initio quantum Monte Carlo, but there are no books, either in Japan or internationally, that focus on its practical simulation operations—especially in a tutorial format. A few specialized books on ab initio quantum Monte Carlo exist in English, but no such books are available in Japanese. Despite nearly 20 years having passed since the availability of user-friendly general-purpose packages, the absence of a tutorial book stands in stark contrast to the extensive resources available for DFT. This book is designed to introduce ab initio quantum Monte Carlo as a natural extension of "practical DFT applications in materials science." It is written in a way that first captures the interest of materials scientists by addressing **what this method can deal with**. Moreover, there are no other books that allow readers to learn independently using their own PCs. The theoretical background of ab initio quantum Monte Carlo serves as a **showcase of many-body electronic theory**, making it of interest even to fundamental scientists working on quantum many-body problems. These topics are arranged in the latter half of the book in the style of a "school trip for grown-ups," where we *revisit* the procedures practiced in the first half and delve deeper into their underlying principles.

This book is intended for the same readership as my previous work, *"ab initio Calculation Tutorial: For Materials Analysis, Informatics and Design"*,[1] namely, beginners from the field of materials science. As in the previous book, I have structured the content to prioritize hands-on experience and intuitive understanding, placing detailed theoretical discussions in the latter half. The book is written in a straightforward manner, avoiding assumptions of prior knowledge in many-body electronic theory or simulation science. It is specifically designed for beginners in simulation, including those in experimental fields and industry, allowing them to set up and practice on their own PCs for self-study. The intended readership includes professionals involved in electronic structure calculations for materials science, graduate students in materials science, condensed matter physics, or quantum chemistry research groups related to electronic structure calculations, as well as undergraduate students studying computational physics or quantum chemistry. The book is especially suited for those interested in many-body electronic methods that go **beyond DFT** and **beyond MO**.

This book reflects my own experiences as a beginning researcher, coming from a background in model theory and facing challenges when entering the unfamiliar field of ab initio calculations. It also incorporates insights gained from explaining this method to others, particularly the parts that were difficult for people to understand or prone to misinterpretation. Additionally, my experience in teaching this subject to students who are not primarily focused on theory has influenced the way the material is presented.

This book is based on the content of the Japanese edition (published in 2025 by Morikita Publishing). The figures are adapted from those used in the above Japanese edition.

Tutorial Environment

This book is structured as a self-study guide for learning ab initio quantum Monte Carlo through hands-on simulation exercises. A key philosophy throughout this book is to ensure that **beginners do not lose motivation during the learning process**. Providing a tutorial under the assumption that **a pre-configured environment is available in a university lab** or that **prior experience with Linux is required** would be of no use to beginners who have only ever used Windows for their studies.

In this book, we carefully guide beginners through the process of setting up their own simulation environment, allowing them to develop familiarity with command-line operations in an engaging way. It would be unrealistic to assume that self-learners can **set up a Linux server** as a prerequisite. For this reason, we have

[1] "ab initio Calculation Tutorial: For Materials Analysis, Informatics and Design" (Springer Nature, 2023), R. Maezono.

long provided tutorials that focus on building a simulation environment directly on personal computers, such as Windows or Mac.

Acknowledgments As a beginner in ab initio electronic structure calculations, I was fortunate to find a solid footing in this field thanks to Professor Richard Needs' group at the Cavendish Laboratory, University of Cambridge, where I spent my postdoctoral years, from 2000 for two years. I am deeply grateful to Dr. Mike Towler and the other group members during my time there for their generous support. In particular, I owe special thanks to Dr. Neil Drummond (now a professor at Lancaster University) for his invaluable assistance in providing access to the tutorial package **CASINO**.

For over 20 years, I have dedicated my research to ab initio quantum Monte Carlo. Throughout this time, I have been fortunate to have long-standing research partners who have supported me. I would like to express my gratitude to Dr. Kenta Hongo (Associate Professor, JAIST) for his unwavering collaboration over the years, as well as to Dr. Tomohiro Ichiba (Specially Appointed Assistant Professor, JAIST) and Dr. Kosuke Nakano (NIMS), who have meticulously carried forward the research foundation. Additionally, I have learned a great deal from ongoing discussions with the following researchers who have worked on ab initio quantum Monte Carlo within my group: Dr. Keishu Uchimura, Dr. Abhishek Raghav, Dr. Abdul Ghaffar, and Dr. Rohit Dahule.

Ookayama, Meguro, Tokyo
June 2025

Ryo Maezono

Competing Interests The author has no competing interests to declare that are relevant to the content of this manuscript.

Contents

Acronyms

The frequently used acronyms in this book are as follows:

DFT	Density Functional Theory
DMC	Diffusion Monte Carlo
QMC	Quantum Monte Carlo
VMC	Variational Monte Carlo
XC	Exchange-Correlation

Regarding Linux operations, the conventions used in this book are as follows:

- **Variables in command descriptions**
  When square brackets `[...]` appear in a command example, they indicate variables. For example:

  ```
  % cp [file01] [file02]
  ```

  This means that these placeholders should be replaced with actual file names when executing the command. For instance:

  ```
  % cp maezono hongo
  ```

  Be careful not to input `[file01]` literally as it appears in the example.
- **Notation for commands that do not fit in a single line**
  When a command is too long to fit within the page width, it is often written as follows:

  ```
  [A too long statement impossible to be accommodated]
  ```

  is instead represented as:

  ```
  [A too long statement impossible \
    to be accommodated]
  ```

  The trailing \ should be interpreted educationally as an instruction to *"remove it and write the command on a single line"*. However, if you enter the command as written, the system will still recognize it correctly.

- **Indicating the working directory using the pwd command//**

```
% pwd
   /home/student/work/setupMaezono
```

When this command is shown, it serves as an indication to the reader that *"The following operations should be performed in this directory"*. This convention is particularly important when providing remote assistance or giving operational instructions, as it helps ensure clarity and prevent errors.

Most of the errors that beginners report as *"It's not working"* are due to executing commands in the wrong directory. For example, if given the instruction *"Check the contents of* `scf.in` *using the* `cat` *command"*, it must be executed in the directory where the file `scf.in` actually exists. Beginners often run

```
% cat scf.in
```

in a directory where `scf.in` is not present, receiving an error message stating *"No such file exists"* and then reporting *"It's not working"*.

Version Information

The contents of this book have been verified to work under the environment available at the time of writing. However, due to circumstances such as future updates of the packages used, it is possible that the examples may not work exactly as described in this book.

At the time of writing this edition, the operating system versions used for verification were the latest OS environments in 2025:

- Macintosh (M1/2020), Ventura13.0.1
- Windows (ARM/2022), Windows11

As noted in Sect. 5.1, the feature "trial wavefunction generation using QuantumEspresso" appears to be no longer supported in the latest versions of QuantumEspresso from v7.4 onward. Since the source code for the v7.3 generation is still available, please use that version instead.

Chapter 1
Introduction

In this chapter, we first describe how the ab initio quantum Monte Carlo method, which you are about to be explained, is positioned relative to other methods, what distinguishing features it has, and what applications it is best suited for. Additionally, we provide guidelines on the tutorial design and the technical aspects of the tutorial exercise environment used in this book.

1.1 Positioning of the Ab Initio Quantum Monte Carlo Method

1.1.1 Avoiding Misconceptions from the Term "Monte Carlo Method"

At the outset, it is important to clarify that the ab initio quantum Monte Carlo method covered in this book is a Monte Carlo method for ab initio applications, which **samples the many-body wave function** of **real materials**. This book discusses a general-purpose method for materials science applications that is considered to be **more reliable than density functional theory** (DFT), which has been the conventional approach for ab initio calculations.

The term "Monte Carlo method" is often associated with well-known approaches such as

- **Dynamics simulations** based on Monte Carlo sampling, which handle or spin orientation fluctuations at finite temperatures.
- **Quantum Monte Carlo methods** that deal with quantum spin system models.

Since many researchers work in these areas, the term "Monte Carlo" is often met with the preconceived notion of "Ah, that method" we frequently encounter. To

R. Maezono, *Ab initio Quantum Monte Carlo Tutorial*,
https://doi.org/10.1007/978-981-95-4134-8_1

complicate matters further, the primary method explained in this book is called **diffusion** quantum Monte Carlo.

For this reason, when the authors begin explaining the "ab initio quantum Monte Carlo method ...," practitioners working in **ab initio electronic structure calculations** often respond with the following misconceptions:

- It must be a classical simulation dealing with diffusion phenomena. Since I am interested in ab initio methods, this is outside my field (misconception).
- It may be useful for dynamics, but it is less reliable than our DFT (misconception).
- It must be related to spin systems or model-based approaches, which are different from the ab initio simulations we perform on real materials (misconception).

These **hasty misunderstandings** often prevent the conversation from progressing. No matter how much we attempt to explain "that is not the case," people simply do not listen.[1] The term "diffusion" in diffusion quantum Monte Carlo is used because, as explained later in Sect. 3.2.2, the **mathematical correspondence** arises from the fact that the "evolution equation that leads to the exact solution of the many-body Schrodinger equation" takes the same form as the "diffusion equation for classical particles."

To reiterate, it is important to emphasize from the beginning that this method is

- Not a Monte Carlo method for classical dynamics ,
- Not a quantum Monte Carlo method for model calculations
- Not a method for handling diffusion phenomena.

To reiterate, the ab initio quantum Monte Carlo method described in this book is an approach that achieves "the most reliable electronic structure calculations" by evaluating the many-body wave function through Monte Carlo sampling [1, 2].

Dispelling Preconceptions Is Extremely Difficult

About 20 years ago, when density functional theory was beginning to gain popularity in Asian countries, explaining ab initio quantum Monte Carlo often led to persistent misunderstandings. Despite repeated explanations over the years, some senior professors would continuously respond with comments like, "It seems to be a powerful method in classical dynamics," making it difficult to steer the discussion back on track without causing offense. Additionally, some researchers specializing in DFT would insist, "Since it's a Monte Carlo method, it must be less reliable than DFT." No matter how much

(continued)

[1] Even within the research community for national grant focused on ab initio methods, there was an instance where we were mistakenly assigned to a group dealing with macroscopic analysis of diffusion phenomena.

we explained, they refused to reconsider their stance. A common trait among researchers is that pride prevents them from backing down once they have taken a position. Even when everyone else understands the concept, there are always those who stubbornly refuse to change their views, perhaps believing that such persistence is a virtue.[2] Once someone has asserted, "As an expert, I know this," it becomes quite difficult to simply admit, "Oh, I didn't know that," and humbly acknowledge the mistake.

1.1.2 Various Approaches in Ab Initio Calculations

A large number of researchers work on **ab initio simulations** for **real materials**, leading to the coexistence of various methodologies. In this section, we will organize these different approaches, gradually clarifying the positioning and distinctive features of ab initio quantum Monte Carlo, which is the main subject of this book.

In device development and materials industry applications, fundamental simulation research is often centered on the rather unglamorous task of **determining the energy values of the compounds themselves** that serve as the base materials.[3] For example, if we calculate the energy value of a system where certain elements in a parent material are substituted with oxygen or hydrogen, the difference from the parent material's energy provides an estimate of **how easily hydrogen or oxygen can penetrate the material**. This, in turn, is useful for studying material degradation at the microscopic level, such as oxidation or embrittlement. For a more detailed explanation of **how computing energy values contributes to materials science**, please refer to my previous book [3].

The energy value of a material can be **computed by solving equations** once the microscopic atomic configuration within the material is specified. Specifically, this is done by numerically solving the many-body Schrödinger equation, which takes the form of a partial differential eigenvalue equation:

$$\left[-\frac{1}{2}\sum_{j=1}^{N}\nabla_j^2 + V\left(\mathbf{r}_1,\cdots,\mathbf{r}_N\right)\right]\cdot\Psi\left(\mathbf{r}_1,\cdots,\mathbf{r}_N\right) = E\cdot\Psi\left(\mathbf{r}_1,\cdots,\mathbf{r}_N\right) \tag{1.1}$$

[2] During a casual conversation at a dinner gathering, a Japanese professor mistakenly conflated the terms "British," which refers to the entire island of Great Britain, and "English," which is specific to England. He stated, "Wales and Scotland are not British, right?". The British attendees gently explained the correct distinction, providing reasons in a tactful manner to avoid causing offense. However, he stubbornly refused to change his view, repeatedly insisting on his misunderstanding. It was quite exasperating.

[3] At first, I also struggled to understand the significance of simply knowing energy values, wondering, "What purpose does this serve?"

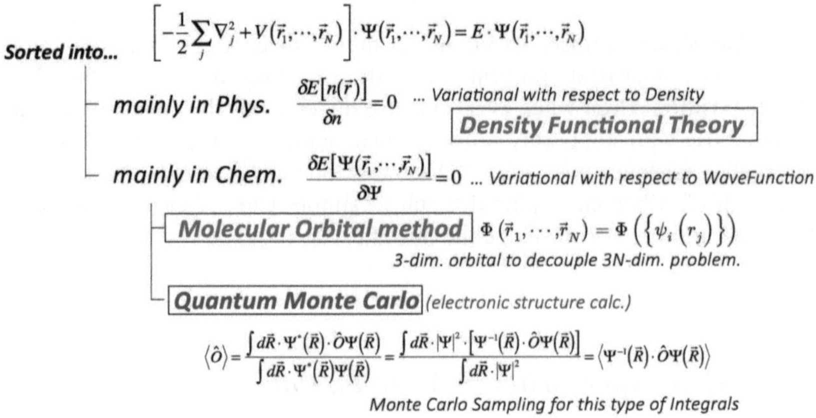

Fig. 1.1 Several major numerical approaches coexist as solutions to the fundamental equations of ab initio calculations, including density functional theory (DFT) and molecular orbital (MO) methods. The ab initio quantum Monte Carlo method covered in this book represents a **third branch** among these approaches

By solving this equation numerically, we can determine the energy value of a material. The research field known as the **ab initio approach** is based on the idea that **by solving this fundamental equation numerically** on a computer, we can elucidate and design material properties from a microscopic perspective. Since the late 1990s, the rapid proliferation and cost reduction of computers have accelerated the adoption of ab initio calculation tools in industry and experimental research. These tools gained popularity because they allow researchers to **predict material properties simply by specifying atomic positions and types**.

The fundamental ab initio equation, Eq. (1.1), is a differential equation. Several major numerical approaches exist for solving it (Fig. 1.1). Solving a differential equation means finding a function that satisfies the equality between the left-hand side and the right-hand side. The common approach is to make an educated guess about the form of the solution, *"it is probably something like this"*, and substitute it into the equation to refine the result. The accuracy of the solution depends on how well the assumed functional form captures the true solution. A more complex assumption retains generality and yields a more accurate answer, but such complexity can make the problem impractically difficult to solve. One strategy, **molecular orbital theory** , seeks to gradually refine an initial guess while keeping the problem computationally feasible. On the other hand, density functional theory (**DFT**) takes an inverse approach by asking whether it is possible to transform the equation itself (mapping the equation system) so that even a simpler and solvable equation form yields an exact solution.

For ab initio simulations of real materials, two major methodologies have emerged:

- **Density functional theory (DFT)**, which has seen explosive adoption in large-scale practical applications, including solid-state systems.

- **Molecular orbital (MO) methods**, which have long been established as highly accurate approaches in molecular science.

Both of these methodologies now have widely available commercial and open-source packages, making them convenient for industrial applications. As these conventional methods have become more widely used, their limitations have become clearer. Some of the most challenging problems include surfaces and interfaces, spin magnetism, and intermolecular forces—key factors in next-generation electronics. To address these challenges, **ab initio quantum Monte Carlo** , the subject of this book, has gained attention as a **third major approach** (Fig. 1.1).

One of the greatest challenges in electronic structure calculations has been *how to handle many-body interactions.* Although the problem itself is clearly defined, "the equation is known, and solving it on a computer should yield the answer," accurately capturing the complex physics arising from electron-electron interactions remains extremely difficult. It is no exaggeration to say that methodologists in the ab initio field have been striving to solve this issue for nearly a century, and the effort continues to this day. To address this problem, DFT takes the approach of *mapping the problem to an effective single-particle form.* However, there exist multiple variations of the exchange-correlation potential used to approximate many-body interactions, leading to differences in predictions depending on the choice of functional [3]. As DFT applications expanded beyond the initial success in covalent solids to more complex systems, such as intermolecular interactions, surfaces, impurities, and magnetism, it became evident that these problems are *highly sensitive to the balance and accuracy of exchange and correlation descriptions.* As a result, concerns have been raised about significant biases introduced by the choice of exchange-correlation potential [3, 10]. These challenges have motivated ongoing research within the DFT framework to *improve the accuracy and descriptive power of exchange-correlation functionals.*

On the other hand, an alternative approach that does not rely on an exchange-correlation potential but instead directly tackles the description of the many-body wave function exists alongside DFT. One such approach is **molecular orbital theory** [3, 27], and another is **ab initio quantum Monte Carlo** [1] (Fig. 1.1). Molecular orbital theory employs a single-particle orbital framework while aiming for a more refined description of the many-body wave function. On the other hand, **ab initio quantum Monte Carlo** does not rely on a single-particle orbital framework. Instead, it takes the approach of **directly sampling the many-body wave function numerically** (Fig. 1.2).[4] As discussed in the next section, this approach was initially slow to gain practical traction due to its high computational cost. However, because of its **exceptionally high parallelization efficiency** (scaling almost linearly, where 1000 cores yield nearly a 1000-fold speedup), it has become increasingly practical in the era of massive-parallel computing, which emerged in

[4] Since high-dimensional spaces like many-body wave functions exhibit the **"curse of dimensionality"** [5], it is often more practical to **sample** rather than **explicitly describe** them.

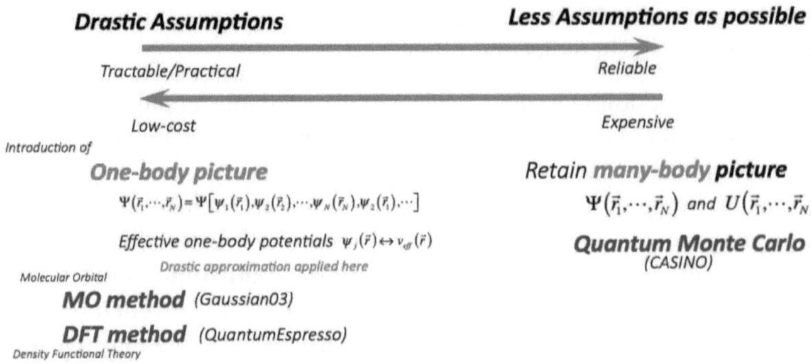

Fig. 1.2 The relationship between **ab initio quantum Monte Carlo** and conventional methods such as **DFT** and **molecular orbital theory** is one of contrasting approaches. Unlike these traditional methods, ab initio quantum Monte Carlo does not rely on a single-particle framework. Instead, it adopts an approach that *treats the many-body wave function directly, preserving its many-body nature*

the 2000s. As a result, it has established itself as **the most reliable general-purpose ab initio method** [1].

1.1.3 How to Solve Equations in Ab Initio Quantum Monte Carlo

In the previous section, we discussed that solving a differential equation requires making an initial assumption about the solution: *"If we make a highly complex assumption without losing generality, we might get an accurate result, but ... "*. Exploring such assumptions rigorously is the approach of molecular orbital theory. However, it is difficult to assume an appropriate analytic functional form in general. Instead, QMC employs random numbers to generate a large number of possible *variable values* for the function, effectively expressing the function through numerical samples and solving the equation by enforcing numerical equality. The essence of the QMC approach lies in integrating these many numerical solutions obtained for individual random samples to infer a solution for the equation in the form of a function. The advantage of *solving the equation more accurately without losing generality* comes at a cost:

- A **large number of numerical samples** must be generated.
- The solution is obtained as a **statistical estimate** , expressed as *"(mean value) ± (statistical error)."*

These aspects present key challenges for the method.

In statistical computations, if we take N samples, the statistical error decreases only as $1/\sqrt{N}$, meaning that it converges very slowly [7]. Achieving sufficiently high predictive accuracy requires a vast computational cost due to this slow convergence. Because of this limitation, ab initio QMC remained impractical for a long time. However, since statistical sampling calculations using random numbers have exceptionally **high parallelization efficiency** and are relatively easy to implement, the practicality of the method began to improve as parallel computing became more accessible in the late 1990s, moving beyond the limited domain of *computing specialists* so far. By the 2010s, as massive-parallel computing with tens of thousands of processors became widespread, QMC attracted significant attention as a practical method capable of surpassing conventional approaches. This provides the most entry-level explanation of the method. You should now have a clearer understanding of *why and where* **massive-parallel computing** is essential for ab initio QMC.

Next, let us move toward a more precise understanding by incorporating mathematical expressions. In DFT and molecular orbital theory, the many-body form of Eq. (1.1) is reformulated into an equation for single-particle orbitals. In DFT, the *exactly* equivalent problem in a single-particle orbital framework is expressed as follows:

$$\left[-\frac{1}{2}\nabla^2 + v_{\text{ext}}(\mathbf{r}) + \hat{U}^{(\text{eff})}_{\text{ele.}}\right]\cdot\psi_j(\mathbf{r}) = \varepsilon_j\cdot\psi_j(\mathbf{r}) \quad (j = 1,\cdots,N). \quad (1.2)$$

Solving this equation, we obtain[5] the single-particle energy levels $\left\{\varepsilon_j\right\}_{j=1}^{N}$, and the single-particle orbital functions $\left\{\psi_j(\mathbf{r})\right\}_{j=1}^{N}$. From the single-particle orbitals, the electron density is obtained as $n(\mathbf{r}) = \sum_{j=1}^{N}\left|\psi_j(\mathbf{r})\right|^2$, and the ground-state energy of the system is given by[6]

In contrast, in the ab initio QMC, the original many-body problem, Eq. (1.1), is directly transformed into the following expression:[7]

$$E=\frac{\int d\mathbf{r}_1\cdots d\mathbf{r}_N\cdot\Psi^*(\mathbf{r}_1,\cdots,\mathbf{r}_N)\left[-\frac{1}{2}\sum_j\nabla_j^2+V(\mathbf{r}_1,\cdots,\mathbf{r}_N)\right]\Psi(\mathbf{r}_1,\cdots,\mathbf{r}_N)}{\int d\mathbf{r}_1\cdots d\mathbf{r}_N\cdot\Psi^*(\mathbf{r}_1,\cdots,\mathbf{r}_N)\cdot\Psi(\mathbf{r}_1,\cdots,\mathbf{r}_N)}.$$

[5] Pay attention to the difference from Eq. (1.1). In Eq. (1.2), the **single-particle orbital function** ψ is written in lowercase, with its argument as the spatial vector $\mathbf{r}$. On the other hand, in Eq. (1.1), the **many-body wave function** Ψ is written in uppercase, with its argument being the snapshot of N electrons, $(\mathbf{r}_1,\cdots,\mathbf{r}_N)$.

[6] $G[n]$ is a functional that can be evaluated once $n(\mathbf{r}) = \sum_{j=1}^{N}\left|\psi_j(\mathbf{r})\right|^2$ is determined. The detailed expression is given in Eq. (16.9).

$$E_{GS} = \sum_{j=1}^{N}\varepsilon_j + G[n].$$

[7] This equation is obtained by multiplying both sides of Eq. (1.1) by $\Psi^*(\mathbf{r}_1,\cdots,\mathbf{r}_N)$ and integrating over all spatial coordinates.

This approach directly evaluates the energy E by performing the above multidimensional integration. This integral is known as a **variational functional** . The integration is carried out using the **Monte Carlo numerical integration method** , which is explained in Sect. 2.2. This is why the method is referred to as a "Monte Carlo method."

Since explicitly writing out multiple variables takes up the entire page width, we introduce the compact notation $\mathbf{R} = (\mathbf{r}_1, \cdots, \mathbf{r}_N)$ and rewrite the variational functional as

$$E = \frac{\int d\mathbf{R} \cdot \Psi^* (\mathbf{R}) \left[-\frac{1}{2} \sum_j \nabla_j^2 + V (\mathbf{R}) \right] \Psi (\mathbf{R})}{\int d\mathbf{R} \cdot \Psi^* (\mathbf{R}) \cdot \Psi (\mathbf{R})} \tag{1.3}$$

Always keep in mind that a single $\mathbf{R}$ represents a **snapshot** of an electron configuration, i.e., an arrangement of N electrons, $(\mathbf{r}_1, \cdots, \mathbf{r}_N)$. We refer to $\mathbf{R}$ as a **many-body configuration** . Similarly, the integral over $d\mathbf{R}$ represents integration over all **possible configuration snapshots** of the N electrons.

1.2 Elementary Introduction to the Ab Initio Quantum Monte Carlo Method

1.2.1 Rough Explanation via the Variational Principle

In the author's previous book, the process of computing energy in ab initio calculations was likened to "fermenting and baking bread dough" [3]. This book will also use the same analogy. In the context of the ab initio quantum Monte Carlo method, the counterpart to the "bread dough" is the many-body wave function $\Psi (\mathbf{R})$ in Eq. (1.3). The counterpart to "baking" is the process of "computing the integral in Eq. (1.3)." Now, recall that the dough-equivalent $\Psi (\mathbf{R})$ was originally something that is "obtained by solving an equation." This raises the question: "How can we 'assign' $\Psi (\mathbf{R})$ when it is something we are supposed to determine by solving it?".

To address this question, we introduce a parameter dependence in $\Psi (\mathbf{R})$ by writing it as $\Psi_\alpha (\mathbf{R})$ in Eq. (1.3), leading to

$$E (\alpha) = \frac{\int d\mathbf{R} \cdot \Psi_\alpha^* (\mathbf{R}) \left[-\frac{1}{2} \sum_j \nabla_j^2 + V (\mathbf{R}) \right] \Psi_\alpha (\mathbf{R})}{\int d\mathbf{R} \cdot \Psi_\alpha^* (\mathbf{R}) \cdot \Psi_\alpha (\mathbf{R})}. \tag{1.4}$$

Since it is possible to **assign a trial form** to the many-body wave function, we introduce an index $\boldsymbol{\alpha}$ to distinguish different trial forms and write it as $\Psi_\alpha (\mathbf{R})$. The energy value E computed using $\Psi_\alpha (\mathbf{R})$ depends on $\boldsymbol{\alpha}$, so we denote it as $E (\boldsymbol{\alpha})$.

An important theorem learned in undergraduate quantum mechanics is the **variational principle** , which states that "the exact solution of Eq. (1.1) minimizes the integral expression in Eq. (1.4)" ([4]). Using this theorem, we arrive at the idea that "by testing various trial functions $\Psi_\alpha(\mathbf{R})$ and selecting the one that minimizes $E(\boldsymbol{\alpha})$, we can obtain the best approximation to the exact solution." This process of optimizing the trial function $\Psi_\alpha(\mathbf{R})$ to minimize $E(\boldsymbol{\alpha})$ is called **variational optimization** . Thus, in response to the question posed at the beginning of this section, "How can we assign $\Psi(\mathbf{R})$ when it is something we are supposed to determine?", the answer is "We assign it by variationally optimizing the trial function $\Psi_\alpha(\mathbf{R})$."

The strength of the approach of "using variational optimization to obtain the exact E as the minimum of $E(\boldsymbol{\alpha})$" lies in its objectivity. By implementing **numerical optimization** to minimize the value of the variational functional $E(\boldsymbol{\alpha})$, we can achieve the exact solution once the minimization is realized. Ultimately, physical intuition is not required. In some cases, numerical optimization can lead to discoveries that surpass human physical insight [8].[8]

To obtain a "highly reliable result" that is as close as possible to the exact solution, it is crucial to allow $\Psi_\alpha(\mathbf{R})$ to have as much flexibility as possible and to compare $E(\boldsymbol{\alpha})$ over a wide range. In other words, the reliability of the result depends on "how flexibly and smoothly the trial function $\Psi_\alpha(\mathbf{R})$ can be relaxed." Using the concept of function space, this can be rephrased as "how broadly the search can sweep through the function space." If the entire function space can be explored, then by minimizing $E(\boldsymbol{\alpha})$, we can find the exact solution hidden somewhere within that space. As a result, the energy value E can be obtained as a **numerically exact solution**.

1.2.2 Procedure in Ab Initio QMC Using Metaphor

The process of relaxing the trial function $\Psi_\alpha(\mathbf{R})$ can be compared to "fermenting bread dough." In a conventional procedure, the steps are "Prepare the dough $\rightarrow$ Let it ferment $\rightarrow$ Bake it". However, in the ab initio QMC, there exists a "skilled baking technique" where the dough continues to ferment even as it is being baked. This "fermenting while baking" technique enables "ultra-fermentation," which is not achievable using the traditional method of fermenting before baking. What exactly this "ultra-fermentation" corresponds to in terms of the many-body wave function $\Psi_\alpha(\mathbf{R})$ will be explained in detail in later chapters. For now, focus on understanding the procedural differences between these computational approaches while familiarizing yourself with the overall workflow.

The conventional approach of "*fermenting before baking*" corresponds to the *variational Monte Carlo* (**VMC**) method. On the other hand, the approach of

[8] See the column in Sect. 1.3.2.

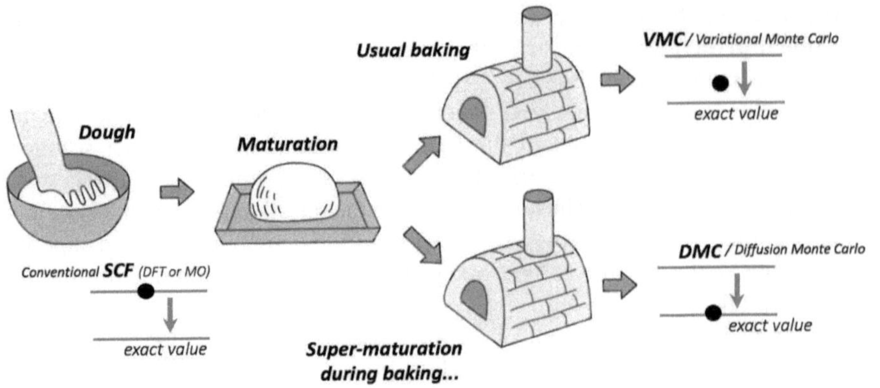

Fig. 1.3 The conventional approach of "*fermenting before baking*" corresponds to the *variational Monte Carlo* (**VMC**) method. In contrast, the approach of "*continuing to ferment the dough while baking it*" corresponds to the *diffusion Monte Carlo* (**DMC**) method

"*continuing to ferment the dough while baking it*" corresponds to the *diffusion Monte Carlo* (**DMC**) method (Fig. 1.3). This book aims to develop an understanding of and practical proficiency in the DMC method.

1.3 What Are Its Characteristics and Applications?

1.3.1 No Restriction on Target Systems

A common question is "To what kinds of material systems can the ab initio quantum Monte Carlo method be applied? Is it only for isolated molecular systems? Or is it limited to solid-state systems?" The answer, for the general-purpose package covered in this book, is that it can be applied to "isolated molecular systems, solid periodic systems, or even model systems."

The reason behind such questions lies in the "constraints imposed by the choice of basis set functions" [3]. In DFT packages, the assumption about whether the method is designed for extended periodic systems or isolated molecular systems depends on the choice of basis set functions used to expand the orbitals, whether plane-wave basis sets or Gaussian basis sets are adopted. When introducing an ab initio QMC package as a post-processing to a DFT package, it naturally leads to the question, "So, which basis sets does this package use?" This, in turn, often leads to further inquiries like, "Does that mean it can only handle solid periodic systems?".

In practical applications, as described in this book, ab initio QMC packages are typically used as a "post-processing step" following the output of SCF-based

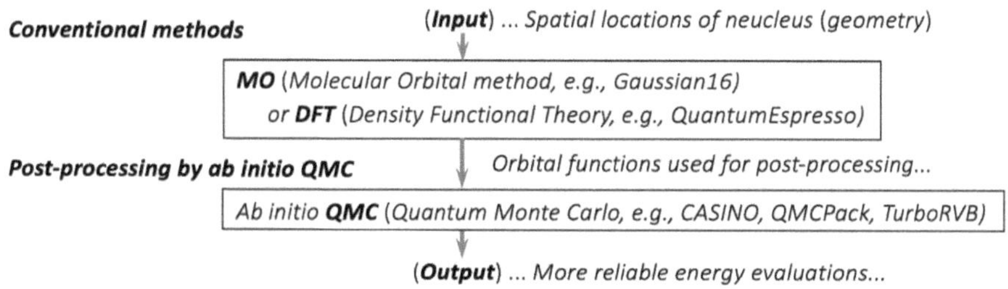

Fig. 1.4 In principle, as shown in Fig. 1.1, quantum Monte Carlo (QMC) electronic structure calculations constitute a "third, independent method." However, in practical applications, QMC is typically used as a post-processing step layered on top of conventional methods such as molecular orbital theory or density functional theory. This setup provides an ideal stage for a workflow that involves "invoking and processing multiple kernel computations" (Sect. 9.2.2)

methods, as illustrated in Fig. 1.4.[9] The term "preceding step" refers to which package is used to generate the trial function $\Psi_\alpha(\mathbf{R})$ in Eq. (1.4). In general implementations of ab initio QMC, both Gaussian and plane-wave basis sets can be used in the preceding SCF stage, and the QMC package is designed to seamlessly handle either basis sets.

The *greatest advantage* of the ab initio QMC is that its fundamental principles are not significantly constrained by the specific form of the trial function $\Psi_\alpha(\mathbf{R})$. As a result, it can accommodate not only plane-wave and Gaussian basis sets but also more general numerical grid representations (numerical basis sets). Additionally, it supports advanced trial function configurations for model systems, such as electron gases and electron-hole gases, in addition to ab initio calculations targeting real (or potentially real) materials.[10]

1.3.2 Where Is the Demand?

1.3.2.1 The Strength of the Variational Method

In Sect. 1.1.2, we discussed the approaches taken by conventional methods such as molecular orbital theory and DFT. When considering "how to approach an unknown exact solution," these conventional methods rely on

- Improving the functional form of the assumed many-body wave function (molecular orbital theory),

[9] SCF-based methods refer to conventional approaches such as DFT and molecular orbital methods. Since both are formulated within the self-consistent field (SCF) framework (*s*elf-*c*onsistent *f*ield), we refer to them collectively as SCF methods.

[10] Beyond ab initio calculations for real (or potentially real) materials, another research approach involves **model calculations**, where simplified models are used to investigate the underlying mechanisms of physical phenomena.

- Refining the exchange-correlation functional, which serves as an effective potential (density functional theory).

In contrast, QMC take a different approach: they refine the approximated solution toward the exact solution based on the **variational principle**.

The refinement strategies in conventional methods rely on the researcher's *physical or chemical insight*. For example, a methodology researcher's mission is to enhance descriptive accuracy based on subjective insight, such as "This level of approximation fails to capture a certain physical process. If we extend the function with additional degrees of freedom in this manner, we should be able to capture that process." In contrast, in QMC, the mission of "approaching the exact solution" is reformulated as a numerical minimization problem for the functional integral in Eq. (1.4). Regardless of the researcher's level of insight, as long as $\Psi_\alpha(\mathbf{R})$ is adjusted to lower the integral value, it is guaranteed to be closer to the exact solution. This provides a remarkably **objective** framework. Unlike conventional methods, the way the many-body wave function $\Psi_\alpha(\mathbf{r}_1, \cdots, \mathbf{r}_N)$ is constructed does not necessarily have to be expressed in terms of orbital functions $\psi(\mathbf{r})$. This allows for the assumption of more flexible and unconstrained functional forms, enabling the search for even lower values of the functional integral.

Furthermore, when advancing from VMC (variational Monte Carlo) to DMC (diffusion Monte Carlo) (Fig. 1.3), the trial function $\Psi_\alpha(\mathbf{R})$ is numerically and **automatically** deformed in a way that lowers the functional integral value. Within this framework of QMC, the question of "why and with what purpose a particular functional form was chosen" becomes secondary. As long as the functional integral value decreases, it is objectively guaranteed to be "more correct." This advantage is not specific to QMC but rather a fundamental strength of the variational principle itself. Variational approaches have often provided breakthroughs in difficult problems where conventional physical intuition and insight have reached their limits.

The Strength of Variational Trial Functions in Many-Body Problems
The "Cooper's trial wave function" in superconductivity and the "Laughlin's trial wave function" in the quantum Hall effect are both examples where the many-body wave function approach and variational methods provided breakthroughs in understanding new mechanisms in condensed matter physics. These achievements were significant enough to be recognized with Nobel Prizes.

1.3.2.2 The Problem of Electronic Correlation

In the case of ab initio QMC, the demand for the method arises from its ability to leverage its objectivity to achieve superior descriptive power over conventional methods, particularly in addressing electronic correlation problems. Electronic correlation is discussed in detail in the author's previous book [3] as well as in Chap. 10 of this book. Broadly speaking, it refers to **higher-order effects arising from wave function deformations due to electron-electron interactions**. Corrections that cannot be captured by bold approximations of electron-electron interactions constitute electronic correlation effects. Although the energy corrections due to electronic correlation are small, they exhibit a diverse range of behaviors depending on the material system.[11] Since capturing material property trends is a key mission of ab initio simulations, the ability to accurately describe electronic correlation becomes a major concern. If electronic correlation is inadequately captured, the predicted material property trends from simulations often become entirely unreliable.

The most straightforward solution as an ab initio method is to separate variables in the many-body wave function $\Psi\left(\mathbf{r}_1, \cdots, \mathbf{r}_N\right)$ from Eq. (1.1) by expressing it as a product of orbital functions $\psi\left(\mathbf{r}\right)$.[12] Since this approximation fails to capture electronic correlation, the most straightforward initial approach to addressing this limitation is to approximate $\Psi\left(\mathbf{r}_1, \cdots, \mathbf{r}_N\right)$ using a more complex expression constructed from orbital functions $\psi\left(\mathbf{r}\right)$. However, continuing along this path leads to an increasingly complex theoretical framework, resulting in significantly higher computational costs.[13]

To address this issue, DFT takes an alternative approach: rather than abandoning the framework obtained through simple variable separation, it seeks to approximate the exact solution by carefully choosing an effective potential $\hat{U}_{\text{ele.}}^{(\text{eff})}$ in Eq. (1.2). This allows for improvements in accuracy without incurring excessive computational costs or introducing significant complexity in the evaluation procedure. The existence of an effective potential that reproduces the exact solution[14] is guaranteed by a proof by contradiction.[15]

This leads to the question: what kind of "effective potential" should be chosen to reproduce the exact solution? In technical terms, this "effective potential" is called the exchange-correlation potential, making this issue the "problem of selecting the exchange-correlation potential" (Fig. 1.5). As discussed in the author's previous book [3], DFT guarantees the existence of an exchange-correlation potential that reproduces the exact solution through a proof by contradiction. However, it does not provide a concrete method for constructing such a potential, and no one

[11] In terms of correction magnitude, relativistic effects are known to be significantly larger [9]. However, because relativistic corrections remain almost constant across different material systems, electronic correlation is more crucial when discussing trends in material properties.

[12] In technical terms, this refers to the Hartree method or the Hartree-Fock method.

[13] A well-known example is the computational scaling of $\sim N^7$ with system size N (such as the number of electrons).

[14] The proper term is "exchange-correlation potential." A detailed explanation is provided in the author's previous book [3] as well as in Chap. 10 of this book. In this chapter, to avoid overwhelming beginners, we provide only a rough conceptual explanation.

[15] The theoretical foundation for this guarantee is provided by a framework known as density functional **theory**.

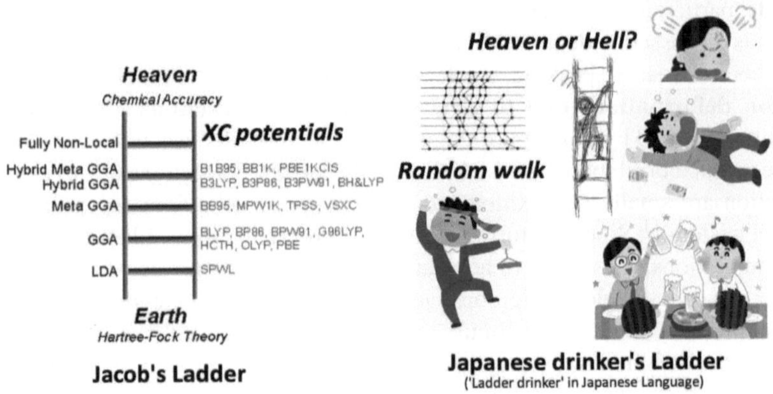

Fig. 1.5 The process of refining exchange-correlation functionals to improve the predictive accuracy of density functional theory is often likened to "climbing Jacob's ladder," in reference to the biblical narrative. In contrast, quantum Monte Carlo electronic structure calculations employ a random walk (drunken walk) approach to achieve higher accuracy, making it more akin to "climbing the ladder" (a Japanese idiom that expresses the act of bar hopping and getting progressively drunk)

knows its explicit form. Historically, extensive efforts have led to the development of practical exchange-correlation potentials. However, there is no universally applicable exchange-correlation potential. In most cases, the selection is based on empirical assurance, such as "This exchange-correlation potential would work reasonably well for covalent systems." It is well known that for certain target systems, the descriptive accuracy may be insufficient. As discussed below, problems involving **intermolecular interactions, magnetism, surfaces, and defects** are particularly sensitive to the choice of exchange-correlation potential and require careful consideration.

A key factor in intermolecular forces is **dispersion forces** (van der Waals forces) [11, 12]. These forces arise when quantum fluctuations cause a neutral charge source to develop a transient charge distribution, which then induces a similar charge distribution in a distant neutral charge distribution. The resulting electrostatic interaction between these fluctuating charge distributions leads to dispersion forces. This is a highly **nonlocal** phenomenon, where an event occurring at one location depends on an event at a distant location. Moreover, since it involves "fluctuations inducing fluctuations," it represents a higher-order quantum fluctuation effect. As a result, conventional exchange-correlation potentials, which are typically based on local approximations with systematic refinements, struggle to accurately describe these interactions [10] (Fig. 1.6).

Magnetic properties and exotic phenomena in strongly localized systems, such as high-T_c superconducting oxides, generally emerge from a delicate balance between exchange and correlation effects (Sect. 10.5.1). When handling such systems with DFT, careful consideration must be given to the construction of the exchange-correlation functional. Otherwise, achieving an accurate description

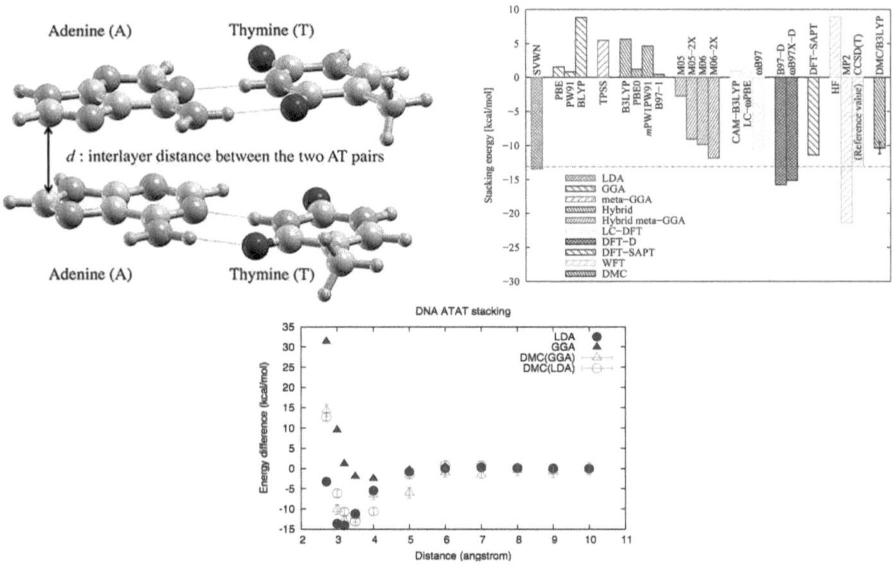

Fig. 1.6 A study evaluating the binding energy between DNA bases using various exchange-correlation potentials [10] showed that when the binding energy (vertical axis) takes a negative value, it correctly predicts the experimental fact that the bases bind together. However, some DFT calculations fail to correctly predict binding. In contrast, ab initio QMC have been shown to correctly predict binding regardless of which DFT method was used to prepare the trial function. A detailed explanation of this figure is provided in Sect. 8.2.4

becomes extremely challenging. Another class of challenges arises in systems involving surfaces, interfaces, and local defects. In such systems, defect sites and interfaces exhibit sharp changes in charge density, resembling delta functions or step functions. Since the assumption of "smoothly varying charge density that can be locally approximated" becomes less valid in these cases, it is intuitively clear that conventional exchange-correlation potentials, which are based on local approximations, struggle to provide an accurate description.

In summary, when conventional methods based on single-electron orbital descriptions attempt to improve the accuracy of electronic correlation, they face limitations in making discoveries beyond subjective insights. Additionally, they encounter the following difficulties:

- In molecular orbital methods, improvements significantly increase computational costs.
- In density functional theory, the selection criteria for exchange-correlation potentials remain unclear.

In contrast, in ab initio QMC, as long as numerical optimization is carried out according to the variational principle, it is objectively guaranteed to approach the exact solution. Furthermore, computational costs scale approximately as $\sim N^3$ with system size N [1], making the cost increase less severe compared to molecular orbital methods.

1.3.3 *What Are the Drawbacks?*

In the previous section, we explained the advantages of ab initio QMC, particularly for problems where conventional approaches struggle to provide accurate descriptions. Given these strengths, one might expect this method to be widely adopted. However, in practice, its utilization remains limited. In this section, we discuss the factors that act as drawbacks to its broader adoption.

1.3.3.1 Computationally Expensive

One major drawback is that **the computational cost is expensive**, and for a long time, the use of supercomputers was inevitable. As discussed in Sect. 1.1.3, this method, in simple terms, generates a large number of random samples and determines statistical equality based on them. However, statistical convergence is slow, scaling as $\sim 1/\sqrt{N}$ with the number of samples N. As a result, a large number of samples must be generated and processed for reliable calculations.

However, since each individual sample calculation can be performed independently on separate processors without communication between them, the **parallelization efficiency** is extremely high. For large-scale computations involving solid-state systems with large unit cells or large molecular systems, this method has been widely adopted on supercomputers exceeding one million parallel processes. On the other hand, for conventional system sizes, as demonstrated in this book's tutorials, modern personal computers are now powerful enough to run these calculations efficiently. With the advancement of processor performance, multicore parallel processing has become standard even in laptops. As a result, lab-scale simulation servers can now perform substantial levels of ab initio QMC calculations.

1.3.3.2 Computable Physical Quantities Are Limited

Compared to DFT, ab initio QMC have significantly higher computational costs and inherent statistical errors in the results. As a result, computing **derived material properties** that require secondary processing of energy values, such as forces or spectra, which are commonly obtained in DFT, is still not a practical option.[16]

[16] Consider statistical quantities expressed as $X_1 = A_1 \pm \delta_1$ and $X_2 = A_2 \pm \delta_2$. When computing their difference, $\Delta X = (A_1 - A_2) \pm \delta_{\mathrm{tot}}$, the mean value is simply the difference $(A_1 - A_2)$.

A particularly significant limitation is that force calculations are practically infeasible.[17] This means that, on its own, ab initio QMC methods **cannot perform structure optimization**. Unlike DFT, which enables "new crystal structure predictions to explore desired material properties," QMC is still far from offering such practical usability. At present, the only application at a practical level is the computation of ground-state energy values.

Another drawback is that **band structures cannot be plotted**. In a quasiparticle approach based on a single-particle picture, band diagrams provide valuable insights and guidance for understanding material properties. However, since ab initio QMC deliberately avoid adopting a single-particle picture to maintain their objective accuracy (Fig. 1.2), they inherently discard the concept of "describing electronic states using bands" from the outset. As a result, calculations of material properties that rely on band structures, such as spectral functions [3], are fundamentally infeasible.

There are methodological developments aimed at reconstructing **single-particle orbitals that reflect many-body interactions** from the many-body wave function treated in ab initio QMC.[18] However, this methodology has not yet reached a level suitable for practical materials science calculations.

1.3.3.3 All Difficulties Concentrated in Numerical Optimization

In ab initio QMC, all the complexities of the many-body problem are encapsulated within the variational principle, ensuring the objective guarantee that "a lower variational functional value indicates a closer approximation to the exact solution." As a result, the challenge of the many-body problem is entirely concentrated in the numerical optimization process, which aims to adjust the many-body wave function to minimize the variational functional value. This numerical optimization has been notoriously difficult, so much so that it has been said to be the cause of "countless graduate students falling in battle."

When the author entered this field around the year 2000, significant progress had already been made in understanding the mechanisms behind the difficulties of numerical variational optimization and in developing improved implementations. As a result, the situation had improved considerably. Even so, tackling methodologically challenging problems often led to struggles with numerical optimization, where minimizing the variational functional value proved extremely difficult. It was a grueling process, one that could leave a researcher feeling discouraged wondering where to start among the vast array of "adjustment knobs." However, advances

However, the total statistical error propagates as $\delta_{\mathrm{tot}} = \sqrt{\delta_1^2 + \delta_2^2}$, which means that statistical errors **accumulate additively**, making precision in derived properties increasingly challenging.

[17] Although fundamental research on methodology is progressing, it has not yet reached a level suitable for practical materials science calculations.

[18] This approach utilizes the concept of natural orbitals [9].

in numerical optimization continue to progress rapidly. As modern algorithms are incorporated into these methods, students of the next generation face far fewer struggles than their predecessors, achieving optimization with relative ease compared to the past.

This is an advanced topic that will not be covered in this book, but in recent years, implementations that represent the many-body wave function, subject to numerical optimization, using neural networks have rapidly advanced. Numerical optimization of neural networks is the core of artificial intelligence technologies, which directly influence the rise and fall of IT companies. Backed by the financial power of major corporations and the collective intelligence of humanity, high-speed and highly efficient numerical optimization leveraging GPUs has been advancing at an unprecedented pace. This trend can be seen as a strategy to "ride for free" on these developments, accelerating progress in solving quantum many-body problems by taking advantage of these state-of-the-art techniques.

1.3.3.4 Conventional Calculation Must Work First for the Process to Begin

In Sect. 1.1.2, we positioned ab initio QMC as "a third, independent method" (Fig. 1.1). While this is theoretically true, in **practical applications to real materials**, as discussed in this book, the initial input for QMC is almost always generated using conventional methods such as DFT. As a result, in practice, QMC is effectively "demoted" to the role of a **post-processor** for conventional methods (Fig. 1.4).

In this framework, ab initio QMC cannot begin unless conventional methods can first produce reliable results. Here, "conventional methods" refer specifically to DFT or molecular orbital methods capable of generating trial wave functions for QMC. Even within DFT, trial function generation is practically infeasible from highly specialized methodologies, such as

- **FLAPW method**, which does not expand the wave function in a single basis set.[19]
- **KKR method**, which does not explicitly handle wave functions.[20]
- **Ultrasoft pseudopotentials or PAW methods**, which use pseudopotentials incompatible with many-body wave function theories.[21]

are generally unsuitable for preparing trial wave functions.

[19] This method describes core regions with atomic orbitals while expanding valence electron regions with plane waves.

[20] Instead of working with wave functions, this method employs Green's functions.

[21] Since DFT treats charge density as the fundamental quantity, these methods are designed based on the principle that as long as the equivalent charge density is reproduced, the wave function itself does not need to be strictly defined.

In practical applications, researchers often identify cases where conventional methods struggle and consider applying ab initio QMC as a potential solution. However, it is not uncommon for this effort to stall at the question: *"Can we even prepare a suitable trial wave function for this system?"*.

1.3.4 How Is It Actually Used?

ab initio QMC demonstrate tremendous power in describing electronic correlation, where conventional approaches often struggle. However, as discussed in the previous subsection, QMC has cumbersome aspects in terms of usability, and at a practical level, it remains a tool that is primarily limited to ground-state energy calculations.[22] One might think, *"If all it can do is compute ground-state energies, then it has no real use."* However, as discussed below, there are fairly significant practical demands for such calculations.

First, for fundamental materials science questions such as the *origin of magnetism* [13, 14] and the *exploration of new crystal phases in extreme conditions, such as deep within the Earth's crust* [15], even just comparing ground-state energy values can lead to research with significant theoretical impact. In these fundamental scientific fields, ab initio QMC are regarded as an essential research tool. For phase diagram elucidation, recent approaches combine DFT calculations with data-driven techniques for crystal structure prediction [16]. Instead of drawing definitive conclusions at the DFT level, a set of candidate structures is retained. Then, a post-processing step using ab initio QMC refines the ground-state energy values with higher reliability. By comparing these refined energy values, researchers can determine which phase is the most stable. This approach has revealed cases where QMC predicts a different stable phase than what was suggested at the DFT level [15].

A more practical application in materials science is using QMC as a **calibration tool** to determine the most appropriate exchange-correlation potential for a given system in DFT [10]. Previously, the only available calibration method for selecting an exchange-correlation potential in DFT was to compare its predictions with experimental data. However, with ab initio QMC now capable of computing real materials, the highly reliable variational principle can serve as an independent **referee**.[23] The approach involves comparing DFT predictions obtained using different exchange-correlation potentials and identifying the one that best matches the QMC results. Once a particular exchange-correlation potential is validated through this

[22] Methodologically, there have been some successful demonstrations in calculating excitation energies and forces, but as of the time of writing, these applications are still far from being viable for practical materials researches.

[23] Calibration based on experimental agreement is often complicated by concerns over how much trust can be placed in older experimental data.

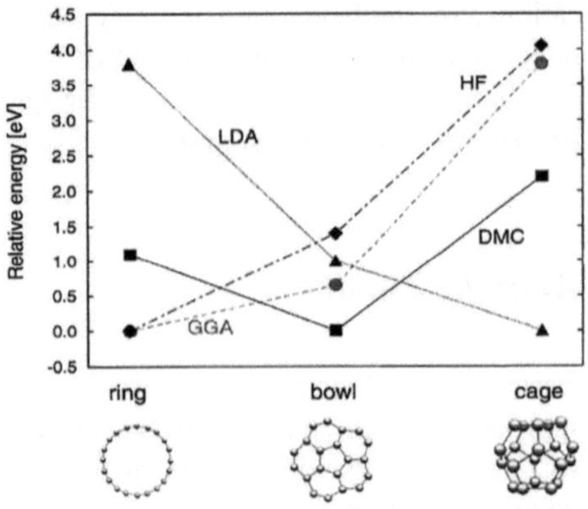

Fig. 1.7 A study evaluating the energy ranking of three structural isomers of C_{20} (ring, bowl, and cage) using different ab initio methods (Figure adapted from [17]). Among the methods "HF/LDA/GGA/DMC," HF and DMC belong to different theoretical frameworks from DFT, while LDA and GGA (For LDA (local density approximation), see Sect. 16.2.2. GGA (generalized gradient approximation) is an improved approximation over LDA, which relies only on local values; GGA incorporates additional information from the gradients of the electron density.) represent two different choices of exchange-correlation potentials within DFT. For each method, the structure with the lowest energy represents *the most stable configuration predicted by that method*. However, the predicted most stable structure varies depending on the method used

process, it is then used as the standard for practical DFT calculations. Performing QMC calibrations for every individual system would be impractical. However, since the descriptive accuracy of an exchange-correlation potential is largely dictated by the nature of bonding (e.g., covalent bonding, intermolecular interactions, etc.), calibration only needs to be conducted for a few **prototype systems**. For systems with similar bonding characteristics, the validated exchange-correlation potential can then be **reliably reused**, assuming that the same physical considerations apply.

When we say *"checking the agreement between ab initio QMC and DFT"*, we do not mean comparing the absolute values of ground-state energies, but rather verifying the **predicted trends in energy variations**. In DFT, the choice of exchange-correlation potential does not follow a variational principle, meaning that, unlike QMC, one cannot assert that *"lower energy values are necessarily more accurate"*. Furthermore, since the energy reference point depends on the specific exchange-correlation potential used, absolute energy values in DFT are not meaningful for direct comparison with QMC results. By **"trend comparison"**, we mean analyzing how the ground-state energy changes when altering system parameters such as composition, structure, or electronic configuration. The agreement between QMC and DFT is assessed by comparing these energy variation trends rather than absolute values (Fig. 1.7).

Using QMC as a calibration tool can be likened to using **a whetstone** for sharpening **the knife** of DFT. For problems where the choice of exchange-correlation potential significantly alters predictions and even leads to academic debates, such calibration studies are of great importance. One such example is the investigation of unknown crystal structures that may become stable in extreme environments, such as **deep within the Earth's crust or in outer space**. Current exchange-correlation potentials have been developed primarily for describing equilibrium phases on Earth. Whether they remain valid under extreme conditions is uncertain. Since experimental validation is not possible, these are cases where we must rely solely on simulations to infer *"what is happening"*. Thus, calibrating the reliability of such predictions is *critical* for the advancement of scientific theories in these fields.

The role of ab initio simulations in the formation of scientific scenario is growing in the field of **life sciences** as well. Traditional exchange-correlation potentials, which have been developed primarily for describing covalent bonding systems, are known to exhibit poor accuracy when predicting the stable structures and reaction pathways of biomolecular systems [10]. For molecular systems, molecular orbital methods, which do not rely on exchange-correlation potentials, have long been regarded as the most reliable ab initio approach. However, biomolecular systems, which play essential roles in life sciences, are typically large in size. Since the computational cost of high-precision molecular orbital methods scales approximately as the seventh power of the system size, applying them to such large biomolecular systems is practically impossible. This is where the advantages of ab initio QMC become particularly noteworthy, as they offer *reliability comparable to or even exceeding that of advanced molecular orbital methods, while maintaining manageable computational costs for large-scale systems*. The predictive reliability of QMC in describing intermolecular interactions is well established [10, 12], making it a promising direction for future research.

For problems closer to industrial applications, evaluating **reaction barrier** energies and **formation enthalpies** is of great importance.[24] In the previous section, we mentioned that ab initio QMC primarily computes ground-state energies. However, when calculating the difference between the ground-state energies before and after a reaction, QMC can achieve sufficient statistical accuracy [1]. Thermodynamic quantities such as formation enthalpy, evaluated in this manner, serve as crucial parameters for computational thermodynamics in phase diagram predictions.[25] Traditionally, parameters for computational thermodynamics were obtained through experiments. However, in recent years, ab initio simulations have increasingly taken on this role. Since the accuracy of parameter evaluations significantly affects the predictive power of computational thermodynamics, the utilization of the **most reliable ab initio method**, namely, ab initio QMC, has become increasingly

[24] These evaluations directly relate to practical challenges such as determining the intermediate structures formed during chemical reactions (which can guide improvements in reaction efficiency) and understanding the ion conduction pathways within crystals (which can help enhance conductivity).

[25] For materials improved through doping or solid solutions, it is essential to determine the maximum solubility percentage that maintains a stable structure contributing to the desired properties. Phase diagrams consolidate such information, making them an invaluable resource.

important for this application. For reaction barrier energies, these can be evaluated by computing ground-state energies while systematically varying the defect positions within a crystal. With the significant acceleration of ab initio QMC, it has recently become feasible to perform evaluations such as nudged elastic band (NEB) calculations, where defect positions within a crystal are systematically varied and computed. These QMC-based evaluations have yielded qualitatively different predictions from DFT, leading to the resolution of discrepancies between experimental results.[26] Studies demonstrating how QMC has resolved such disputes by reconciling theoretical predictions with experimental findings have also been reported [23].

Up to this point, we have discussed the characteristics of ab initio QMC and their expected applications. The underlying principles that enable QMC to achieve high predictive reliability will be explained in detail in the upcoming practical tutorials ($\sim$ Chap. 6) and theoretical background sections (Chap. 7). However, it is worth noting that the fundamental principles of this method were already established as early as the 1970s [24–26]. For a long time, despite the established theoretical foundations, widespread adoption of ab initio QMC in materials science was hindered, primarily by computational resource limitations and the challenges of numerical optimization. However, in the 1990s, efforts to develop general-purpose software packages for QMC accelerated. Notable packages that emerged during this period include **CASINO** from the UK [18],[27] **QWalk** from the USA [22], and **CHAMP**, a joint effort between US and European researchers [21]. These packages enabled users to download and utilize QMC simulations, ushering in a new era of accessibility. All of them have been deployed on flagship-level supercomputers and have contributed significantly to scientific advancements.

From the 2010s onward, the architecture of large-scale supercomputers shifted from relying on general-purpose processors to actively incorporating specialized computing hardware, such as GPUs, which, while powerful, are often challenging to use efficiently. Since ab initio QMC have historically been **supercomputer-native**, their software packages needed to adapt to this evolving landscape. For **large-scale practical computations in materials science** that maximize supercomputer performance, there has been a notable consolidation of efforts around the US-based package **QMCPack** [19].[28] On the other hand, for the development of advanced methodologies, such as force calculations, excitation energy computations, and error correction schemes, a different approach has been maintained. This direction focuses on agile, iterative development in **research lab-scale server environments**, where new techniques can be tested and refined more flexibly. In this context, in addition to the aforementioned **CASINO**, the European package **TurboRVB** [20]

[26] A common pattern in scientific debates arises when experiment A and experiment B report conflicting results, with theory A supporting experiment A and theory B supporting experiment B. Such situations often fuel academic controversies.

[27] During this time, the author was a postdoctoral researcher in the CASINO development group.

[28] Several researchers from the author's group have contributed to QMCPack development.

has been actively maintained and updated over the years.[29] Further discussion on these aspects is provided in Chap. 9.

1.4 Overview of the Following Content

1.4.1 Approach to Content Structure

This book is intended for beginners and practical users. Instead of presenting theoretical concepts before practical applications, we adopt the same "learning by doing" approach as in the author's previous book [3]. Although the explanations may be somewhat redundant, the material is structured to deepen understanding gradually, like *making incremental grooves with a utility knife to cut through cardboard*, ensuring that learners can progress without losing motivation. The more advanced theoretical discussions are placed later in the book, following practical exercises, allowing readers to absorb the theory more effectively in a manner akin to *school trip again by an adult.*

The practical workflow of ab initio QMC is illustrated in Fig. 1.8. Experienced users often design tutorials that guide beginners through this workflow step by step from the beginning. However, forcing beginners to follow the workflow without first digesting the underlying concepts often results in a loss of motivation. In this book, we begin with a functional check, providing an overview of *what the inputs and outputs are* [§2]. We then **work backward** to explore the various concepts necessary to understand and execute these calculations, addressing the question *"How were the components of this computation prepared?"* [§3–§5]. **Finally**, in §6, we consolidate understanding by following the entire workflow in the style of **an "adult school trip"**. The detailed theoretical background is presented in §7–§10, offering insights from the perspective of many-body electron theory to highlight the distinct characteristics of ab initio QMC methods.

I Actually Disliked Tutorial-Based Learning ...
The author began receiving requests to teach electronic structure calculation courses around 2009. At that time, the course structure typically consisted of an extended theoretical background section filled with mathematical formalism, followed by practical exercises. Personally, I was never fond of tutorial-style learning during my student years, where one is encouraged to *"first try running the program at a basic level without fully understanding the*

(continued)

[29] Some members from the author's group have played a key role in the development of TurboRVB.

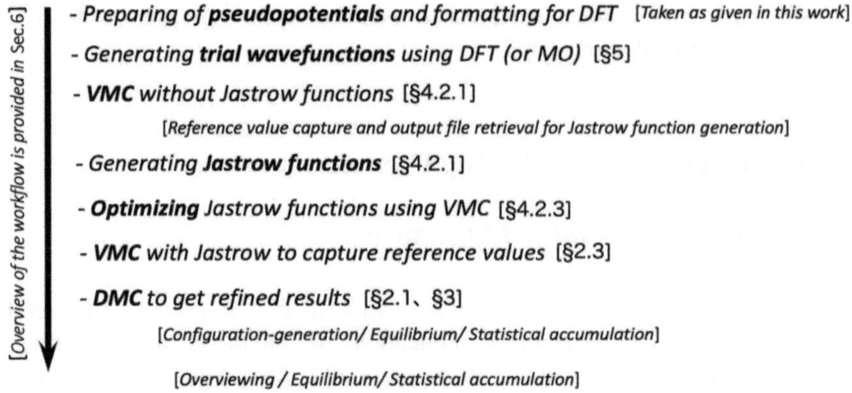

Fig. 1.8 An overview of the overall workflow covered in this book. Rather than forcing readers to follow the practical workflow from the beginning without first understanding the underlying concepts, we start with a functional check, providing an overview of **what the inputs and outputs are** [§2]. Next, we explore the fundamental concepts necessary to understand and execute these calculations [§3–§5]. **Finally**, in §6, we reinforce comprehension by tracing the entire practical workflow in the style of **an "adult school trip"**. The detailed theoretical background is covered in §7–§10, offering insights from many-body electron theory to understand the **unique characteristics** of ab initio QMC methods

details". However, if one prioritizes a writing style that is *"precise and free of unnecessary elaboration"*, the result tends to resemble a mathematics or formal science textbook, where the first chapter is filled with an exhaustive list of definitions—ultimately deterring beginners from engaging with the material.

I began to appreciate the **"learn by doing"** approach when I took on the task of building a PC cluster server as a complete beginner. With guidance from students, I worked through the process but quickly realized that many ICT-related concepts, filled with loanwords and technical jargon, were nearly impossible to grasp simply by reading books or online articles without hands-on experience. For example, the mechanics of budget SIM cards, cloud file services like Dropbox, or server access via public key authentication were all topics that, no matter how clearly explained in writing or speech, remained difficult to understand. However, the moment I actually purchased and used these tools, I immediately thought, *"Ah, so that's how it works!"* It was then that I truly understood what *"learning by doing"* meant.

As in the previous book [3], this work is strictly limited to content that, *"if understood thoroughly, enables self-study for the rest"*. Unnecessary additions were deliberately avoided to prevent an overwhelming *"cover everything"* approach. Specifically, the discussion is focused solely on **ground-state energy calculations**, ensuring that the key concepts of ab initio QMC are explained with clarity and depth. Since solid-state periodic systems can be studied using only free software, the initial conceptual explanations are based on this approach. Tutorials using Gaussian basis sets are presented later in the book. By first introducing the core concepts through solid-state periodic systems, readers, even those who may not have access to Gaussian-based software, can still gain a comprehensive understanding through a concise, terminology-rich explanation in the latter sections.

1.4.2 Assumed Reader Level

Since this book is intended for readers who have taken an interest in its subject matter, it is assumed that they already have some level of experience with electronic structure calculations using either *density functional theory (DFT)* or *molecular orbital methods*. This book assumes that the reader is already familiar with the operation of DFT simulations at the level covered in the author's previous book [3]. Additionally, the fundamental Linux operations required for this book can be learned from the author's previous work [3]. Regarding parallel computing, this book assumes a conceptual understanding at the level explained in Chapter 1 of the author's earlier book *"Building a DIY PC Cluster"* [6].

Even If Somewhat Inaccurate or Redundant
Even when dealing with undefined concepts, providing a rough explanation in context, an introduction that leans more on intuition rather than strict accuracy, can help readers gradually become familiar with the topic before more precise explanations are given in later chapters. This approach is what we liken to "gradually cutting through cardboard." For beginners, aiming from the outset for "perfectly precise wording that no one could possibly dispute" often results in excessive disclaimers, making the explanation difficult to follow. During my undergraduate years, I often encountered professors who were overly fixated on precision. They would stand at the podium, hesitating: "Well, we could say that *xx* is *yy*, but ... no, not quite. Ah, maybe *zz* is ... no, that's not entirely correct either. Hmm ... okay, then, *aa* is ... no, that's not precise enough either ... ." Meanwhile, the students, still new to the subject, were left feeling uncertain and lost. It was a frustrating experience. It is often more effective in the early stages of learning to make bold, straightforward statements, rather than getting caught up in excessive precision. If necessary,

(continued)

a simple footnote such as "This is a slightly imprecise way of putting it" can suffice, allowing for a smoother introduction to the topic.[30]

1.4.3 Tutorial Environment

As in the author's previous book on DFT calculation tutorials [3], this tutorial is designed so that readers can study independently on their own PCs. The exercises are structured to be operated via the command line on either a Windows PC or a Mac. Setup procedures for each OS are provided in the appendices. These instructions describe how to configure each OS to function as a Linux PC, making the exercises in this book fundamentally "tutorials that can be performed on a Linux PC." Additionally, the author's previous book [3] describes a similar tutorial setup using "Ubuntu OS" installed on a Raspberry Pi.

The QMC package used in this tutorial is **"CASINO"** [18].[31] The version of CASINO used in this book is a simplified distribution provided by the development group. This is the same version used in the author's previous book on parallel computing exercises [6]. Readers will download CASINO from the Internet and install it on their personal multicore PCs. Installation instructions are provided in a separate appendix.

The version of **CASINO** used in this book is from around 2010, making it an older implementation compared to the package code currently used at the cutting edge of research in 2025. Furthermore, the usage method described in this book follows a **more outdated approach** rather than the currently recommended usage of CASINO. There are several reasons for deliberately adopting this approach:

- For individual self-study, it is desirable that the package can be *easily installed on personal Windows or Mac machines*. As of 2025, state-of-the-art ab initio QMC packages for professional use are designed to fully exploit *large-scale*

[30] A similar phenomenon can be observed in Japanese language education for international students. I have rarely seen students successfully master Japanese when taught with an excessive focus on precisely conveying particles and verb conjugations.

Students who first attempt to use the language in a more pidgin-like manner, saying "speed-up-wise" instead of "speedy," "setup-should-be" instead of "required to be setup," "non-doable" instead of "not capable," or "do with, figure-make" instead of "by making a figure," and immerse themselves in a large volume of vocabulary tend to acquire Japanese proficiency much faster.

[31] CASINO was developed by Professor Richard Needs' group at the Cavendish Laboratory, University of Cambridge, around the year 2000, when the author was a postdoctoral researcher there. It continues to be maintained and expanded by Professor Drummond (Lancaster University) and others in the UK.

parallelization (on the order of millions of cores) and cutting-edge acceleration technologies such as GPUs. These implementations heavily rely on the latest libraries available on flagship-level supercomputers. Installing and using such software on a personal PC would require modifying libraries and handling technical dependencies, demanding a high level of expertise. For beginners, this is practically infeasible.

- In modern implementations, well-established algorithms are often hidden from input settings to ensure that beginner users do not need to explicitly consider them. As a result, users are unable to finely adjust the conditions of these individual algorithms or observe how different settings affect their behavior.
- In modern practical usage trends, the process is highly automated so that users can reach the final result with just a single click. Many internal procedural steps are carried out behind the scenes, hidden from the user. As a result, it becomes unclear what steps are involved and what specific processing is taking place at each stage.

For these reasons, we deliberately choose an older implementation and usage method to *restore the educational aspect* of "understanding the algorithms while running the calculations."

Thick-Layered Implementations

Since 2010s, trends in large-scale supercomputer usage have seen to be shifted. Previously, packages were developed and maintained with the guiding principle of being *universally operable across different environments, without dependence on specific hardware*. However, in the years that followed, the focus shifted toward *aggressively optimizing for specific hardware architectures* to fully exploit the performance of expensive, cutting-edge processors. Previously, the structure of the physical algorithms could be easily understood just by reading the program code. In contrast, modern implementations are dominated by non-physics-related components, such as "definitions for efficient memory and register usage" and "data transfer handling with GPUs." As a result, a significant portion of the code is now dedicated to hardware optimization rather than the core physics simulation itself. It is as if one must "cut through thick layers of fat before reaching the lean meat." To maximize performance, these programs make extensive use of specialized libraries tailored to specific supercomputers. As a result, such program codes have low portability, making it increasingly difficult to install and run them on personal PCs.

References

1. “Quantum Monte Carlo simulations of solids”, W.M.C. Foulkes, L. Mitas, R. J. Needs, and G. Rajagopal, Rev. Mod. Phys. 73, 33 (2001), https://doi.org/10.1103/RevModPhys.73.33
2. “Quantum Monte Carlo Approaches for Correlated Systems” (Cambridge University Press, 2017), F. Becca, S. Sorella, ISBN: 978-1107129931
3. “Ab initio Calculation Tutorial: For Materials Analysis, Informatics and Design”, (Springer Nature, 2023), R. Maezono, ISBN: 978-98-19909-21-6
4. “Density-functional Theory of Atoms And Molecules” (Oxford University Press, 1989), R.G. Parr, W. Yang, ISBN: 978-01-95092-76-9
5. “The Elements of Statistical Learning: Data Mining, Inference, and Prediction, Second Edition” (Springer, 2009), T. Hastie, R. Tibshirani, J. Friedman, ISBN: 978-03-87848-57-0
6. “How to Build Your Own PC Cluster” (in Japanese), Ryo Maezono, Morikita Publishing (2017/12/14) ISBN:978-46-27818-21-7
7. “An Introduction to Computer Simulation Methods: Applications to Physical Systems, Third Edition” (Amazon Digital Services, 2017), H. Gould, J. Tobochnik, W. Christian, ISBN: 978-19-74427-47-5
8. “Excitons and biexcitons in symmetric electron-hole bilayers”, R. Maezono, Pablo Lopez Rios, T. Ogawa, and R.J. Needs, Phys. Rev. Lett. 110, 216407 (2013), https://doi.org/10.1103/PhysRevLett.110.216407
9. “Introduction to Computational Chemistry, Third Edition” (Wiley, 2017), F. Jensen, ISBN: 978-11-18825-99-0
10. “The Importance of Electron Correlation on Stacking Interaction of Adenine-Thymine Base-Pair Step in B-DNA: A Quantum Monte Carlo Study”, K. Hongo N.T. Cuong, and R. Maezono, J. Chem. Theory Comput. 9, 1081 (2013), https://doi.org/10.1021/ct301065f
11. “The Theory of Intermolecular Forces” (Oxford University Press, 1997), A. Stone, ISBN: 978-01-98558-83-5
12. “Anomalous non-additive dispersion interactions in systems of three one-dimensional wires”, A.J. Misquitta, R. Maezono, N.D. Drummond, A.J. Stone, and R.J. Needs, Phys. Rev. B 89, 045140:1–9 (2014). https://doi.org/10.1103/PhysRevB.89.045140
13. “Interpretation of Hund’s multiplicity rule for the carbon atom”, K. Hongo, R. Maezono, Y. Kawazoe, H. Yasuhara, M.D. Towler, and R.J. Needs, J. Chem. Phys. 121, 7144 (2004), https://doi.org/10.1063/1.1795151
14. “Emergence of a Kondo singlet state with the Kondo temperature well beyond 1,000K in the proton-embedded electron gas”, Y. Takada, R. Maezono, and K. Yoshizawa, Phys. Rev. B. 92, 155140 (2015), https://doi.org/10.1103/PhysRevB.92.155140
15. “Candidate structure for the H_2-PRE phase of solid hydrogen”, T. Ichibha, Y. Zhang, K. Hongo, R. Maezono, F.A. Reboredo, Phys. Rev. B 104, 214111 (2021), https://doi.org/10.1103/PhysRevB.104.214111
16. “Potential high-T_c superconductivity in $YCeH_{20}$ and $LaCeH_{20}$ under pressure”, P. Song, H. Zhufeng, K. Nakano, K. Hongo, R. Maezono, Mater. Today Phys. 28, 100873 (2022), https://doi.org/10.1016/j.mtphys.2022.100873
17. “Structure and Stability of Molecular Carbon: Importance of Electron Correlation”, J.C. Grossman, L. Mitas, K. Raghavachari, Phys. Rev. Lett. 75, 3870 (1995), https://doi.org/10.1103/PhysRevLett.76.1006
18. “Continuum variational and diffusion quantum Monte Carlo calculations”, R.J. Needs, M.D. Towler, N.D. Drummond, P.L. Rios, J. Phys.: Condens. Matter 22, 023201 (2010), 1 https://doi.org/10.1088/0953-8984/22/2/023201
19. “QMCPACK: an open source ab initio quantum Monte Carlo package for the electronic structure of atoms, molecules and solids”, J. Kim, A.D. Baczewski, T.D. Beaudet, A. Benali, M.C. Bennett, M.A. Berrill, N.S. Blunt, E.J.L. Borda, M. Casula, D.M. Ceperley, J. Phys.: Condens. Matter 30, 195901 (2018), https://doi.org/10.1088/1361-648X/aab9c3

20. "TurboRVB: A many-body toolkit for ab initio electronic simulations by quantum Monte Carlo", K. Nakano, C. Attaccalite, M. Barborini, L. Capriotti, M. Casula, E. Coccia, M. Dagrada, C. Genovese, Y. Luo, G. Mazzola, A. Zen, S. Sorella, J. Chem. Phys. 152, 204121 (2020), https://doi.org/10.1063/5.0005037
21. "Energy and variance optimization of many-body wave functions", C.J. Umrigar, C. Filippi, Phys. Rev. Lett. 94, 150201 (2005), https://doi.org/10.1103/PhysRevLett.94.150201
22. "QWalk: A quantum Monte Carlo program for electronic structure", L.K. Wagner, M. Bajdich, L. Mitas, J. Comput. Phys. 228, 3390 (2009), https://doi.org/10.1016/j.jcp.2009.01.017.
23. "Ti interstitial flows giving rutile TiO_2 reoxidation process enhanced in (001) surface", T. Ichibha, A. Benali, K. Hongo, and R. Maezono, Phys. Rev. Mater. 3, 125801 (2019), https://doi.org/10.1103/PhysRevMaterials.3.125801
24. "Monte Carlo Methods In Ab Initio Quantum Chemistry (World Scientific, 1994), B.L. Hammond, W.A. Lester Jr., P.J. Reynolds, ISBN: 978-98-10203-22-1
25. "Monte Carlo Methods in Statistical Physics, Second Edition" (Springer-Verlag, 1986), edited by K. Binder, ISBN: 978-35-40165-14-9
26. "Quantum Monte Carlo Methods in Physics and Chemistry" (Springer, 1998), M.P. Nightingale, C.J. Umrigar, ISBN: 978-07-92355-51-9
27. "Molecular Electronic-Structure Theory" (Wiley, 2013), T. Helgaker, P. Jorgensen, J. Olsen, ISBN: 978-11-18531-47-1

Chapter 2
Running a Calculation First

Using the downloaded tutorial materials, we will first confirm if a calculation could run. Depending on whether your OS is Mac or Windows, proceed to Chap. 11 [MacOS] or Chap. 12 [WinOS] to set up your environment on your PC. Through the tutorials in this chapter, one will learn various simulation operations, Monte Carlo integration for evaluating the variational functional, parallel computing techniques, and statistical processing methods such as reblocking.

In this book, we use aluminum nitride (AlN), an insulating solid, as an example of a solid periodic system, and the water molecule (H_2O) as an example of an isolated molecular system (Fig. 2.1). The goal of the simulations is to accurately calculate the ground-state energy values of these systems. Up to Chap. 6, we will focus on the AlN example to learn the workflow of computational practice and the underlying theory. Then, in Sect. 6.2, we will apply the same procedure to an isolated molecular system, reinforcing the concepts as a review.

2.1 First, Check if It Runs

In this book, the computational workflow is likened to "baking bread to perfection." In this chapter, all the necessary ingredients corresponding to the bread dough have already been prepared by the instructor, and the dough has already been fermented. The reader's first experience will be to "try baking it." The content of this section can be likened to "learning how to operate a bread maker" and "test-tasting the freshly baked bread."

R. Maezono, *Ab initio Quantum Monte Carlo Tutorial*,
https://doi.org/10.1007/978-981-95-4134-8_2

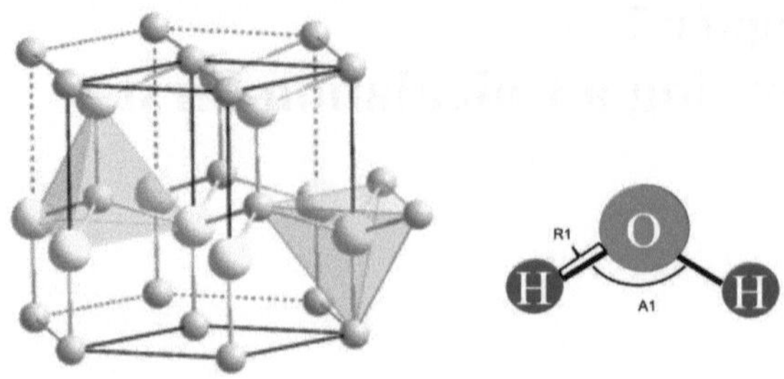

(*a*) **AlN** *crystal* (§2-§6) (*b*) $\mathbf{H_2O}$ *molecule* (§6.2)

Fig. 2.1 The tutorial examples used in this book include the following: For solid periodic systems, aluminum nitride (AlN), an insulating solid; For isolated molecular systems, the water molecule (H_2O)

2.1.1 Navigating to the Working Directory Containing Input/Output Files

Move to the following directory and check the file contents.

```
% pwd
    /Users/maezono/work/qmcTutorials/introByPw
% ls
al_pp.data    input  bwfn.data
    correlation.data   n_pp.data   correlation.init
```

These files correspond to the "fermented bread dough prepared by the master baker," with each file serving a specific role in the analogy as follows:

- **bwfn.data**
  The core part of the many-body wave function $\Psi_\alpha(\mathbf{R})$ in the variational functional of Eq. (1.4),

$$E(\alpha) = \frac{\int d\mathbf{R}\cdot\Psi_\alpha^*(\mathbf{R})\left[-\frac{1}{2}\sum_j \nabla_j^2 + V(\mathbf{R})\right]\Psi_\alpha(\mathbf{R})}{\int d\mathbf{R}\cdot\Psi_\alpha^*(\mathbf{R})\cdot\Psi_\alpha(\mathbf{R})},$$

  which does not depend on the variational degrees of freedom $\boldsymbol{\alpha}$.
- **correlation.data**
  The part of the many-body wave function $\Psi_\alpha(\mathbf{R})$ in the above equation that carries the dependence on the variational degrees of freedom $\boldsymbol{\alpha}$. The file "correlation.data" provided in this directory contains a version where the $\boldsymbol{\alpha}$ dependence has already been optimized. In the analogy, this corresponds to the material that has been *fermented before baking*.[1]

[1] See Sect. 1.2.2.

- **al_pp.data, n_pp.data**
  A part of the potential $V(\mathbf{R})$ in the above equation (numerical data describing the potentials from ionic cores, called **pseudopotential**).
- **input**
  Instructions for baking the bread (adjusting the temperature, timing for stopping the heat, etc.).

At this stage, the master has prepared everything for you. By inputting these instructions into the oven, the bread will be baked according to the specified baking time and heat level, as described in the instructions.

For this initial "test run," we will execute a DMC simulation.[2] This corresponds to a baking method where the fermented dough continues to undergo ultra-fermentation even while being baked. The preparation and optimization of the many-body wave function files, "bwfn.data" and "correlation.data," will be explained in detail in later chapters. Similarly, the specifications of the pseudopotential files "al_pp.data" and "n_pp.data," which numerically represent the atomic potential, will be discussed in later chapters. However, this book assumes prior knowledge of the fundamental concept of pseudopotentials.[3]

2.1.2 Running the Example Files

Now, let's use these materials to "bake the bread," that is, to run the simulation. The prepared files above are the input files for the ab initio QMC package **CASINO**. Before proceeding, confirm that you are in the working directory where these files are located using the **pwd** command below. Then, execute the "CASINO" binary **casino** in the background using the parallel execution command **mpirun**:[4,5]

```
% pwd
    /Users/maezono/work/qmcTutorials/introByPw
% mpirun -np 8 casino &
```

Since this calculation takes more than a minute, execute

```
% top
```

to verify that the CPU cores are indeed running in eightfold parallelization.[6]

---

[2] Diffusion Monte Carlo, a quantum diffusion Monte Carlo calculation.

[3] See the author's previous work [1].

[4] Prior knowledge of the parallel execution command **mpirun** and background execution using "&" is assumed. See the author's previous work [1] for reference.

[5] The "-np 8" option in the **mpirun** command specifies execution with 8-core parallelization. Adjust the value accordingly to match your PC specifications, such as "4-core" or "2-core."

[6] To exit the **top** command, press **Ctrl+C**. For further details, refer to the author's previous work [1].

When the calculation is complete, the terminal will display

```
[1]+  Done     mpirun -np 8 casino
```

indicating that the calculation has finished. After that, use the **ls** command to check that the files in the directory have been updated as follows:

```
% ls
   al_pp.data     config.in          dmc.hist   out
   bwfn.data      config.out         input
   bwfn.data.bin  correlation.data   n_pp.data
```

The newly generated files after the calculation are

- **out**
  Standard output file. This contains a progress report of the simulation. Checking this file allows you to verify whether the simulation ran correctly.
- **dmc.hist**
  A calculation output file that records how the energy values fluctuated during the simulation. This is referred to as a **hist file**. Details on its contents will be explained later.
- **config.in, config.out**
  Intermediate files used for passing data to subsequent calculations. These are referred to as **config files**. The details of their contents will be explained later.
- **bwfn.data.bin**
  A binary version of the input file **bwfn.data**. By converting the text-based **bwfn.data** into machine-readable binary format, subsequent reuse on the same machine is accelerated.

2.1.3 Analyzing the Results

With this, the calculation has been completed, and the results are ready. In our analogy, this corresponds to the state where "the bread has been baked." Next, we proceed to the task of "analyzing the results," which is akin to "test-tasting the freshly baked bread to check if it has been properly baked." Confirm that you are in the working directory where the results were obtained (using the **pwd** command below), and then execute the analysis utility **reblock**:

```
% pwd
    /Users/maezono/work/qmcTutorials/introByPw
% reblock
   ...
```

The program will prompt for user input. Enter "100" when prompted. It will ask for input two more times—enter "1" first and then "32." The meaning of these input values will be explained later. Then you'll get

```
ENERGY COMPONENTS WITH REBLOCKED ERROR BARS
===========================================
                                 Mean (au/prim cell)    Err (au/prim cell)
 Total energy (using Ewald) :  -2.39757137935202E+01  4.70399388013678E-03
Tot pot energy (using Ewald) :  -4.32327183042669E+01  6.10894296804199E-02
         Kinetic energy (K) :   1.92570045107467E+01  6.16509828005772E-02
         Kinetic energy (T) :   1.95116041191771E+01  2.15647247309610E-01
      Kinetic energy (FISQ) :   1.97662037276075E+01  4.16702279393892E-01
          Ewald interaction :  -1.12038902718514E+00  2.70330230509409E-02
           Local e-i energy :  -2.14892798567153E+01  7.97132977197474E-02
        Nonlocal e-i energy :   1.12419315191910E+00  2.45976817228051E-02
            Constant energy :  -2.17472425722856E+01
Program finished.
```

showing the analysis of the result.

The meaning of the input values "100" and "32" will be explained later. However, the reason for entering "1" was in response to the following prompt:

```
Choose one of (1) au per prim cell; (2) eV per prim cell;
(3) kcal per prim cell; (4) au per atom;
(5) eV per atom; and (6) kcal per atom.
```

Thus, the final displayed value corresponds to "(1) au per prim cell," meaning that the result is expressed in atomic units[7] per unit cell.

The output of **reblock** includes many items in addition to the first entry, "Total energy (using Ewald)." The details of these additional items will be explained later. For now, it is sufficient to understand that the value of "Total energy (using Ewald)" represents the **final energy value**. This means that the result corresponds to "Mean value: −2.39757137935202E+01," "Error: 4.70399388013678E–03" which gives the final result as $E = -23.9757 \pm 0.0047$. The reason why the result is a statistical estimate with an error bar will be gradually explained in the next section.

Next, we use the utility command **graphit** to visualize how the energy values evolved during the simulation. This helps verify whether the calculation behaved normally. First, confirm that you are in the same working directory using the **pwd** command, and then execute the **graphit** command:

```
% pwd
    /Users/maezono/work/qmcTutorials/introByPw
% graphit
```

As a result of executing the command, a new window appears displaying a figure similar to Fig. 2.2. Two plots are shown. The horizontal axis represents the step number, indicating the progression of the simulation. The upper plot will be explained later (Sect. 3.2.5). The lower plot illustrates how the computed energy values fluctuate as the simulation progresses. Upon closer inspection, three curves are plotted. These will be explained later. Although the figure may not be very easy to read, focusing on the region where the step number is below 50 reveals that the energy values gradually decrease (a more typical example of this behavior will be

[7] For the atomic units, see the footnote in Sect. 6.1.6.

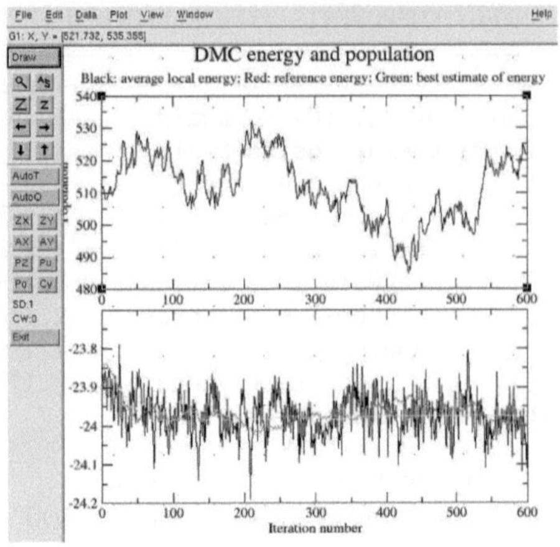

Fig. 2.2 The screen output from executing **graphit**. The horizontal axis represents the step number, indicating the progression of the simulation. The upper plot will be explained later (Sect. 3.2.5). The lower plot illustrates how the computed energy values fluctuate as the simulation progresses

presented later). Recall the variational principle, which states that "lower energy values indicate closer proximity to the exact solution." This means that "as the simulation progresses, it approaches a more accurate solution." This is the essence of what we referred to as "ultra-fermentation," where the dough continues to mature even while being baked.

2.2 Sampling and Evaluation by Statistical Averaging

In the previous section, we observed that "results are obtained as statistical values with error bars" and that "energy values are generated as the simulation progresses through its steps." Let us now explore these aspects in greater detail. Specifically, we aim to understand how the energy value obtained from the integral in Eq. (1.4)

$$E(\alpha) = \frac{\int d\mathbf{R}\cdot\Psi_\alpha^*(\mathbf{R})\left[-\frac{1}{2}\sum_j \nabla_j^2 + V(\mathbf{R})\right]\Psi_\alpha(\mathbf{R})}{\int d\mathbf{R}\cdot\Psi_\alpha^*(\mathbf{R})\cdot\Psi_\alpha(\mathbf{R})} \tag{2.1}$$

is evaluated using statistical sampling and averaging.

From this point on, the content of this section will proceed in the following order:

- Integrals of the form

$$I = \int_\Omega F(x)\cdot p(x)\cdot dx \quad \text{where,} \quad \int_\Omega dx\cdot p(x) = 1 \quad , \quad p(x) \geq 0$$

can be numerically evaluated as sampling averages (Sects. 2.2.1–2.2.3).

- The variational functional in Eq. (2.1) can be transformed into an integral of the above form I (Sect. 2.2.4).

This will be the logical progression of our discussion.

2.2.1 Evaluating Integrals via Statistical Averaging

Consider the sampling average of a function $F(x)$ at M sampled points $\{x_j\}$ along the x-axis

$$I = \frac{1}{M}\sum_{j=1}^{M} F\left(x_j\right)$$

where we assume that M is a very large number. Formally, this can be rewritten using the delta function as

$$\begin{aligned} I &= \frac{1}{M}\sum_{j=1}^{M} F\left(x_j\right) = \frac{1}{M}\sum_{j=1}^{M} \int_{\Omega} F(x)\,\delta\left(x - x_j\right)\cdot dx \\ &= \int_{\Omega} F(x)\,\frac{1}{M}\sum_{j=1}^{M} \delta\left(x - x_j\right)\cdot dx \end{aligned}$$

Now, defining

$$p(x)\cdot dx = \frac{1}{M}\sum_{j=1}^{M} \delta\left(x - x_j\right)\cdot dx \tag{2.2}$$

we can express I as

$$I = \frac{1}{M}\sum_{j=1}^{M} F\left(x_j\right) = \int_{\Omega} F(x)\cdot p(x)\cdot dx =: \langle F(x)\rangle_p \tag{2.3}$$

Here, $\langle F(x)\rangle_p$ represents the "expectation value weighted by $p(x)$."

The right-hand side of Eq. (2.2) corresponds to the operation of "sweeping through all M total samples at a given x and counting only the points where x_j matches x." This mathematically represents the histogram of the distribution $\{x_j\}_j^M$. In the limit where the histogram bin width approaches zero, this expression converges to the probability distribution function of $\{x_j\}$.[8] Indeed, Eq. (2.2) is nonnegative, and

$$\int_\Omega dx \cdot p(x) = \frac{1}{M}\sum_{j=1}^{M}\int_\Omega dx \cdot \delta\left(x - x_j\right) = \frac{1}{M}\sum_{j=1}^{M} 1 = 1$$

Thus, it satisfies the definition of a probability distribution function:

$$\int_\Omega dx \cdot p(x) = 1 \quad , \quad p(x) \geq 0 \tag{2.4}$$

Rewriting Eq. (2.3) by reversing the left-hand and right-hand sides, we obtain

$$I = \int_\Omega F(x) \cdot p(x) \cdot dx = \frac{1}{M}\sum_{j=1}^{M} F\left(x_j\right) \quad , \quad \left\{x_j\right\} \sim p(x) \tag{2.5}$$

This equation provides a **strategy** for evaluating an integral through the sampling average. The notation "$\left\{x_j\right\} \sim p(x)$" means that "the probability of sampling points $\left\{x_j\right\}$ follows the probability distribution $p(x)$." In other words, if we can generate a sequence of sampling points $\left\{x_j\right\}$ according to the probability distribution function $p(x)$, we can use this sequence to **evaluate the integral as a statistical average**.

Curse of Dimensionality

Equation (2.5) presents **Monte Carlo integration** as one option for numerically evaluating an integral by using the sampling average. On the other hand, elementary numerical integration methods typically introduce quadrature rules such as the trapezoidal rule.[9] These quadrature-based integration methods, such as the trapezoidal rule, are known to suffer from increasing errors as the dimensionality increases, a phenomenon referred to as the **curse of dimensionality**. In contrast, Monte Carlo integration has the advantage that its error does not depend on the dimensionality [3], making it overwhelmingly superior for high-dimensional integrals, such as those encountered in many-particle systems.

[8] Since this is an introductory explanation for beginners, we do not discuss rigorous mathematical treatment of elements such as dx.

[9] Related methods include Simpson's rule and Newton-Cotes formulas [2].

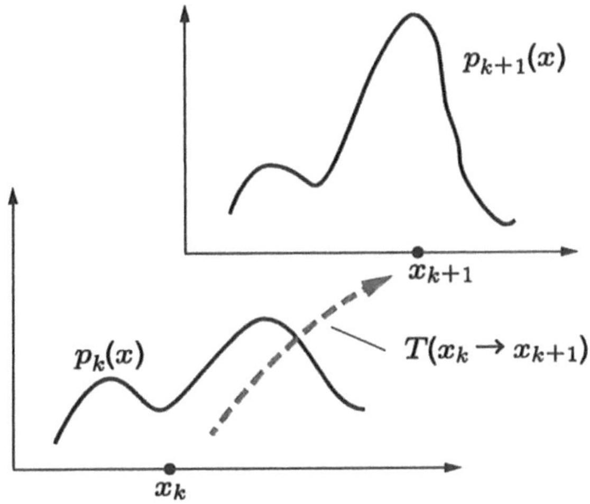

Fig. 2.3 The transition from the sampling point x_k at step k to the sampling point x_{k+1} at step $(k+1)$ occurs according to the probability $T(x_k \to x_{k+1})$. The probability of finding x_k or x_{k+1} along the x-axis is represented as the vertical axis values, such as $p_k(x)$

2.2.2 *Generating Sampling Sequences Following a Desired Distribution*

When evaluating an integral as a sampling statistic using Eq. (2.5), the remaining technical challenge is to establish a method for generating a sampling sequence $\{\mathbf{x}_j\}$ that follows a given probability distribution $\bar{p}(\mathbf{x})$. This can be achieved using the **Markov chain Monte Carlo method** (MCMC).[10]

Consider a scenario where a single sampling point evolves over time, changing its position as $\{x_1, x_2, \cdots, x_k, \cdots\}$. Let $p_k(x)$ represent the probability of the sampling point appearing at position x at time step k. If this probability converges to $p_{k\gg 1}(x) = \bar{p}(x)$, then we have successfully generated a sequence $\{x_j\}$ that follows the desired distribution $\bar{p}(x)$. This satisfies our requirement of obtaining a sequence $\{x_j\}$ that is generated according to a given probability distribution $\bar{p}(x)$.

Now, suppose that at each step, the configuration is updated from x_k to x_{k+1} according to a transition probability $T(x_k \to x_{k+1})$ (Fig. 2.3). The transition probability $T(x \to y)$ is a normalized probability with respect to the destination y, satisfying

$$\int dy \cdot T(x \to y) = 1 \tag{2.6}$$

If we use an update transition probability $T(x_k \to x_{k+1})$ that satisfies the following condition for a given target distribution $\bar{p}(x)$

$$\bar{p}(x_k) \cdot T(x_k \to x_{k+1}) = \bar{p}(x_{k+1}) \cdot T(x_{k+1} \to x_k) \tag{2.7}$$

---

[10] Other methods include approaches based on the Fokker-Planck equation and thermal bath methods. The Fokker-Planck equation approach will be discussed later in Sect. 7.2.2.

then we can show that the probability of a sampling point appearing at x converges as $p(x) \rightarrow \bar{p}(x)$. This means that the generated sequence follows the desired distribution $\bar{p}(x)$. The condition given in Eq. (2.7) is known as the **detailed balance** condition.

By imposing the detailed balance condition, we can demonstrate

- (1) Once the system reaches the desired distribution $\bar{p}(x)$, it will not deviate from it.
- (2) The system will always converge toward the desired distribution $\bar{p}(x)$.

This establishes that the sampling sequence **converges to the desired distribution** $\bar{p}(x)$.

First, (1) above can be demonstrated as follows: If the current distribution is $p_n(x)$ and transitions to $p_{n+1}(x)$ in the next step, we can express this as

$$p_{n+1}(x) = \int dx' \cdot p_n(x') T(x' \rightarrow x) \tag{2.8}$$

Here, $p_{n+1}(x)$ represents the probability that a configuration appears at position x. This probability is obtained by summing over all previous positions x' from which a transition to x occurs with probability $T(x' \rightarrow x)$. The probability that the configuration is at x' in the previous step is given by $p_n(x')$. Thus, the probability of transitioning from x' to x is given by $p_n(x') \cdot T(x' \rightarrow x)$. By summing over all possible previous positions x', we obtain $p_{n+1}(x)$, which represents the total probability of the configuration appearing at x in the next step.

To prove (1), we need to show that if $p_n(x) = \bar{p}(x)$ at step n, then $p_{n+1}(x) = \bar{p}(x)$ at step $n+1$. Evaluating this, we find

$$\begin{aligned}
p_{n+1}(x) &= \int dx' \cdot p_n(x') T(x' \rightarrow x) = \int dx' \cdot \bar{p}(x') T(x' \rightarrow x) \\
&= \int dx' \cdot \bar{p}(x) T(x \rightarrow x') \qquad (2.9) \\
&= \bar{p}(x) \int dx' \cdot T(x \rightarrow x') = \bar{p}(x) \qquad (2.10)
\end{aligned}$$

Thus, we have successfully demonstrated the desired property. In Eq. (2.9), we used the detailed balance condition (2.7), and in Eq. (2.10), we applied the normalization condition of T given in Eq. (2.6).

Next, to prove (2) that the system necessarily converges toward the desired distribution $\bar{p}(x)$, we define the distance measuring how close the current distribution is to the target distribution as

$$D_n = \int dx \cdot |\bar{p}(x) - p_n(x)| \tag{2.11}$$

The goal is then to show that this distance decreases with each iteration, i.e.,

$$D_{n+1} \leq D_n.$$

We proceed with this approach: Let us evaluate

$$D_{n+1} = \int dx \cdot |\bar{p}(x) - p_{n+1}(x)|.$$

Using Eq. (2.10),

$$\bar{p}(x) = \int dx' \cdot \bar{p}(x') T(x' \to x)$$

and Eq. (2.8)

$$p_{n+1}(x) = \int dx' \cdot p_n(x') T(x' \to x)$$

we can express the integrand of Eq. (2.11) as

$$|\bar{p}(x) - p_{n+1}(x)| = \left| \int dx' \cdot (\bar{p}(x') - p_n(x')) T(x' \to x) \right|.$$

Substituting this into Eq. (2.11), we obtain

$$\begin{aligned} D_{n+1} &= \int dx \cdot \left| \int dx' \cdot (\bar{p}(x') - p_n(x')) T(x' \to x) \right| \\ &= \int dx \cdot T(x' \to x) \left| \int dx' \cdot (\bar{p}(x') - p_n(x')) \right| && (2.12) \\ &= \left| \int dx' \cdot (\bar{p}(x') - p_n(x')) \right| && (2.13) \\ &\leq \int dx' \cdot |\bar{p}(x') - p_n(x')| = D_n && (2.14) \end{aligned}$$

Thus, we have successfully demonstrated the desired result. In the derivation,

- In Eq. (2.12), we used the fact that T is a probability and thus nonnegative, allowing us to move it outside the absolute value.
- In Eq. (2.13), we used the normalization condition of T, which ensures that the integral over all possible transitions sums to 1.
- In Eq. (2.14), we applied the inequality stating that the absolute value of an integral is always less than or equal to the integral of the absolute value of its integrand.

2.2.3 Metropolis Method

In the previous section, we established that if there exists a transition probability T satisfying the detailed balance condition

$$\bar{p}(x) \cdot T(x \to y) = \bar{p}(y) \cdot T(y \to x) \tag{2.15}$$

then we can generate a sampling sequence $\{x_k\}$ whose occurrence frequency follows the desired distribution $\bar{p}(x)$. As a concrete implementation of a transition probability T that satisfies the detailed balance condition (2.15), we introduce the **Metropolis method.**[11]

In the Metropolis method, we use a probabilistic process that satisfies the symmetry condition:

$$F(x \to y) = F(y \to x)$$

(the implementation of which will be discussed later) to construct the transition probability $T(x \to y)$ as follows:

$$T(x \to y) = \begin{cases} F(x \to y) \cdot \dfrac{\bar{p}(y)}{\bar{p}(x)} & \cdots [\bar{p}(y) < \bar{p}(x)] \\ F(x \to y) & \cdots [\bar{p}(y) > \bar{p}(x)] \end{cases} = \min\left[1, \frac{\bar{p}(y)}{\bar{p}(x)}\right] \tag{2.16}$$

This ensures that T satisfies the detailed balance condition. This method first proposes an update using the probabilistic process $F(x \to y)$. If the proposed update moves to a region where the distribution function value is higher $[\bar{p}(y) > \bar{p}(x)]$, it is accepted unconditionally. If the update moves to a region where the distribution function value is lower, it is accepted with a probability of

$$\left[\frac{\bar{p}(y)}{\bar{p}(x)}\right] \leq 1.$$

This ensures that the sampling sequence follows the desired probability distribution.

It is straightforward to verify that the implementation of T in Eq. (2.16) satisfies the detailed balance condition. For example, assuming $\bar{p}(y) > \bar{p}(x)$, we have

$$\frac{T(x \to y)}{T(y \to x)} = \frac{F(x \to y)}{F(y \to x) \cdot \dfrac{\bar{p}(x)}{\bar{p}(y)}} = \frac{\bar{p}(y)}{\bar{p}(x)}$$

[11] More precisely, this is known as the *Metropolis-Hastings algorithm* [3]. The original work by Metropolis in 1953 was formulated as a special case for the Boltzmann distribution, and it was later generalized by Hastings in 1970.

By cross multiplying the leftmost and rightmost terms, we confirm that the detailed balance condition is indeed satisfied. A similar argument holds for the case $\bar{p}(y) < \bar{p}(x)$.

Finally, we address the remaining issue: the concrete implementation of the probability process $F(x \to y)$. We generate the configuration update $x \to y$ using a uniform distribution $\xi \sim U(0, 1)$ as follows:

$$y = x + \left(\xi - \frac{1}{2}\right) \times \Delta \tag{2.17}$$

This means that we determine the update target by multiplying a uniform random number ξ, centered around $1/2$ with an amplitude of 0.5, by a suitable stride width Δ chosen based on the problem at hand.[12] The value of Δ is specified by the user depending on system size and other relevant parameters. Since the random number is uniformly distributed, there is no imbalance in the transition between $x \rightleftharpoons y$, ensuring that

$$F(x \to y) = F(y \to x)$$

is satisfied.

The same uniform random number ξ used in implementing the probabilistic process $F(x \to y)$ can also be reused to determine whether the update is accepted in the probabilistic process T from Eq. (2.16), allowing us to complete the entire update process in a single step. Defining the probability T from Eq. (2.16) as

$$\pi = \min\left[1, \frac{\bar{p}(y)}{\bar{p}(x)}\right]$$

we can visualize this as dividing the interval [0,1] into two regions: the interval $[0,\pi]$ is "black," and the interval $[\pi,1]$ is "white." If we throw a "pin" (a uniformly distributed random number ξ from [0,1]), the update is accepted if it lands in the black region and rejected if it lands in the white region. This corresponds exactly to the procedure described by Eq. (2.16). Thus, we generate $\xi \sim U(0, 1)$, determine a trial update y using Eq. (2.17), and implement the update acceptance rule as

$$\xi < \frac{\bar{p}(y)}{\bar{p}(x)}$$

If this condition is satisfied, the update is accepted. This ensures the realization of a **random walk that satisfies the detailed balance condition**.

[12] There are also interesting discussions regarding the choice of pseudo-random numbers, particularly in the context of eliminating autocorrelation, which will be discussed later [4, 5].

The discussion so far has been illustrated using a one-dimensional distribution $\bar{p}(x)$ for simplicity. However, as mentioned in the column on the *curse of dimensionality*, the true strength of Monte Carlo integration lies in sampling from high-dimensional spaces, such as $\bar{p}(x_1, x_2, \cdots, x_N)$. In this case, there are two possible implementation approaches:

- Updating all N components simultaneously and then deciding whether to accept or reject the update.
- Updating the components sequentially, one at a time, deciding on acceptance or rejection at each step.

The latter approach is known as **Gibbs sampling.**[13]

2.2.4 Evaluating the Variational Functional via Sampling

The variational functional in Eq. (1.4),

$$E(\alpha) = \frac{\int d\mathbf{R} \cdot \Psi_\alpha^*(\mathbf{R}) \left[-\frac{1}{2}\sum_j \nabla_j^2 + V(\mathbf{R}) \right] \Psi_\alpha(\mathbf{R})}{\int d\mathbf{R} \cdot \Psi_\alpha^*(\mathbf{R}) \cdot \Psi_\alpha(\mathbf{R})}$$

contains the operator

$$\hat{H} = -\frac{1}{2}\sum_j \nabla_j^2 + V(\mathbf{R}) \tag{2.18}$$

which is called the **Hamiltonian**. For convenience, we will use the notation $\hat{H}$ in subsequent expressions. The variational functional can then be rewritten as

$$\begin{aligned} E(\alpha) &= \frac{\int d\mathbf{R} \cdot \Psi_\alpha^*(\mathbf{R}) \cdot \hat{H}\Psi_\alpha(\mathbf{R})}{\int d\mathbf{R} \cdot \Psi_\alpha^*(\mathbf{R}) \cdot \Psi_\alpha(\mathbf{R})} = \int d\mathbf{R} \cdot \frac{\Psi_\alpha^* \cdot \underline{\Psi_\alpha \cdot \Psi_\alpha^{-1}} \cdot \hat{H}\Psi_\alpha}{\int d\mathbf{R} \cdot \Psi_\alpha^*(\mathbf{R}) \cdot \Psi_\alpha(\mathbf{R})} \\ &= \int d\mathbf{R} \cdot P(\mathbf{R}) \cdot \left[\underline{\Psi_\alpha^{-1}} \cdot \hat{H}\Psi_\alpha \right] \end{aligned} \tag{2.19}$$

This reformulation makes it clear how the functional can be evaluated via statistical sampling. In the transformation above, we introduced

$$P(\mathbf{R}) = \frac{\Psi_\alpha^*(\mathbf{R}) \cdot \underline{\Psi_\alpha(\mathbf{R})}}{\int d\mathbf{R} \cdot \Psi_\alpha^*(\mathbf{R}) \cdot \Psi_\alpha(\mathbf{R})} \tag{2.20}$$

[13] Originally, this method emerged in the context of heat bath algorithms.

Since

$$\Psi_\alpha^* (\mathbf{R}) \cdot \Psi_\alpha (\mathbf{R}) = |\Psi_\alpha (\mathbf{R})|^2 ,$$

$P (\mathbf{R})$ is nonnegative. Furthermore, integrating over all possible $\mathbf{R}$ results in cancellation between the numerator and denominator, yielding

$$\int d\mathbf{R} \cdot P (\mathbf{R}) = 1 \quad , \quad 0 \le P (\mathbf{R}) \le 1.$$

Thus, $P (\mathbf{R})$ satisfies the necessary properties of a probability distribution function, as discussed in Eq. (2.4). Thus, we can rewrite the variational functional as

$$\begin{aligned} E (\alpha) &= \int d\mathbf{R} \cdot P (\mathbf{R}) \cdot \left[\Psi_\alpha^{-1} \hat{H} \Psi_\alpha \right] = \langle \Psi_\alpha (\mathbf{R})^{-1} \cdot \hat{H} \Psi_\alpha (\mathbf{R}) \rangle_{P(\mathbf{R})} \\ &= \frac{1}{M} \sum_{j=1}^{M} \left[\Psi_\alpha^{-1} (\mathbf{R}_j) \cdot \hat{H} \Psi_\alpha (\mathbf{R}_j) \right] \\ &= \frac{1}{M} \sum_{j=1}^{M} E_L (\mathbf{R}_j) \quad , \quad \{\mathbf{R}_j\} \sim P (\mathbf{R}) \end{aligned} \tag{2.21}$$

This demonstrates that the variational functional value $E (\alpha)$ can be evaluated as a sampling average over M configurations $\{\mathbf{R}_j\}$. It is important to recall that each sampled configuration $\mathbf{R}_j$ is a single point in a $3N$-dimensional space, representing a snapshot of the configuration of N electrons:

$$\mathbf{R}_j = (\mathbf{r}_1, \cdots , \mathbf{r}_N)_j .$$

The quantity being sampled is

$$E_L (\mathbf{R}) = \Psi_\alpha^{-1} (\mathbf{R}) \hat{H} \Psi_\alpha (\mathbf{R}) , \tag{2.22}$$

which is referred to as the **local energy**.

2.2.5 *Variational Monte Carlo for Many-Body Wave Functions*

Summarizing the discussion so far, the variational functional value of the ground-state energy, $E (\alpha)$, can be evaluated as follows:

1. Given a trial wave function $\Psi_\alpha (\mathbf{r}_1, \cdots , \mathbf{r}_N)$, the probability distribution function in Eq. (2.20) is determined.

2. A sequence of electron configurations $\left\{\mathbf{R}_j = \left(\mathbf{r}_1^{(j)}, \cdots, \mathbf{r}_N^{(j)}\right)\right\}_{j=1}^{M}$ is generated according to this probability distribution function using a Markov chain Monte Carlo method that satisfies the detailed balance condition. Note that the number of sampled configurations M is independent of the number of electrons N.
3. After a sufficient number of steps, the sampled sequence $\left\{\mathbf{R}_j\right\}$ follows the probability distribution of Eq. (2.20). Therefore, by computing the sampling statistical average in Eq. (2.21), the value of $E\left(\boldsymbol{\alpha}\right)$ can be numerically evaluated.

2.3 Input Specification for Variational Monte Carlo (Entry-Level)

The numerical optimization of $E\left(\boldsymbol{\alpha}\right)$ with respect to $\boldsymbol{\alpha}$ (referred to as *fermenting the dough*), the preparation of $\Psi_{\alpha}\left(\mathbf{R}_j\right)$ (*preparing the dough*), and the principle and method of *super-fermentation while baking* will be explained gradually in later chapters. In this section, we first describe how the concepts explained so far relate to the simulation input files.

2.3.1 *Isolating the VMC Part*

The calculation in Sect. 2.1 was performed to verify whether the computational environment was properly set up. For this purpose, the input was configured so that the entire sequence from VMC to DMC would automatically proceed. In the CASINO package, such a setup allows for the automatic transition from *VMC → DMC* for practical convenience. However, from an educational standpoint, it is beneficial to separate these calculations to clearly distinguish *this part is VMC* and *this part is DMC*. This separation allows us to explain the reasoning behind the transition from *VMC → DMC*.[14]

First, set up the working directory as follows:

```
% pwd
    /Users/maezono/work
% cp -r qmcTutorials/introByPw 01check
% rm 01check/correlation.init
% cp -r 01check/ 02vmcOnly
% pwd
    /Users/maezono/work
% ls
    01check/ 02vmcOnly/
```

[14] Separating each computational process is also important for handling large-scale computational tasks. When performing VMC calculations that take several days to complete, it is necessary to divide the computation into distinct steps and gradually submit them to a supercomputer.

From this point onward, all simulation exercises will be conducted within the `work` directory. The directory `01check` is a copy of the results from Sect. 2.1, where the initial simulation run was performed. This was further copied to `02vmcOnly` to migrate the calculation setup. Then, execute the following commands:

```
% pwd
    /Users/maezono/work
% cd 02vmcOnly/
% ls
    al_pp.data      config.out  n_pp.data
    bwfn.data       correlation.data  out
    bwfn.data.bin   dmc.hist
    config.in       input
% clearup
    Clearing up after CASINO in directory: .
    Finished.
% ls
    al_pp.data  bwfn.data.bin     input
    bwfn.data   correlation.data  n_pp.data
```

By executing the **clearup** command, all previous calculation output files (such as `out`, `dmc.hist`, and `config.out`) that were left in `02vmcOnly` have been deleted, ensuring that the directory is now cleared for a new simulation run.

Next, to separate VMC and DMC and run only the VMC calculation, edit the `input` file as instructed below:

```
% pwd
    /Users/maezono/work/02vmcOnly
% more input
....
# RUN
runtype            : vmc_dmc #*! Change to vmc!
...
# VMC
vmc_equil_nstep    : 1000    #*! Explained later
vmc_nstep          : 5000    #*! Explained later
vmc_nblock         : 1       #*! Explained later
vmc_nconfig_write  : 512     #*! Change to 0!
...
```

These modifications ensure that only the VMC calculation is executed, without transitioning to DMC. For confirmation, after modifying these parameters as instructed above, the updated values should appear as follows:

```
% pwd
    /Users/maezono/work/02vmcOnly
% more input
....
# RUN
runtype            : vmc
# VMC
vmc_equil_nstep    : 1000
```

```
vmc_nstep          : 5000
vmc_nblock         : 1
vmc_nconfig_write : 0
...
```

Previously, the **runtype** parameter was set to **vmc_dmc**, which specified that the calculation should automatically continue from VMC to DMC. By changing this setting to vmc, the calculation is now restricted to VMC only.

In Eq. (2.21),

$$E(\alpha) \approx \frac{1}{M}\sum_{j=1}^{M} \Psi_\alpha^{-1}\left(\mathbf{R}_j\right) \cdot \hat{H}\Psi_\alpha\left(\mathbf{R}_j\right)$$

the number of samples, M, corresponds to the parameter **vmc_nstep** = 5000. The preceding parameter, **vmc_equil_nstep** = 1000, specifies that before starting the actual sampling, the system will undergo 1000 steps of pre-equilibration (thermalization) to stabilize the sampling process. In Sect. 2.2.2, we mentioned that the sampling sequence converges to the distribution $\bar{p}(x)$. However, only after performing a sufficient number of configuration updates under the detailed balance condition does the sampling start to produce configurations with the desired frequency distribution. This process of reaching equilibrium is called **equilibration**. The parameter **vmc_equil_nstep = 1000** specifies the number of steps required for this equilibration before starting the actual sampling.

After editing the input file as described above, execute the following command to run the calculation:

```
% pwd
    /Users/maezono/work/02vmcOnly
% mpirun -np 8 casino &
```

Once the calculation is complete, verify that it has **terminated successfully** by checking the standard output file **out**:

```
% pwd
    /Users/maezono/work/02vmcOnly
% tail out
    ...
    Total CASINO real time : : :          3.1050
    ...
    Ends 2023/09/16 15:14:28.676
```

If the above message appears in **out**, it confirms that the simulation has completed normally.

Let's check the contents of the directory after the calculation:

```
% pwd
    /Users/maezono/work/02vmcOnly
% ls
    al_pp.data     config.out          out
```

```
bwfn.data        correlation.data  vmc.hist
bwfn.data.bin    input             n_pp.data
```

Unlike the case in Sect. 2.1, the files `config.in` and `dmc.hist` are absent, but instead, a new file named `vmc.hist` has been generated. The `vmc.hist` file contains the sampling history of the local energy values $\left\{E_L\left(\mathbf{R}_j\right)\right\}$, as given in Eq. (2.22). Additionally, `config.out` stores the last snapshot of the simulation:

$$\mathbf{R}_M = \left(\mathbf{r}_1^{(M)}, \cdots, \mathbf{r}_N^{(M)}\right) .$$

These pieces of information are useful for performing a **resume calculation**, where the simulation is continued by extending the number of steps further.

2.3.2 Understanding the Efficiency of Parallelization

So far, we have executed simulations using the command

```
% mpirun -np 8 casino &
```

where the option `-np 8` specifies that the calculation is performed using eight parallel cores. Now, let's conduct an exercise by running the calculation with different levels of parallelization, using four cores, two cores, and a single core (nonparallel execution), and examine how the results compare. This subsection is also related to understanding the specification of the accumulation step count, specifically whether `vmc_nstep = 5000` refers to the number of steps per computational core or the total number of steps across all cores.

Execute the following:

```
% pwd
    /Users/maezono/work/02vmcOnly
% mv out out08
% clearup
% mpirun -np 4 casino &
```

This process moves the most recent eight-core parallel computation results to `out08` for safekeeping, clears all previous calculation results, and then initiates a new computation using four-core parallelization. After the calculation completed, execute the following:

```
% mv out out04
% clearup
% mpirun -np 2 casino &
% mv out out02
% clearup
% casino &
% mv out out01
```

After verifying the normal completion of each computation as described in the previous section, the standard output file is moved for safekeeping before proceeding to the next lower parallelization level. This process repeats by first running the calculation with two cores, then moving the output to out02, clearing previous results, and finally running a nonparallelized (single-core) calculation, saving its output as out01. For a single-core, nonparallel calculation, execute **casino** without using the **mpirun** command. Once all the out files are prepared,

```
% grep MPI out0*
out01: Sequential run: not using MPI.
out02: Running in parallel using 2 MPI processes.
out04: Running in parallel using 4 MPI processes.
out08: Running in parallel using 8 MPI processes.
```

to confirm that each calculation was performed using the intended number of parallel cores.

A detailed explanation will be provided later, but the values captured from the standard output below correspond to the energy estimates from the VMC calculation.

```
% grep 'On-the-fly' out*
    out01: -23.836750123164 +/- 0.019116659098 On-...
    out02: -23.857483619815 +/- 0.016530828020 On-...
    out04: -23.827592443012 +/- 0.018860097942 On-...
    out08: -23.844945586056 +/- 0.022192063664 On-...
```

It can be observed that the results remain consistent within the error bars regardless of the number of parallel cores used for the calculation. When interpreting **statistical estimates** obtained from Monte Carlo calculations, it is essential to pay attention to whether **the values agree within the error bars**. Even if the mean values differ slightly, as long as they overlap within the error bars, they should be regarded as "the same result." Developing proficiency in this interpretation is necessary.

As inferred from the fact that the error bars in the above results are approximately the same size,[15] the specification "vmc_nstep = 5000" means that *the total number of samples is 5000, regardless of the number of parallel cores used*. If the calculation is performed on a single core, each core accumulates 5000 configurations. For two cores, each core accumulates 2500 configurations. For four cores, each core accumulates 1250 configurations, and for eight cores, each core accumulates 625 configurations.[16]

Next, we capture the computation time as follows:

```
% grep 'real time' out*
    out01: Total CASINO real time : 3.1050
    out02: Total CASINO real time : 1.6430
    out04: Total CASINO real time : 0.8870
    out08: Total CASINO real time : 0.7700
```

[15] The magnitude of the error bars is determined by the number of statistical samples M.

[16] Therefore, it is natural to set **"vmc_nstep"** as a multiple of the number of cores used.

It can be observed that doubling the number of cores approximately halves the computation time. When increasing from four cores to eight cores, the time reduction is not exactly half. This discrepancy arises because the PC used in this case does not actually have eight physical cores but instead relies on hyper-threading for parallelization, which influences the performance.

2.3.3 Continuation Calculation and Block Length

Next, we explain the concept of **blocking** in statistical accumulation. A naive approach would be to straightforwardly accumulate 5000 samples if $M = 5000$ samples are required. However, for certain systems, accumulating 5000 samples in a single run could take several days, especially for large-scale systems. When specifying **vmc_nstep = 5000**, the program accumulates 5000 steps without outputting any intermediate files or progress reports. During this period, there is no way to verify whether the computation is proceeding correctly. Moreover, if an unexpected event such as a power outage occurs, the entire calculation must be discarded and restarted.

To address this issue, the total 5000 steps can be divided into blocks, for example, 1000 steps per block, allowing intermediate files to be output at each block. This way, even if the calculation is interrupted, it can be resumed from the intermediate files, progressing step by step. Such continuation from intermediate files is called a **resume calculation**. On large-scale supercomputers, job restrictions such as "within 24 hours per job" are common. By adjusting the number of steps per block to fit within this time limit, calculations can proceed efficiently.

Now, let's accumulate statistics again with 5000 steps:

```
% pwd
   /Users/maezono/work/02vmcOnly
% rm out0*
% clearup
% mpirun -np 8 casino &
```

Checking the result:

```
% grep 'On-the-fly' out
   out: -23.844945586056 +/- 0.022192063664 On-...
```

The error bar is approximately 0.02. To reduce this error bar by half, we need to further accumulate statistics. The error in statistical sampling decreases as $1/\sqrt{M}$ with respect to the sample size M [6]. Thus, to halve the error, four times the number of steps is required. In this case, we need a total of 20,000 steps, but since we already have 5000 steps accumulated, we can resume the calculation and add only 15,000 more steps.

To execute this resume calculation, set the input as follows. First, back up the existing input file:

```
% pwd
    /Users/maezono/work/02vmcOnly
% cp input input_org
```

Then, edit the input file as follows:

```
# RUN
...
newrun      : F     #*! Changed from T to F
...
# VMC
...
vmc_nstep  : 15000 #*! Remaining 15,000 steps
vmc_nblock : 3     #*! Block size set to 5,000
...
```

The **newrun** parameter is changed from "T" to "F," indicating that this is not a new calculation but a continuation. The **vmc_nstep** parameter is set to "15,000 remaining steps" to reach the desired total of 20,000 steps. Since the previous stage has already accumulated 5000 steps, it is necessary to maintain "5000 steps per block" to ensure statistical consistency and avoid issues (details to be explained later). Therefore, to divide the additional 15,000 steps into "blocks of 5000 steps each," the **vmc_nblock** (number of blocks) is set to 3.

Now, let's conduct the resume calculation tutorial. First, we will perform again a new calculation with 5000 steps and then continue by resuming this calculation to complete the remaining 15,000 steps. To begin, the edited input file for resuming is saved as **input_res**, and the original input file for the new calculation, **input_org**, is restored as **input**.

```
% pwd
    /Users/maezono/work/02vmcOnly
% cp input input_res
% cp input_org input
```

Next, we execute the calculation using eight parallel processes:

```
% pwd
    /Users/maezono/work/02vmcOnly
% clearup
% mpirun -np 8 casino &
```

After the calculation finishes, verify that **config.out** and **vmc.hist** have been generated. These files will be used for the next resume calculation, so be mindful of their role in the process.

Next, proceed with the following commands:

```
% pwd
    /Users/maezono/work/02vmcOnly
% cp input_res input
```

```
% cp vmc.hist vmc.hist_1st
% mv config.out config.in
% mv out out_1st
```

Here, the resume calculation input file **input_res** is replaced as **input**, and the previously generated configuration file **config.out** is renamed to **config.in** for continuation. By doing this, the final configuration of the previous step,

$$\mathbf{R}_{M=5000} = \left(\mathbf{r}_1^{(M)}, \cdots, \mathbf{r}_N^{(M)}\right)$$

is used as the initial configuration for the next step, ensuring that the calculation continues from $\mathbf{R}_{M+1}$. The file **vmc.hist** was renamed as **vmc.hist_1st** to keep a record for later comparison. Since we used the **cp** command, the original **vmc.hist** file containing the first 5000 steps remains intact. The **out** file should be renamed to **out_1st** for safekeeping. Otherwise, the next standard output will be appended to this file.

In the next stage of the calculation, since we are continuing from the previous computation, **do not execute clearup** before running the calculation.

```
% pwd
    /Users/maezono/work/02vmcOnly
% mpirun -np 8 casino &
```

By doing this, the subsequent sampling history will be appended to the previously generated **vmc.hist** file, resulting in a total history covering 20,000 steps recorded in this file.

Once the calculation is complete, compare the results using

```
% grep fly out*
out:     -23.849821905216 +/- 0.009844648148 On-the...
out_1st: -23.844945586056 +/- 0.022192063664 On-the...
```

This confirms that the final result recorded in the **out** file corresponds to an accumulation of four times the number of steps as in **out_1st**, resulting in the error bar being reduced by half. Additionally, checking the file sizes

```
% ls -lh
...
... 389K  9 16 19:56 vmc.hist
...  97K  9 16 19:56 vmc.hist_1st
```

we see that the sampling history file has increased in size by a factor of 4, confirming that it now contains results from 20,000 steps instead of just 5000 steps.

2.4 Reblocking Analysis

In Sect. 2.1, we used the **reblock** command to analyze the calculation results. This command is important for accurately computing statistical errors. In this section, we will explain the reasoning behind this need for "careful analysis."

In practical QMC calculations,

$$\bar{f} = \langle f_j \rangle = \frac{1}{M}\sum_{j=1}^{M} f_j \quad , \quad \sigma_{\bar{f}}^2 = \left\langle \left(f_j - \bar{f}\right)^2 \right\rangle = \frac{1}{M}\sum_{j=1}^{M} \left(f_j - \bar{f}\right)^2 \tag{2.23}$$

the mean value $\bar{f}$ and variance $\sigma_{\bar{f}}^2$, computed as above, are used to determine the consistency of results.[17]

For example, consider performing statistical processing with M =1000 samples. Suppose we divide these 1000 samples into subgroups of "sample size $m = 10$" and take the mean value within each subgroup first. For this case, the number of subgroups becomes $N = M/m = 100$. Intuitively, it should be understandable that "the mean value calculated using individual samples from $M = 1000$ samples" and "the mean value calculated using the $N = 100$ mean values from the subgroups" are identical. Expressing this mathematically, setting $M = N \cdot m$,

$$\bar{f} = \frac{1}{N \cdot m}\sum_{j=1}^{N \cdot m} f_j = \frac{1}{N}\left(\frac{1}{m}\sum_{j=1}^{m} f_j + \frac{1}{m}\sum_{j=m+1}^{2m} f_j + \cdots\right) = \frac{1}{N}\sum_{k=1}^{N} \bar{f}_k \tag{2.24}$$

Here,

$$\bar{f}_k = \frac{1}{m}\sum_{j=(k-1)m+1}^{km} f_j \tag{2.25}$$

is the "mean value within a subgroup." Taking N such subgroup means, the mean value obtained is

$$\bar{f}_{(N)} = \left\langle \bar{f}_k \right\rangle$$

Since this corresponds to the rightmost term of (2.24),

$$\bar{f} = \bar{f}_{(N)}$$

Thus, we conclude that "the mean value remains unchanged even when the data is block-averaged into subgroups."

[17] If $f = \bar{f} \pm \sqrt{\sigma_{\bar{f}}^2}$ overlaps within the "error bar range," it is considered to be the "same value."

The reason we obtained the result that "the mean value remains unchanged even when block-averaged in advance" is that the target mean value

$$\bar{g} = \langle g(x) \rangle$$

possesses the linearity property

$$g(x_1 + x_2) = g(x_1) + g(x_2).$$

The variance value

$$\sigma_f^2 = \left\langle \left(f_j - \bar{f}\right)^2 \right\rangle$$

does not follow the same reasoning as the mean value, since $g(x)$ is a quadratic function and thus nonlinear. As a result, "pre-blocking alters the variance value." In mathematical form, the variance of the block-averaged values $\left\{\bar{f_k}\right\}$ from Eq. (2.25) is given by

$$\sigma_{(N)}^2 = \frac{1}{M} \sum_{k=1}^{N} \left(\bar{f_k} - \bar{f}\right)^2.$$

This does not match the original variance σ_f^2 before blocking, leading to

$$\sigma_f^2 \neq \sigma_{(N)}^2.$$

In practice, when the variance is computed while varying the reblocking size m, a typical reblocking size dependence, as shown in Fig. 2.4, emerges. As the reblocking size increases, the variance increases, reaches a flat plateau region, and then starts oscillating erratically.

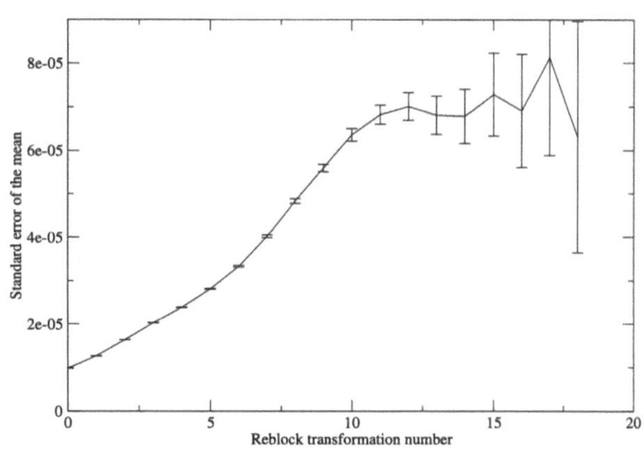

Fig. 2.4 The variance value changes depending on the block processing size. The vertical axis represents the variance value, while the horizontal axis represents the reblocking size. Error bars indicate the "variance of the variance value." (The figure is reproduced from the CASINO manual [7])

When observing the dependence shown in Fig. 2.4, which reblocking size's variance value should we trust and adopt? First, we do not adopt variance values that exhibit oscillatory behavior for large reblocking sizes. The oscillatory fluctuations occur because increasing the reblocking size reduces the number of block samples N, making the variance value increasingly uncertain.[18] The variance values in such an uncertain region lack statistical reliability. Indeed, the "error bars of the error bars" depicted in Fig. 2.4 also become significantly large.

Now, what about the region with small reblocking sizes? Between the variance value at reblocking size $m = 1$ and that at $m = 2$, which one is more reliable? The answer is that the variance value at $m = 2$ is more reliable. The reason is that "correlation between sample points is more effectively removed": The variance evaluation formula in Eq. (2.23) operates correctly under the assumption that each sampling point f_j is independent of the others. However, in Monte Carlo sampling, **pseudo-random numbers** are used for sampling. Additionally, as discussed in Sect. 2.2.2, a **Markov process**-based implementation is employed. In these cases, the mechanism inherently determines the next sampling point based on the current one, inevitably introducing **correlations between consecutive sampling points**. In the case of a reblocking size $m = 1$, the evaluation is directly affected by this correlation. However, when reblocking with $m = 2$, the representative points after reblocking are spaced further apart, reducing the correlation. Thus, the variance evaluation at $m = 2$, where the correlation is weaker, is considered to capture a more accurate value.

With $m = 4$, the correlation weakens even further. As the reblocking size increases, the computed variance is expected to converge toward the true value. On the other hand, if the reblocking size becomes too large, the sample size decreases, leading to increased uncertainty in the variance, as mentioned earlier. As a result, as shown in Fig. 2.4, the variance exhibits a "convergence to the true value" in a flat plateau region, followed by oscillatory fluctuations. Therefore, the variance captured in the plateau region is adopted as the "more accurate variance." The **reblock** utility allows users to observe this behavior and specify the "appropriate reblocking size" for estimation.

When executing **reblock** in the working directory containing the hist file, the following information is displayed after selecting the display unit system:

```
REBLOCKING ANALYSIS OF TOTAL-ENERGY DATA
 ========================================
 (Energy data in units of au/prim cell.)
  RTN   Blk leng   Std error in mean       Error in std error
    0          1   1.672595274850299E-02   4.734603042829402E-04
```

[18] Variance, that is, the degree of data dispersion, has meaning only when the total number of data points is sufficiently large. If there are only two data points, the concept of "degree of dispersion" loses significance. Ultimately, when there is only one data point, "dispersion" has no meaning. Formally evaluating this, the variance for a single data point becomes infinite, leading to the result that "anything can be said about the degree of dispersion" (any value can be considered as dispersion without contradiction).

```
1         2   1.681163409763426E-02   6.735438976599983E-04
2         4   1.703165591318651E-02   9.665538883823915E-04
3         8   1.641518809575703E-02   1.321701178278133E-03
4        16   1.550597190679065E-02   1.777196052730665E-03   *** BEST ***
5        32   1.678223943729000E-02   2.756655893926293E-03
6        64   1.561307666122193E-02   3.728911280979427E-03
7       128   1.527116632710172E-02   5.480043425227117E-03
8       256   1.216568095271472E-02   7.165198471189222E-03
```

The "Blk leng" column represents the reblocking size m that appeared in Eq. (2.25), and the corresponding standard error (the square root of the variance) is given in the "Std error in mean" column. The "*** BEST ***" label indicates the reblocking size determined by the program package as the optimal choice. Since this decision is made using a simple algorithm, users should examine the data and determine the most appropriate reblocking size themselves. Specifically, by observing the "Std error in mean" values from top to bottom, one can identify where the values initially increase and then fluctuate, helping to determine the best "Blk leng" value to adopt. While the program suggests "16" as the best reblocking size, the author would prefer selecting a slightly smaller value, such as "(Blk leng) = 8".[19] By entering this reblocking length, the final result, as obtained in Sect. 2.1, is displayed.

References

1. "Ab initio Calculation Tutorial: For Materials Analysis, Informatics and Design", (Springer Nature, 2023), R. Maezono, ISBN: 978-98-19909-21-6
2. "Computational Physics: Problem Solving with Python" (Wiley-VCH, 2015) R.H. Landau, M.J. Paez, C.C. Bordeianu ISBN: 978-35-27413-15-7
3. "The Elements of Statistical Learning: Data Mining, Inference, and Prediction, Second Edition" (Springer, 2009), T. Hastie, R. Tibshirani, J. Friedman, ISBN: 978-03-87848-57-0
4. "Random number generators tested on quantum Monte Carlo simulations", K. Hongo, R. Maezono, K. Miura, J. Comput. Chem. 31, 2186 (2010), https://doi.org/10.1002/jcc.21509
5. "Quantum Monte Carlo method for electronic structure calculations", K. Hongo, R. Maezono, Nucl. Sci. Tech. 2, 51 (2011).
6. "An Introduction to Computer Simulation Methods: Applications to Physical Systems, Third Edition" (Amazon Digital Services, 2017), H. Gould, J. Tobochnik, W. Christian, ISBN: 978-19-74427-47-5
7. "Continuum variational and diffusion quantum Monte Carlo calculations", R.J. Needs, M.D. Towler, N.D. Drummond, P.L. Rios, J. Phys.: Condens. Matter 22, 023201 (2010), https://doi.org/10.1088/0953-8984/22/2/023201

[19] It is often safer to overestimate the standard error rather than underestimating conservatively to avoid overstating precision.

Chapter 3
Diffusion Monte Carlo Method

The quantum diffusion Monte Carlo (**DMC**) method, which we have metaphorically described as "baking while achieving ultra-maturation to produce high-quality bread," implements a mechanism whereby statistical accumulation is performed while the sampled trial function gradually approaches the exact solution. In this chapter, we will explain the principle behind this process of "converging toward the exact solution" (Sect. 3.1). Next, we will explain how this principle is implemented on a computer (Sect. 3.2). With an understanding of these principles, we will conduct practical DMC calculations to comprehend how and at which computational stages "maturation" progresses (Sects. 3.4, 3.3).

3.1 Projection Method Using Imaginary-Time Evolution

3.1.1 Time Evolution of the Wave Function

Let us recall the origin of the many-body Schrodinger equation in (1.1):

$$\hat{H} \cdot \Psi(\mathbf{R}) = E \cdot \Psi(\mathbf{R}) \tag{3.1}$$

By applying the optical-mechanical analogy to the Hamilton-Jacobi equation obtained from classical mechanics, the time-dependent Schrodinger equation was derived as the fundamental equation of wave mechanics [3] :[1]

[1] In the context of this book, $\mathbf{R}$ is a $3N$-dimensional vector, but at the time when quantum mechanics was constructed using this logic, $\mathbf{R}$ was considered as a three-dimensional vector, and discussions were developed in the form of a single-particle problem.

R. Maezono, *Ab initio Quantum Monte Carlo Tutorial*,
https://doi.org/10.1007/978-981-95-4134-8_3

$$i\frac{\partial \Phi(\mathbf{R},t)}{\partial t} = \hat{H}\cdot\Phi(\mathbf{R},t) \tag{3.2}$$

By separating the variables in space and time as

$$\Phi(\mathbf{R},t) = \Psi(\mathbf{R})\cdot\chi(t)$$

we obtain

$$\frac{i\partial_t\chi(t)}{\chi(t)} = \frac{\hat{H}\Psi(\mathbf{R})}{\Psi(\mathbf{R})} = E.$$

Since the left-hand side and the right-hand side have different dependent variables, if this equation holds identically, both sides must be constant. We denote this constant as E. This separation constant E has the same dimension as the Hamiltonian $\hat{H}$, which corresponds to the dimension of energy, and thus it is represented as E.

Through this variable separation, the equation is separated into

$$\hat{H}\Psi(\mathbf{R}) = E\cdot\Psi(\mathbf{R}),$$
$$\chi(t) = C\cdot\exp[-iEt].$$

The partial differential eigenvalue problem concerning the spatial part corresponds to the many-body Schrodinger equation presented at the beginning of this book, namely, Eq. (1.1).

Solving the partial differential eigenvalue problem for the spatial part, given by Eq. (3.1), we obtain a set of eigenvalues E and eigenfunctions Ψ:

$$(E_0,\Psi_0),(E_1,\Psi_1),\cdots,(E_k,\Psi_k),\cdots.$$

Thus, for the k-th eigenvalue level (E_k,Ψ_k),

$$\Psi_k(\mathbf{R})\cdot\exp[-iE_kt]$$

is a solution of Eq. (3.2). Due to the linearity of the equation, the superposition over all k,

$$\Phi(\mathbf{R},t) = \lambda_0\cdot\Psi_0(\mathbf{R})\cdot\exp[-iE_0t]+\lambda_1\cdot\Psi_1(\mathbf{R})\cdot\exp[-iE_1t]+\cdots \tag{3.3}$$

serves as the general solution.

3.1.2 Concept of Utilizing Imaginary-Time Evolution of the Wave Function

We consider an arbitrary many-body trial wave function $\tilde{\Psi}_\alpha^{\text{trial}}(\mathbf{R})$ and set it as the initial condition $\Phi(\mathbf{R},0) = \tilde{\Psi}_\alpha^{\text{trial}}(\mathbf{R})$.[2] Using the time-dependent Schrodinger equation,

$$i\frac{\partial \Phi(\mathbf{R},t)}{\partial t} = \hat{H}\cdot\Phi(\mathbf{R},t) \quad , \quad \Phi(\mathbf{R},0) = \tilde{\Psi}_\alpha^{\text{trial}}(\mathbf{R}) \tag{3.4}$$

this function evolves in real time. The solution takes the general form of Eq. (3.3),

$$\Phi(\mathbf{R},t) = \Psi_0(\mathbf{R})\cdot\exp[-iE_0 t] + \lambda_1\cdot\Psi_1(\mathbf{R})\cdot\exp[-iE_1 t] + \cdots$$

as it evolves in time (with normalization $\lambda_0 = 1$). Here, it is important to note that the initial trial function $\tilde{\Psi}_\alpha^{\text{trial}}(\mathbf{R})$ is an **approximate** function. However, the time-evolving function $\Phi(\mathbf{R},t)$, which takes this as an initial condition, already contains, as one of its components, the **exact ground-state wave function** $\Psi_0(\mathbf{R})$. The exact solution $\Psi_0(\mathbf{R})$ that we seek is "already present" within the function. The reason why the trial approximate function fails to accurately capture the exact solution is that it is contaminated by various excited state components $\{\Psi_{j\neq 0}(\mathbf{R})\}$. If we regard these excited state components as unwanted noise and successfully eliminate them, we can extract only the exact solution $\Psi_0(\mathbf{R})$.

Next, we consider how to extract only the exact ground-state component $\Psi_0(\mathbf{R})$. The key idea lies in the difference in time dependence between the ground-state component and the excited-state components. Since $E_0 < E_1 < E_2 < \cdots$, the ground-state component that we seek exhibits the slowest time oscillation, while the excited-state components, which act as unwanted noise, oscillate at higher frequencies. By exploiting this difference in frequency and filtering out only the fast oscillating components, we can **extract only the exact ground-state solution**.

The goal of "removing only the high-frequency components" can be achieved by transforming the real-time variable t in the oscillation $\exp\left[-iE_j t\right]$ into imaginary time as $it \to \tau$. With this transformation, the oscillation $\exp\left[-iE_j t\right]$ is converted into decay $\exp\left[-E_j\tau\right]$, causing the excited-state components with larger energy E_j to decay and vanish more quickly. Reorganizing this idea, we can express it as follows. The real-time dynamics given by Eq. (3.4) is transformed by the change of variable $it \to \tau$ into

$$-\frac{\partial \Phi(\mathbf{R},\tau)}{\partial \tau} = \hat{H}\cdot\Phi(\mathbf{R},\tau)\,, \tag{3.5}$$

[2] We denote the trial function $\Psi_\alpha(\mathbf{R})$ from Eq. (1.4) as $\tilde{\Psi}_\alpha^{\text{trial}}(\mathbf{R})$. As explained in Eq. (1.4), the index α represents a set of variational parameters. This notation is adopted to avoid confusion with $\Psi_k(\mathbf{R})$ in Eq. (3.3). The function Ψ_k is an exact eigenfunction, and k is an integer labeling the eigenfunction.

where the imaginary-time dynamics of $\Phi(\mathbf{R}, \tau)$ is governed by

$$\Phi(\mathbf{R}, \tau) = \Psi_0(\mathbf{R}) \cdot \exp[-E_0\tau] + \lambda_1 \cdot \Psi_1(\mathbf{R}) \cdot \exp[-E_1\tau] + \cdots .$$

Here, the term containing Ψ_0 exhibits the slowest decay, while the undesired excited-state components Ψ_1 and beyond decay first and disappear.

However, in this form, even the desired ground-state component Ψ_0 would eventually vanish. To prevent this, we introduce an energy offset E_T and modify Eq. (3.5) as

$$-\frac{\partial \Phi(\mathbf{R}, \tau)}{\partial \tau} = \left(\hat{H} - E_T\right) \cdot \Phi(\mathbf{R}, \tau) \tag{3.6}$$

With this modification, the corresponding time evolution results in a shift of each energy E_j by E_T, leading to

$$\begin{aligned}\Phi(\mathbf{R}, \tau) = \Psi_0(\mathbf{R}) \cdot \exp[-(E_0 - E_T)\tau] + \lambda_1 \cdot \Psi_1(\mathbf{R}) \cdot \exp[-(E_1 - E_T)\tau] \\ +\lambda_2 \cdot \Psi_2(\mathbf{R}) \cdot \exp[-(E_2 - E_T)\tau] + \cdots\end{aligned} \tag{3.7}$$

By doing this, if we can adjust E_T to match E_0, then the desired ground-state component Ψ_0 alone will remain as a constant term under imaginary-time evolution, while all other excited-state components will decay and vanish due to $\left(E_j - E_T\right) \geq 0$ for $j \geq 1$. Consequently, we obtain

$$\Phi(\mathbf{R}, \tau \to \infty) = \Psi_0(\mathbf{R}) . \tag{3.8}$$

This shows that the desired exact solution component $\Psi_0(\mathbf{R})$ can be successfully extracted. This handling is what we have been metaphorically referring to as "further maturing while baking."

The Magic of Transforming Real Time into Imaginary Time

The transformation from real-time oscillations to imaginary-time decay is a dramatic change, but formally, it is merely a variable transformation. This corresponds to considering not only the real axis (horizontal axis) but also the imaginary axis (vertical axis) in the complex plane. It demonstrates the power of analytic continuation [4]. In many-body perturbation theory, the technique of "transforming a two-time Green's function into a thermal Green's function through analytic continuation, making computations dramatically more feasible" (Sect. 17.2.7) can also be attributed to this kind of magic.

As described above, if the distribution that has converged to $\Psi_0(\mathbf{R})$ is used for the variational functional in Eq. (2.21),

$$E(\boldsymbol{\alpha}) \approx \frac{1}{M}\sum_{j=1}^{M}\left(\tilde{\Psi}_{\alpha}^{\text{trial}}(\mathbf{R}_j)\right)^{-1} \cdot \hat{H}\tilde{\Psi}_{\alpha}^{\text{trial}}(\mathbf{R}_j) = \frac{1}{M}\sum_{j=1}^{M} E_L(\mathbf{R}_j)$$

$$\{\mathbf{R}_j\} \sim P(\mathbf{R}) = \frac{\tilde{\Psi}_{\alpha}^{\text{trial}*}(\mathbf{R})\,\tilde{\Psi}_{\alpha}^{\text{trial}}(\mathbf{R})}{\int d\mathbf{R}\cdot\tilde{\Psi}_{\alpha}^{\text{trial}*}(\mathbf{R})\cdot\tilde{\Psi}_{\alpha}^{\text{trial}}(\mathbf{R})} \tag{3.9}$$

then the computed value of $E(\boldsymbol{\alpha})$ will match the exact eigenvalue corresponding to the exact solution, that is, the exact ground-state energy value.

The above is an outline of the DMC method, which has been metaphorically described as "further refining the dough while baking." In the DMC method, a calculation is performed to drive the imaginary-time evolution given by Eq. (3.6). During this process, the initially prepared wave function $\Phi(\mathbf{R}, 0) = \tilde{\Psi}_{\alpha}^{\text{trial}}(\mathbf{R})$ gradually evolves into $\Phi(\mathbf{R}, \tau) \to \Psi_0(\mathbf{R})$. This corresponds to the metaphor of the "dough" $\Phi(\mathbf{R}, \tau)$ undergoing "further refinement," approaching the exact solution. In practical applications, the initial trial function $\Phi(\mathbf{R}, 0) = \tilde{\Psi}_{\alpha}^{\text{trial}}(\mathbf{R})$ is typically chosen from a function that has already undergone some degree of refinement.[3] This corresponds to the idea of "further refining the dough" to bring it closer to the exact solution after it has already been partially matured before baking. Note that the black line in Fig. 2.2 represents the evolution of the local energy values $E_L(\mathbf{R}_j)$ as they change during the "super-refinement" process. Meanwhile, the dark gray plot shows the evolution of the offset energy E_T introduced in (3.6).[4] At this stage, note that two energy estimates, E_L (local energy) and E_T (offset energy), have been introduced.

Readers may have already noticed that the above framework works only under the assumption that "E_T can be adjusted to E_0." However, the exact ground-state energy E_0 is unknown, and the purpose of the calculation is to determine it. So, one might question how it is possible to adjust E_T to E_0. A clever strategy, which will be discussed later (Sect. 3.2.5), makes this adjustment feasible, and this mechanism forms the core of the DMC method.

How to Introduce Diffusion Monte Carlo Method

There are various ways to introduce diffusion Monte Carlo (DMC). In addition to the approach taken in this chapter, which starts from the real-time oscillation of many-body wave functions, there are also more sophisticated explanations using Green's functions, projection operators, and path integrals

(continued)

[3] The procedure for this will be described later in (Sect. 4). For now, we proceed with the assumption that the instructor has prepared the already refined dough using `correlation.data`.

[4] On a PC display with color, this plot is shown in red. Additionally, Fig. 2.2 includes another light gray, hollow plot (displayed in green on a PC), which represents a different energy estimate appearing in the detailed implementation of the DMC method [2]. For now, you can proceed without concern for this value.

(Sect. 7.3.2). However, when explaining why this method converges to the exact solution in a way that is accessible to beginners, starting with such advanced mathematical tools tends to lengthen the discussion unnecessarily. That said, for understanding and discussing advanced improvements such as the "T-move scheme" (Sect. 8.3.4) and "lattice-regularized Monte Carlo methods" (Sect. 8.2.3), it is essential to eventually refine the elementary understanding presented here by upgrading to a more rigorous comprehension. The theoretical foundations covered in Sect. 7 provide a gateway to such an upgrade (Sect. 7.3.2).

3.2 Implementation of Imaginary-Time Evolution

We have explained that by driving the imaginary-time evolution of Eq. (3.6), one can obtain the sampling distribution $\Phi(\mathbf{R}, \tau)$, which gradually approaches the exact solution. The next question to be addressed is "How can the imaginary-time evolution described by the differential equation (Eq. (3.6)) be implemented on a computer?" Some readers might speculate, "Since this involves solving a differential equation numerically, perhaps it is evaluated by discretization." However, rather than using discretization, we take the approach of viewing Eq. (3.6) as a **diffusion equation**, allowing for an equivalent implementation via a **random walk** simulation of diffusing particles. By understanding this implementation method, we can also address the question raised in the previous section: "How can the reference energy E_T be adjusted to approach the exact ground-state energy E_0?"

3.2.1 *Diffusion Equation*

We will discuss the **equivalence between the diffusion equation and random walks** in more detail in the following sections. However, to ensure that we do not lose track amidst long mathematical derivations, this section provides an overview of what we aim to explain.

Let us first review the diffusion equation. Consider the number of particles in a closed region enclosed by a surface S, denoted as $N(t)$. The conservation of particle number in terms of inflow and outflow is expressed as

$$\frac{\partial N(t)}{\partial t} = -\int_S d\mathbf{S} \cdot \mathbf{j}(\mathbf{r}, t)$$

Here, $\mathbf{j}(\mathbf{r}, t)$ represents the particle flux, and the surface integral above accounts for the total inflow across the surface S.[5] This equation expresses the balance between the inflow of particles and the increase in particle count within the closed region.

The number of particles $N(t)$ can be expressed as the volume integral of the particle density $\rho(\mathbf{r}, t)$:

$$N(t) = \int_V dV \cdot \rho(\mathbf{r}, t).$$

Substituting this into the previous equation and converting the surface integral into a volume integral using Gauss's theorem [5], we obtain

$$\begin{aligned}\frac{\partial}{\partial t}\int_V dV \cdot \rho(\mathbf{r}) &= -\int_S d\mathbf{S} \cdot \mathbf{j}(\mathbf{r}) \\ &= -\int_V dV \cdot [\nabla \cdot \mathbf{j}(\mathbf{r})].\end{aligned}$$

By focusing on the integrand of the volume integral, we derive the **local description for particle conservation**:[6]

$$\frac{\partial \rho(\mathbf{r}, t)}{\partial t} + \nabla \cdot \mathbf{j}(\mathbf{r}, t) = 0.$$

When no external manipulation is applied, the particles naturally flow from regions of high concentration to regions of low concentration, moving toward equilibrium. Such a flow can be expressed using the negative gradient of the density $\nabla\rho(\mathbf{r}, t)$:[7]

$$\mathbf{j}(\mathbf{r}, t) = -D \cdot \nabla\rho(\mathbf{r}, t).$$

Substituting this **gradient flow** into the above conservation law, we obtain:[8]

$$\frac{\partial \rho(\mathbf{r}, t)}{\partial t} = D \cdot \nabla^2\rho(\mathbf{r}, t). \tag{3.10}$$

[5] The negative sign indicates inflow direction. Since $d\mathbf{S}$ is an outward-facing vector, the integral would otherwise represent outward flux without this sign.

[6] This describes a quantitative relation that holds locally at each point.

[7] The negative sign indicates flow from high to low concentration. A gradient is positive when moving from lower to higher values.

[8] At the macroscopic level, it is customary to derive the diffusion equation in a *simplified manner* by seemingly assuming the gradient flow relation (Fick's law) just as a prerequisite, shown here. However, as will be discussed in the microscopic derivation (Sect. 7.1), the gradient flow is not a prerequisite for the derivation. It is not that "the diffusion equation is derived because there is a gradient flow" but rather that the diffusion equation is derived as a consequence of random particle motion, and the gradient flow emerges as a result of it.

This equation governs the time evolution of the density distribution $\rho(\mathbf{r}, t)$ of a diffusing particle system moving from high to low concentration toward equilibrium. It is known as the **diffusion equation**.

By the way, the density distribution $\rho(\mathbf{r}, t)$ represents a continuous formulation of a histogram that describes the spatial distribution of individual particle positions $\{\mathbf{r}_j\}$.[9] When particles move toward equilibrium driven by the diffusion, each particle undergoes **Brownian motion**, continuously colliding with others. Brownian motion is driven by random forces, leading to position updates described as a **random walk**:

$$\mathbf{r}_j(t + \Delta t) = \mathbf{r}_j(t) + \left[\text{random positional fluctuation; } \Delta\mathbf{r}_j\right]$$

As will be discussed in detail in the following sections, this "random positional update" follows a Gaussian distribution:

$$\mathbf{r}_j(t + \Delta t) = \mathbf{r}_j(t) + \boldsymbol{\chi}_{\text{Gauss}}^{(3)}[0, 2D\Delta t], \tag{3.11}$$

where $\boldsymbol{\chi}_{\text{Gauss}}^{(N)}\left[\mu, \sigma^2\right]$ represents an N-dimensional vector where each component is a random variable following a Gaussian distribution with mean μ and variance σ^2. Here, note that the coefficient D appearing in the diffusion equation (3.10) is related to the variance of the Gaussian distribution in (3.11), that is, the "step size of the random walk." This coefficient D is referred to as the **Einstein coefficient** due to its contribution to the theoretical framework of Brownian motion.

Summarizing the above results, the time evolution of the density distribution $\rho(\mathbf{r}, t)$ driven by the diffusion equation

$$\frac{\partial \rho(\mathbf{r}, t)}{\partial t} = D \cdot \nabla^2 \rho(\mathbf{r}, t)$$

corresponds to the time evolution of the "histogram of particle positions," where the positions are updated via the random walk

$$\mathbf{r}_j(t + \Delta t) = \mathbf{r}_j(t) + \boldsymbol{\chi}_{\text{Gauss}}^{(3)}[0, 2D\Delta t].$$

This equivalence allows us to interpret the time evolution of the distribution function driven by the diffusion equation (3.10) not by directly solving a partial differential equation but instead by performing a random walk simulation using (3.11) and constructing histograms at successive time steps. In other words, as illustrated in Fig. 3.1, the key insight is to regard the time evolution of the density distribution $\rho(\mathbf{r}, t)$ as the time evolution of a histogram and reinterpret it in terms of the time evolution of the particle positions that compose the histogram.

---

[9] A description based on individual particle positions is called the **Lagrangian description**, whereas a field-based description using the density $\rho(\mathbf{x}, t)$ is known as the **Eulerian description**.

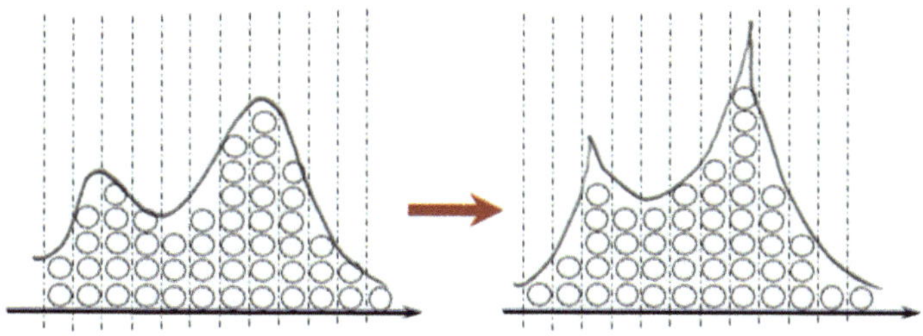

Fig. 3.1 The image of the many-body wave function $\Phi(\mathbf{R}, \tau \to \infty)$ being "refined" through DMC calculation, gradually approaching the exact $\Psi_0(\mathbf{R})$. The meaning of the cusp (sharp point) depicted on the side of $\Psi_0(\mathbf{R})$ is explained in detail in Sect. 8.1. As illustrated, the idea of interpreting the many-body wave function as a "histogram of particles" is discussed in Sect. 3.2.1

Let us recall our mission in ab initio QMC. Our goal is to determine the time evolution of the distribution $\Phi(\mathbf{R}, \tau)$ governed by the imaginary-time Schrodinger equation (3.6):

$$\frac{\partial \Phi(\mathbf{R}, \tau)}{\partial \tau} = \left(\frac{1}{2}\sum_{j=1}^{N} \nabla_j^2 + E_T - V(\mathbf{R})\right) \Phi(\mathbf{R}, \tau).$$

Ignoring the term $[E_T - V(\mathbf{R})]$, this equation takes almost the same form as the diffusion equation (3.10), effectively describing diffusion in a $3N$-dimensional space. Therefore, instead of solving the partial differential equation directly, we can equivalently determine the time evolution of $\Phi(\mathbf{R}, \tau)$ by performing a "random walk simulation in $3N$-dimensional space," analogous to (3.11):

$$\mathbf{R}_j(\tau + \Delta\tau) = \mathbf{R}_j(\tau) + \chi_{\mathrm{Gauss}}^{(3N)}[0, 2D\Delta\tau]. \tag{3.12}$$

By iteratively constructing histograms of particle positions at successive time steps, we can obtain the evolution of $\Phi(\mathbf{R}, \tau)$ in an equivalent manner (Fig. 3.2).

In this manner, the idea discussed in the previous section that implementing imaginary-time evolution allows us to approach the exact solution is realized through a random walk simulation in diffusion Monte Carlo (DMC). The term "diffusion" originates from the analogy to the diffusion equation, as described here.

In this subsection, we discussed the correspondence between the diffusion equation and random walks. The diffusion equation is more generally treated as the **Fokker-Planck equation**, while the corresponding "time evolution equation of random walks" is known as the **Langevin equation**. In the following sections, we will explain the generalized **equivalence between the Fokker-Planck equation and the Langevin equation**. The mathematical derivations will be lengthy, but their fundamental outline has already been presented in this subsection.

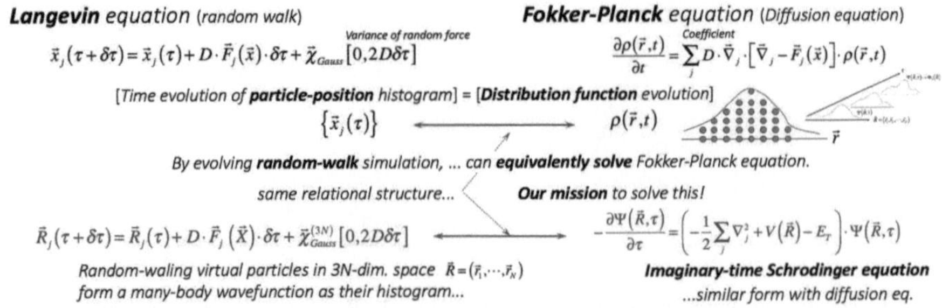

Fig. 3.2 The imaginary-time Schrodinger equation can be regarded as a "higher-dimensional version (3N-dimensional)" of the three-dimensional Fokker-Planck equation. In the case of the three-dimensional Fokker-Planck equation, there was a corresponding "particle random walk (in three-dimensional space)" whose histogram followed the Fokker-Planck equation. Therefore, for the higher-dimensional Fokker-Planck equation, which corresponds to the imaginary-time Schrodinger equation, we can consider a corresponding "particle random walk (in 3N-sional space) whose histogram follows the imaginary-time evolution of $\Psi(\mathbf{R}, \tau)$"

3.2.2 Random Walk and Diffusion Equation

Let us denote the position of the j-th particle as $\mathbf{x}_j(t)$. Then, we consider a system where the particle position evolves over time according to

$$\mathbf{x}_j(t+\Delta t) = \mathbf{x}_j(t) + D\cdot\mathbf{F}(\mathbf{x})\cdot\Delta t + \chi_{\text{Gauss}}^{(3)}[0, 2D\Delta t]\,. \tag{3.13}$$

$\chi_{\text{Gauss}}^{(N)}\left[\mu, \sigma^2\right]$ is, as mentioned in the previous section, an N-dimensional vector of random variables generated from a Gaussian distribution with mean μ and variance σ^2. $\mathbf{F}(\mathbf{x})$ represents the "flow experienced by a particle at position $\mathbf{x}$." By rewriting the above equation, we obtain $\Delta\mathbf{x}_j/\Delta t = D\cdot\mathbf{F}(\mathbf{x}) + \cdots$, indicating that $\mathbf{F}(\mathbf{x})$ serves as a "velocity orientation" applied to each particle at a given location. This type of flow is referred to as **drift flow**. The third term, $\chi_{\text{Gauss}}^{(3)}[0, 2D\Delta t]$, represents a random value generated by Gaussian noise, resulting in the random updating of particle positions, forming a **random walk**. This description of particle motion under the influence of random forces is known as the **Langevin equation**.

Representing the positions $\left\{\mathbf{x}_j(t)\right\}$ of N_w particles[10] using a **distribution function** $\rho(\mathbf{x})$. That is, the value of $\rho(\mathbf{x})$ expresses the density of particles at position $\mathbf{x}$, indicating regions with high or low concentration. Suppose each particle updates its position according to the random walk equation (3.13). Then, the overall distribution of particles also changes over time, leading to a time-dependent distribution function $\rho(\mathbf{x}, t)$.

The fact that "each particle position $\left\{\mathbf{x}_j(t)\right\}$ changes from moment to moment" and that "the particle density distribution function $\rho(\mathbf{x}, t)$ varies over time"

[10] The subscript w originates from "Walker." Since these particles undergo random "walks," they are referred to as "Walkers."

represents **two different ways of describing the same situation.**[11] Given this, we pose the question "How does the time evolution of the distribution function that describes position updates governed by the Langevin equation (3.13) behave?" The derivation is lengthy, so we defer it to the theoretical explanation Sect. 7 after gaining proficiency in the operation of the DMC method. However, the conclusion is as follows:

$$\frac{\partial \rho (\mathbf{x}, t)}{\partial t} = D \cdot \nabla \cdot [\nabla - \mathbf{F}(\mathbf{x})] \cdot \rho (\mathbf{x}, t) \tag{3.14}$$

This equation, which describes the time evolution of the distribution function, is called the **Fokker-Planck equation**. One key point to note is the quantitative correspondence: "The constant D appearing in (3.13) corresponds to the coefficient D in (3.14)."

When the drift flow $\mathbf{F} = 0$ in Eq. (3.14), it reduces to

$$\frac{\partial \rho (\mathbf{x}, t)}{\partial t} = D \cdot \nabla^2 \rho (\mathbf{x}, t)$$

which corresponds to the diffusion equation discussed in the previous section. Note that the "random force acting on the particle position" $\chi^{(3)}_{\text{Gauss}} [0, 2D\Delta t]$ in the Langevin equation (3.13) has been reinterpreted as the diffusion phenomenon.

The correspondence between the Langevin equation and the Fokker-Planck equation described above can be interpreted as follows: Suppose we are given the mission to "evolve the distribution function $\rho (\mathbf{x}, t)$ over time based on the Fokker-Planck equation (3.14)." Instead of solving the differential equation directly, we can perform a diffusion simulation of particles using the corresponding Langevin equation (3.13), construct the distribution histogram step by step, and thereby achieve the same time evolution prescribed by the Fokker-Planck equation **equivalently**. We apply this idea to our mission of "evolving the wave function in time using the imaginary-time many-body Schrodinger equation."

3.2.3 Implementing Imaginary-Time Evolution by Random Walk

Our mission is to drive the imaginary-time many-body Schrodinger equation (3.6). Substituting the Hamiltonian (2.18) into this equation, we obtain

$$\frac{\partial \Phi (\mathbf{R}, \tau)}{\partial \tau} = \left(\frac{1}{2} \sum_{j=1}^{N} \nabla_j^2 + E_T - V(\mathbf{R}) \right) \Phi (\mathbf{R}, \tau) \tag{3.15}$$

[11] This corresponds to the Lagrangian and Eulerian perspectives mentioned in the previous footnote.

Our goal is to implement the imaginary-time evolution of this equation.

Applying the "equivalent correspondence" discussed in the previous section, the imaginary-time evolution of $\Phi(\mathbf{R}, \tau)$ in the above equation can be equivalently obtained as the time evolution of the histogram distribution function of particles undergoing position updates via a random walk:

$$\mathbf{R}_j(\tau + \Delta\tau) \approx \mathbf{R}_j(\tau) + \chi^{(3N)}_{\text{Gauss}}[0, \Delta\tau] \tag{3.16}$$

Here, the coefficient is set as $D = 1/2$. However, since $\mathbf{R} = (\mathbf{r}_1, \cdots, \mathbf{r}_N)$, the diffusing particles (**walkers**) exist as single points in a $3N$-dimensional space. Accordingly, the random force $\chi^{(3N)}_{\text{Gauss}}[0, \Delta\tau]$ also has a dimensionality of $3N$. The reason why " $\approx$" is used instead of "=" in Eq. (3.16) is that, for now, we are ignoring the term "$E_T - V(\mathbf{R})$" in Eq. (3.15). Without this term, based on the discussion in the previous section, it should be clear that the " $\approx$" in Eq. (3.16) would turn into an "=."

The treatment of the term "$E_T - V(\mathbf{R})$" requires further preparation and will be discussed later in Sect. 3.2.5. Note that in Eq. (3.15), the first term involving ∇^2 represents particle diffusion as described above, while the second term, "$E_T - V(\mathbf{R})$," corresponds to the right-hand side of a first-order reaction equation in chemical reaction kinetics. Therefore, these terms are sometimes referred to as the **diffusion term** and the **reaction term** (or the rate term, as it corresponds to the reaction rate).

First Encounter with the Sign Problem

The discussion here, which leverages the formal analogy between the imaginary-time many-body Schrodinger equation and the Fokker-Planck equation, hinges on the assumption that "$\Psi_\alpha(\mathbf{R}, \tau)$ can be regarded as a probability distribution". However, the wave function describing electrons, which are fermions, can take both positive and negative values and can have real and imaginary components. A careful reader might notice the issue: "Is it not impossible to regard it as a probability distribution since it lacks positivity?" This point will be discussed in Sect. 8.2.

3.2.4 Sampling by Random Walk

Instead of "solving" the imaginary-time many-body Schrodinger equation (3.15) as a differential equation, we construct a histogram through a random walk of a walker ensemble in $3N$-dimensional space, which corresponds to something like

(3.16).[12] After sufficient time has elapsed, the particle positions $\{\mathbf{R}_j\}$ generated successively by this random walk should exhibit a distribution following the histogram $\Phi(\mathbf{R}, \tau \gg 1)$. As mentioned in Eq. (3.8), this distribution $\Phi(\mathbf{R}, \tau \gg 1)$ approaches the exact ground-state solution. If we adopt this as the sampling distribution $\Psi_\alpha(\mathbf{R})$ in the variational functional of Eq. (3.9),

$$E(\alpha) \approx \frac{1}{M}\sum_{j=1}^{M} \Psi_\alpha^{-1}\left(\tilde{\mathbf{R}}_j\right) \cdot \hat{H}\Psi_\alpha\left(\tilde{\mathbf{R}}_j\right) = \frac{1}{M}\sum_{j=1}^{M} E_L\left(\tilde{\mathbf{R}}_j\right)$$

$$\left\{\tilde{\mathbf{R}}_j\right\} \sim P(\mathbf{R}) = \frac{\Psi_\alpha^*(\mathbf{R})\,\Psi_\alpha(\mathbf{R})}{\int d\tilde{\mathbf{R}} \cdot \Psi_\alpha^*(\mathbf{R}) \cdot \Psi_\alpha(\mathbf{R})} \tag{3.17}$$

then, as discussed around Eq. (3.9), we can obtain the exact energy value E.

Here, note the intention behind using the tilde notation $\left\{\tilde{\mathbf{R}}_j\right\}$ to represent the sampling points in the above equation. As can be seen from the expression $P(\mathbf{R}) = \cdots$ in Eq. (3.17), $\left\{\tilde{\mathbf{R}}_j\right\}$ corresponds to the particle positions appearing with a probability distribution that follows the histogram of the **squared** wave function $|\Phi|^2$. On the other hand, the random walk generated by Eq. (3.16) produces $\{\mathbf{R}_j\}$, which appears with a probability distribution corresponding to the histogram of the wave function Φ itself, **without squaring**. Roughly speaking, it might seem acceptable to use $\{\mathbf{R}_j\}$ as a substitute for $\left\{\tilde{\mathbf{R}}_j\right\}$, but the two are indeed distinct. This substitution becomes valid by mapping the quantity subjected to imaginary-time evolution, $\Phi(\mathbf{R}, \tau)$, onto a different function $f(\mathbf{R}, \tau)$, which will be discussed later. In the actual implementation of ab initio QMC, simulations are performed using this mapping. The details will be deferred to the theoretical explanation in Sect. 7, while for now, the focus will be on mastering the operational procedures. With this broad understanding, the goal is to develop an intuitive grasp of how the "convergent solution of the Fokker-Planck equation" can be equivalently obtained through the "convergent distribution obtained by random walks governed by the Langevin equation" and how the sampling in Eq. (3.17) is carried out to yield more accurate results.

3.2.5 Maintaining a Constant Total Number of Walkers

In Sect. 3.2.2, the total number of walkers was denoted as N_w. Recalling that the histogram formed by the walker distribution corresponds to $\Phi(\mathbf{R}, \tau)$ (see Fig. 3.1), it follows that its integral value scales with the total number of walkers N_w. Here,

[12] The treatment of the term "$E_T - V(\mathbf{R})$" has yet to be addressed, so we tentatively refer to it as "something like" (3.16), but the overall structure remains unchanged.

recalling Eq. (3.7),

$$\Phi(\mathbf{R}, \tau) = \Psi_0(\mathbf{R}) \cdot \exp[-(E_0 - E_T)\tau] + \lambda_1 \cdot \Psi_1(\mathbf{R}) \cdot \exp[-(E_1 - E_T)\tau] + \lambda_2 \cdot \Psi_2(\mathbf{R}) \cdot \exp[-(E_2 - E_T)\tau] + \cdots . \tag{3.18}$$

At the end of Sect. 3.1.2, it was stated that if the reference energy E_T is adjusted to match the exact ground-state energy E_0, only the desired ground-state component Ψ_0 remains as a constant term, ensuring convergence to the exact solution, $\Phi(\mathbf{R}, \tau) \to \Psi_0(\mathbf{R})$. When this convergence is achieved, $\Phi(\mathbf{R}, \tau)$ remains constant, and thus its integral value should also remain constant. Consequently, the total number of walkers N_w also becomes constant.

When adjusting E_T during the calculation process, if E_T is too small and lower than E_0, the term Ψ_0 will also decay in the form of $\exp[-\lambda\tau]$ $(\lambda > 0)$, causing both the integral value of $\Psi_\alpha(\mathbf{R})$ and the total number of walkers N_w to vanish. Conversely, if E_T is too large and higher than E_0, the term Ψ_0 behaves as $\exp[+\lambda\tau]$ $(\lambda > 0)$, leading to an increase and eventual divergence in both the integral value of $\Psi_\alpha(\mathbf{R})$ and the total number of walkers N_w. Thus, by adjusting E_T to maintain N_w at a constant value throughout the simulation, convergence to the exact solution $\Psi_\alpha(\mathbf{R}, \tau) \to \Psi_0(\mathbf{R})$ is achieved. Moreover, at this stage, **the value of E_T, adjusted to keep N_w constant, converges to the lowest eigenvalue E_0**. In this way, the exact ground-state energy E_0 can be estimated from the converged value of E_T, which is an adjustable parameter in the simulation. The energy value estimated in this manner is called the **growth estimator**.

Now, what exactly causes changes in N_w during the simulation process, that is, during the imaginary-time evolution described by Eq. (3.15)? Looking at Eq. (3.16), we see that it simply represents the initial N_w walkers individually updating their positions without altering the total number N_w. Here, we must recall that we have postponed addressing how to handle the term $E_T - V(\mathbf{R})$. Previously, we referred to this term as the **rate term**, stating that it corresponds to the right-hand side of the first-order reaction equation found in chemical reaction kinetics. If we ignore the first term (the diffusion term) in Eq. (3.15), we obtain

$$\frac{\partial \Phi(\mathbf{R}, \tau)}{\partial \tau} \sim -(V(\mathbf{R}) - E_T) \cdot \Phi \quad , \quad \therefore \quad \Phi(\mathbf{R}, \tau) \sim \exp[-(V(\mathbf{R}) - E_T) \cdot \tau]$$

This means that the presence of the potential $V(\mathbf{R})$ affects the integral value of $\Phi(\mathbf{R}, \tau)$, and consequently, the number of walkers N_w increases or decreases. Examining the variable $\mathbf{R}$ in the above equation, we see that it suggests "in regions where the potential is high, the function value $\Phi(\mathbf{R}, \tau)$ decreases." Recalling that $\mathbf{R} = (\mathbf{r}_1, \cdots, \mathbf{r}_N)$, this implies that "snapshots of electron configurations corresponding to higher potential energy become less likely to occur." Here, we can also see the role of the tuning parameter E_T. Since the number of walkers is controlled based on whether the potential value is higher or lower than the offset E_T, if the total number of walkers decreases too much, increasing E_T makes the potential appear relatively lower, thereby increasing the number of walkers. Through this mechanism, E_T functions as an adjustment parameter to maintain a constant number of walkers.

Note that the variation in the number of walkers, N_w, is what was displayed in the upper plot of the "graphit" output screen (Fig. 2.2). Users monitor this variation to check whether N_w remains roughly constant throughout the simulation.

Avoid Confusing the Number of Walkers with the Number of Electrons
It is important to remember that we represent $\Phi(\mathbf{R}, \tau)$ using a histogram of N_w walkers. For instance, even when the number of electrons is $N = 1$, we still express the single-electron wave function using a histogram of N_w walkers. Naturally, the number of walkers N_w and the number of electrons N are unrelated. The value N_w corresponds to the "**resolution** of the discretized wave function representation." A single-electron wave function ($N = 1$) can be represented with $N_w = 512$ walkers or with $N_w = 1024$ walkers. During a seminar at a prestigious university, a "Professor from the Physics department" mistakenly confused the number of walkers N_w with the number of electrons N. Despite multiple careful and polite explanations, which the younger researchers around him understood and acknowledged, this "Professor" stubbornly insisted that "if the number of particles is 1, then the number of walkers must also be 1." This frustrating experience ultimately led me to miss an important career opportunity. Some researchers, particularly those in closely related fields, are unfortunately too preoccupied with pride and competition to genuinely listen to others.

3.3 Input Settings for DMC Simulations

A prolonged focus on theoretical background through mathematical formulations can be exhausting. So, at this point, let's take a break from theory and return to hands-on practice. We will now conduct quantum diffusion Monte Carlo (DMC) simulations to experience firsthand how the mathematical formulations correspond to actual computational procedures.

3.3.1 Preparing the Working Directory

Prepare the working directory for the exercise using the following steps:

```
% pwd
    /Users/maezono/work
```

```
% ls
    01check/ 02vmcOnly/ qmcTutorials/
```

After confirming the above structure, proceed with

```
% cp -r 02vmcOnly 03congen
% cd 03congen/
```

This way allows us to reuse previously prepared files to set up the working directory efficiently. Proceed by checking the contents with "ls" at each step, and then remove unnecessary files using

```
% rm out*
% clearup
% rm vmc.hist_1st input_res
```

After cleanup, prepare the input file with

```
% mv input_org input
```

Ensure that the working directory is correctly set up by verifying its contents:

```
% pwd
    /Users/maezono/work/03congen
% ls
    al_pp.data  bwfn.data.bin     input
    bwfn.data   correlation.data  n_pp.data
```

3.3.2 *DMC Procedure and Its Correspondence with Input Specifications*

Reviewing the previous section, the steps of the DMC calculation are as follows:

1. Prepare N_w walkers generated with a frequency given by the probability distribution of the many-body wave function $\Psi_\alpha(\mathbf{R})$.[13] This step is referred to as the **configuration generation process (congen)**.
2. Perform random walks on these walkers for $T_{\text{eq}} = N_{\text{eq}} \cdot \Delta t$. As a result, the initial trial function $\Phi(\mathbf{R}, \tau = 0) = \Psi_\alpha(\mathbf{R})$ undergoes histogram evolution over time, leading to the distribution convergence $\Phi(\mathbf{R}, \tau) \to \Psi_0(\mathbf{R})$. This step is referred to as the **equilibration process (equil)** (see Fig. 3.3).
3. Once convergence is achieved, continue the random walk for an additional $T_{\text{stat}} = N_{\text{stat}} \cdot \Delta t$ to generate configurations, and execute the sampling described in (3.17) using N_{stat} samples. This step is referred to as the **statistical accumulation process (stats)**.

[13] Visualize this as preparing N_w snapshots of multiparticle configurations.

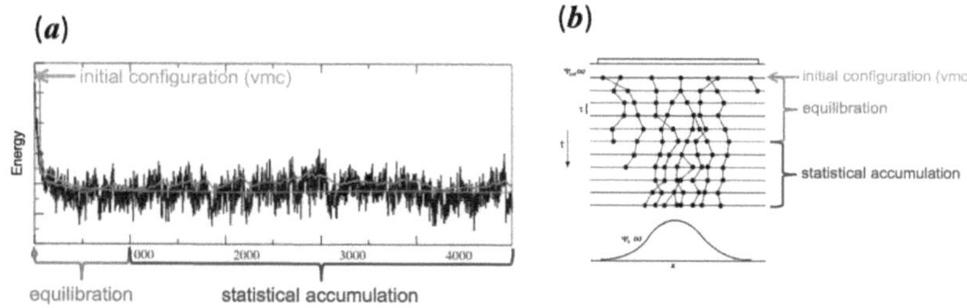

Fig. 3.3 Before calculating the mean value, a certain number of sampling iterations are performed without accumulating statistics (equilibration process). Once the distribution has equilibrated, the averaging process begins (statistical accumulation process). The equilibration of the distribution corresponds to the convergence of the many-body wave function toward the exact solution. Consequently, the sampled local energy values decrease in accordance with the variational principle (panel (**a**)). During this equilibration process, walkers with high local energy values are eliminated, while walkers with lower local energy values are generated. This results in the stabilization and convergence of the sampling distribution (panel (**b**))

From here, we will proceed step by step through the configuration preparation process $\rightarrow$ equilibration process $\rightarrow$ statistical accumulation process, explaining each step while performing practical exercises.[14]

Configuration preparation process (congen) is performed using the VMC calculation described in Sect. 2.3. The configurations $\{\mathbf{R}_j\}$ generated there are produced with frequencies that follow the distribution histogram of "$\Psi_\alpha(\mathbf{R})$ as defined by 'bwfn.data' and 'correlation.data'," implemented via methods such as the Metropolis algorithm. The last N_w configurations output from this VMC calculation are saved as an output file for use as the initial walkers in the next-stage DMC calculation.

Modify the following items in the input file as necessary to match the values shown here.

```
% pwd
    /Users/maezono/work/03congen
% more input
...
runtype             : vmc     #*!
newrun              : T       #*!
...
vmc_equil_nstep     : 1000    #*! Sec.2.3.1
vmc_nstep           : 5000    #*! Sec.2.3.1
vmc_nblock          : 1       #*! Sec.2.3.1
vmc_nconfig_write : 512       #*! Sec.3.3
...
dmc_equil_nstep     : 100     #*! Sec.3.3
dmc_equil_nblock    : 5       #*! Sec.3.3
```

[14] "congen" stands for "configuration generation," "equil" stands for "equilibration," and "stats" stands for "statistical accumulation."

```
dmc_stats_nstep    : 500    #*! Sec.3.3
dmc_stats_nblock   : 25     #*! Sec.3.3
dmc_target_weight : 512.d0 #*! Sec.3.3
dtdmc              : 0.01   #*!
...
```

Here, **vmc_nconfig_write** specifies the number of walkers N_w output from VMC, set to 512. The same number appears in **dmc_target_weight** as "512.d0" in real number format, which specifies the N_w used in the DMC simulation. The reason for specifying the former as an integer and the latter as a real number will be explained later. At this stage, understand that

$$[\text{vmc\_nconfig\_write}] = [\text{dmc\_target\_weight}] = N_w$$

This setting applies regardless of the number of parallel computation cores; it always refers to "a total of 512 walkers." Therefore, it is natural to choose **the number of walkers N_w as a multiple of the number of cores** so that the workload is evenly distributed across cores.

Next, we describe the equilibration process "$T_{\text{eq}} = N_{\text{eq}} \cdot \Delta t$" and the statistical accumulation process "$T_{\text{stat}} = N_{\text{stat}} \cdot \Delta t$." The parameter **dtdmc** represents Δt and is specified as "0.01." The values of N_{eq} and N_{stat} are defined as

$$[\text{dmc\_equil\_nstep}] = N_{\text{eq}}$$

$$[\text{dmc\_stats\_nstep}] = N_{\text{stat}}$$

In this case, the settings indicate that "100 steps are used for equilibration, followed by 500 steps for statistical accumulation." At this point, the approximate number of samples for statistical accumulation is given by "$N_w \times N_{\text{stat}} = 512 \times 500$".[15]

As described in Sect. 2.3.3, we discuss the blocking of statistical accumulation. Here, the specification that the 100 steps of the equilibration process are divided into five blocks is written as **dmc_equil_nblock = 5**. Therefore, the number of steps per block is $100/5 = 20$.

Similarly, the block size for the accumulation process is specified by **dmc_stats_nblock**. For statistical processing as discussed in Sect. 2.3.3, it is necessary to ensure that the "number of steps per block" remains consistent between the equilibration and accumulation processes. Following this rule, the 500 steps in the accumulation process are divided into m blocks so that each block contains 20 steps. Since "$500/m = 20$," we obtain $m = 25$, which results in the specification **dmc_stats_nblock = 25**.

[15] During the DMC simulation, even in the statistical accumulation process, the "initial configuration count of 512" fluctuates over time (Sect. 3.2.5), so it does not remain constant at 512. Hence, this is stated as an approximation.

3.4 DMC Calculation Tutorials

3.4.1 Sequence Leading to Accumulation Process

After preparing the input as described in the previous section, execute

```
% pwd
    /Users/maezono/work/03congen
% mpirun -np 8 casino &
```

to run the **configuration generation process**. Once the calculation is complete, use the **ls** command to confirm that out and vmc.hist have been generated. The end of the out file should display

```
% tail out
    ...
    Total CASINO CPU time  : : :       0.0000
    Total CASINO real time : : :       0.8190
    ...
    Ends 2023/09/26 21:39:17.145
```

Verify that this output confirms the calculation has been completed successfully.

Next, store these results safely.

```
% pwd
    /Users/maezono/work/03congen
% mkdir congen
% mv input out vmc.hist config.out congen
```

As shown above, create a directory named "congen" and move the four files "input, out, vmc.hist, and config.out" into it. These files serve as important calculation logs.

Next, proceed to the **equilibration process**.

```
% cp congen/input .
% cp congen/config.out .
% mv config.out config.in
```

Here, we utilize the configuration output from the **configuration generation process (congen)**, namely, "config.out," as the initial walkers for the subsequent DMC calculation ("config.in") (Sect. 3.3.2). The roles of the "vmc.hist" and "config.out" files were discussed at the end of Sect. 2.3.1. If your memory is unclear, refer back to that section to refresh your understanding.

Next, edit the input file for the equilibration process as follows:

```
...
runtype : dmc_equil
newrun  : T
testrun : F
...
```

The parameter **runtype**, which specifies the calculation mode, is changed from **vmc** to **dmc_equil** (see Fig. 3.4). Also, note that **newrun** is set to **T**. This indicates that the DMC calculation is a new run.

<table>
<tr><th>Config. generation</th><th>Equillibration</th><th>Statistical Accmulation</th></tr>
<tr><td>vmc</td><td>dmc_equil</td><td>dmc_stats</td></tr>
<tr><td>vmc</td><td colspan="2">dmc</td></tr>
<tr><td colspan="3">vmc_dmc</td></tr>
</table>

Fig. 3.4 The "runtype" setting in the input file is fundamentally divided into "vmc," "dmc_equil," and "dmc_stats." Additionally, there are options such as "dmc" and "vmc_dmc" that allow multiple calculations to be executed in one go

Then, execute the calculation with

```
% mpirun -np 8 casino &
```

Once the calculation is completed, check the output by running

```
% tail -n 40 out
```

This should display output similar to the following:

```
...
=========================================================
In block : 4

Number of moves in block                 : 20
Load-balancing efficiency (%)            : 99.354
Number of config transfers               : 30
Acceptance ratio (%)                     : 99.604
New best estimate of DMC energy (au)     : -23.96888519
Max no of attempts before accept move    : 3

Time taken in block    : : :          0.0000
=========================================================
In block : 5
...
Total CASINO real time : 5.1830
...
Ends 2023/09/26 21:50:19.799
```

As specified in Sect. 3.3.2, the calculation completes after five blocks of equilibration.

Once the calculation is completed, use the **ls** command to check that the following three output files have been generated: "config.out," "out," and "dmc.hist." As explained at the end of Sect. 2.3.1, the file dmc.hist[16] contains the evolution of the sampled values, while config.out stores the position information of the N_w walkers. Then, archive these results with the following commands:

[16] For VMC, this was vmc.hist.

```
% mkdir equil
% mv input out dmc.hist config.out equil/
```

Next, prepare for the **statistical accumulation process**.

```
% rm config.in
% cp equil/config.out config.in
```

This step transfers the walker configuration (config.out) obtained in the equilibration process to serve as the initial distribution "config.in" for the next stage. The purpose of the statistical accumulation process is to further random walk this equilibrated distribution for sampling.

As mentioned in Sect. 2.1, we used "graphit" to visualize the process, including the equilibration phase. Since we want to keep a record of the sampling process, we append it to the "equilibration phase record" as follows:

```
% cp equil/dmc.hist .
```

This copies "dmc.hist" obtained from the equilibration process into the working directory, allowing new data to be appended to it.

Next, execute the following command:

```
% cp equil/input .
```

This copies the previously used "input" file. Then, edit its content as follows:

```
...
runtype : dmc_stats
newrun  : F
testrun : F
...
```

The parameter **runtype**, which specifies the calculation mode, has now changed from **dmc_equil** to **dmc_stats**. Note that **newrun** is set to F. This indicates that the DMC calculation is not new but rather a continuation from the equilibration process.

Next, execute the following command:

```
% mpirun -np 8 casino &
```

to run the calculation. Once the calculation is complete, use

```
% tail -n 20 out
```

to check that the computation has reached the 25th block, as specified in Sect. 3.3.2. Additionally, execute the following commands:

```
% ls -lh equil/dmc.hist
    -rw-r--r--@ 1 maezono staff  28K  9 26 21:50 equil/dmc.hist
% ls -lh dmc.hist
    -rw-r--r--@ 1 maezono staff 163K  9 26 21:54 dmc.hist
```

to confirm that "dmc.hist" has been appended with new data and that its file size has increased. Running **graphit** from this working directory will read "dmc.hist" and produce a result like Fig. 3.5a.

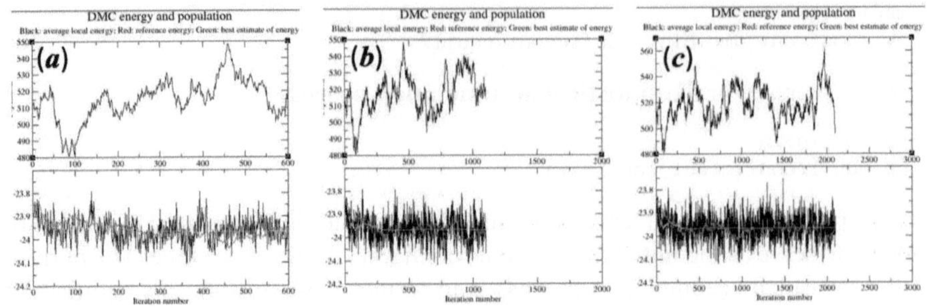

Fig. 3.5 Visualization of the DMC accumulation process using **graphit**. The accumulation is shown up to 600 steps (panel (**a**)), up to 1100 steps (panel (**b**)), and up to 2100 steps (panel (**c**))

3.4.2 Continuation of Accumulation Process

Next, let's learn how to continue this calculation further to increase the statistical accumulation count. By increasing the accumulation count M, the error bar decreases as $\sim 1/\sqrt{M}$ [1]. In practical applications, the statistical accumulation is continued until the required statistical accuracy (small enough error bars) is achieved.

First, save the results obtained so far as follows.

```
% pwd
    /Users/maezono/work/03congen
% mkdir stats01
% mv input out dmc.hist config.out stats01/
% ls stats01/
    config.out    dmc.hist    input    out
```

Then, execute the following commands:

```
% cp stats01/* .
% rm config.in out
% mv config.out config.in
```

After this, use the **ls** command to confirm that **dmc.hist** and **config.in** are present in the working directory, ensuring that they are carried over from the previous step. The **input** file remains unchanged from the previous step. By doing so, the continuation calculation will accumulate another 500 steps, just as in the previous stats run.

Then, execute the calculation with

```
% mpirun -np 8 casino &
```

Once the calculation has finished, verify completion by executing

```
% tail -n 20 out
```

and confirming that the calculation has proceeded up to the 25th block as specified in Sect. 3.3.2. Running **graphit** in this working directory will yield a plot similar to Fig. 3.5b. Due to the appended records in **dmc.hist**, the statistical accumulation now includes 100 equilibration steps, 500 steps from the previous section, and an additional 500 steps from this section, bringing the total to 1000 accumulated steps.

Next, let's perform a reblocking analysis on the results. Executing **reblock** will display

```
% reblock
...
 There are 1100 lines of data in total.
 There are 100 lines marked as equilibration data.
 Please enter the number of initial lines of data that you wish to discard,
 or enter -1 to discard the data marked as equilibration data.
```

This prompt asks how many initial steps should be excluded from statistical analysis as part of the equilibration process. For our case, we enter "100." Next, the prompt

```
Please select units for your energy data.
Choose one of (1) au per prim cell; (2) eV per prim cell;
(3) kcal per prim cell; (4) au per atom; (5) eV per atom; and (6) kcal per
atom.
```

asks for the unit selection. We choose "1," which corresponds to atomic units per primitive cell. The following output appears:

```
REBLOCKING ANALYSIS OF TOTAL-ENERGY DATA
========================================
(Energy data in units of au/prim cell.)
 RTN   Blk leng   Std error in mean       Error in std error
   0          1   1.718793568034981E-03   3.845262369498652E-05
  ...
   5         32   3.639385605792452E-03   4.678971347651916E-04
   6         64   4.381840749148847E-03   8.102026411268567E-04
   7        128   2.821208929624061E-03   7.643055167311876E-04
  ...
```

As in Sect. 2.1, we select "32." The final output appears:

```
Chosen block length: 32.
ENERGY COMPONENTS WITH REBLOCKED ERROR BARS
                              Mean (au/prim cell)   Err (au/prim cell)
Total energy (using Ewald): -2.39681486488763E+01   3.639385605792452E-03
  ...
Program finished.
```

From this, we determine that the result is $E = -23.9681 \pm 0.0036$. The error bar has been reduced from the previous value in Sect. 2.1, where $E = -23.9757 \pm 0.0047$. Although the numerical values appear different, they agree within the error bars.

As previously mentioned, increasing the statistical accumulation count M reduces the error bar as $\sim 1/\sqrt{M}$. Therefore, to halve the error bar, four times the statistical accumulation is required. In Sect. 2.1, the error bar was obtained with 500 steps of accumulation. To halve this error bar, an estimated 2000 steps (four times the original) are necessary. So far, through "stats02," we have accumulated

$500 \times 2 = 1000$ steps. We still need an additional 1000 steps. Let's perform a computation that accumulates this remaining 1000 steps in a single run.

First, store the results of the previous computation with

```
% mkdir stats02
% mv input out dmc.hist config.out stats02
```

Then, check that the size of **dmc.hist** has increased, confirming that accumulation is being appended to previous calculations:

```
% ls -lh stats0*/dmc.hist
-rw-r--r--@ 1 maezono staff 163K  9 26 21:54 stats01/dmc.hist
-rw-r--r--@ 1 maezono staff 298K  9 26 22:01 stats02/dmc.hist
```

Next, prepare for further accumulation calculations by carrying over the configuration file from "stats02":

```
% cp stats02/* .
% rm config.in out
% mv config.out config.in
```

Next, edit the following section of the **input** file:

```
dmc_stats_nstep  : 1000
dmc_stats_nblock : 50
```

Here, the objective is to add another 1000 steps, so **dmc_stats_nstep** is set to 1000. For statistical analysis, the number of steps per block must be kept consistent. In the previous runs, the settings were

```
dmc_stats_nstep = 500
dmc_stats_nblock : 25
```

Since **500/25 = 20 = 1000/50**, setting **dmc_stats_nblock = 50** ensures that the number of steps per block remains the same while accumulating 1000 additional steps.

Next, execute the following command:

```
% mpirun -np 8 casino &
```

Since this calculation takes time, periodically check the progress using

```
% tail -n 20 out
```

Monitor the last displayed block number; it should gradually increase. Once it reaches 50 blocks, the calculation is complete. After completion, execute **graphit** to obtain a plot like Fig. 3.5c, confirming that the accumulated statistical sampling has reached 2000 steps.

Next, perform the reblocking analysis. As before, enter "100/1/32," and you will obtain

```
% reblock
 ...
 Chosen block length: 32.
                              Mean (au/prim cell)    Err (au/prim cell)
 Total energy (using Ewald) : -2.39741870330965E+01  3.11904958312324E-03
   ...
```

The error bar is now 0.0031. Compared to the previous result from §2.1, where the statistical accumulation was 500 steps ("$E = -23.9757 \pm 0.0047$"), we see that accumulating four times more samples has approximately halved the error. The reason why it is not exactly halved is that in DMC calculations, the target probability distribution evolves as the simulation progresses.

Finally, execute the following commands:

```
% mkdir stats03
% mv input out dmc.hist config.out stats03
```

After archiving the results, verify that "dmc.hist" has increased in size as a result of additional accumulation by executing

```
% ls -lh stats*/dmc.hist
  -rw-r--r--@ 1 maezono  staff  163K  9 26 21:54 stats01/dmc.hist
  -rw-r--r--@ 1 maezono  staff  298K  9 26 22:01 stats02/dmc.hist
  -rw-r--r--@ 1 maezono  staff  570K  9 26 22:08 stats03/dmc.hist
```

This confirms that "dmc.hist" has been appended, increasing its file size.

3.4.3 Reviewing the Overall Process

For educational purposes, we have explained the computational procedures by dividing them into "configuration preparation phase," "equilibration phase," and "statistical accumulation phase."[17] These steps can be executed all at once by specifying "**runtype** = vmc_dmc" as we did in Sect. 2.1 (Fig. 3.4). Now, let's try performing this process independently while checking the specifications in the input file.[18]

First, prepare the working directory as follows:

```
% pwd
    /Users/maezono/work
% cp -r 03congen/ 04total
% cd 04total/
```

[17] Even in practical applications, when dealing with large systems at the forefront of research or using shared supercomputers with job time limits set within 24 hours, computations are often divided into phases and executed sequentially.

[18] Additionally, specifying "**runtype** = dmc" enables both the "equilibration phase (dmc_equil)" and the "statistical accumulation phase (dmc_stats)" to be executed in a single run.

```
% clearup
% rm -r stats0* equil
% cp congen/input .
% rm -r congen
```

Since the input file was copied from the configuration preparation phase, edit it as follows:

```
runtype             : vmc_dmc
newrun              : T
...
dmc_equil_nstep     : 100
dmc_equil_nblock    : 5
dmc_stats_nstep     : 2000
dmc_stats_nblock    : 100
...
```

This time, since 2000 steps are accumulated in one go, if the "block size is set to 20 steps," then the specified number of blocks should be as indicated above. Ensure that you understand this setting correctly.

Next, use the **mpirun** command to submit the calculation. Once the calculation is complete, verify the file size of **dmc.hist** by executing

```
% pwd
    /Users/maezono/work/04total
% ls -lh dmc.hist
  -rw-r--r--@ 1 maezono staff 570K 9 26 22:18 dmc.hist
```

Ensure that the file size of **dmc.hist** matches the one obtained when "the statistical accumulation was built up to 2000 steps over multiple runs" in the previous section.

Use **graphit** to verify that "2000 steps of accumulation have been performed." Next, perform the reblocking analysis. The output should be

```
Chosen block length: 128.
  Total energy (using Ewald): -2.39768109024146E+01  3.07111557796641E-03
  ...
```

Confirm that this result "matches the previous section's outcome within the statistical error range."

References

1. "An Introduction to Computer Simulation Methods: Applications to Physical Systems, Third Edition" (Amazon Digital Services, 2017), H. Gould, J. Tobochnik, W. Christian, ISBN: 978-19-74427-47-5
2. "Continuum variational and diffusion quantum Monte Carlo calculations", R.J. Needs, M.D. Towler, N.D. Drummond, P.L. Rios, J. Phys.: Condens. Matter 22, 023201 (2010), https://doi.org/10.1088/0953-8984/22/2/023201

3. “Schrödinger’s Mechanics” (World Scientific, 1989), D.B. Cook, ISBN: 978-99-71507-60-2
4. “Visual Complex Analysis” (Clarendon Press, 1999), T. Needham, ISBN: 978-01-98534-46-4
5. “Mathematical Methods for Physics and Engineering, Third Edition” (Cambridge University Press, 2006), K.F. Riley, M.P. Hobson, S.J. Bence, ISBN: 978-05-21679-71-8

Chapter 4
Variational Optimization of Many-Body Wave Functions

In the previous chapters, we explained how to run VMC and DMC simulations using the "well-matured dough $\Psi_\alpha(\mathbf{R})$ prepared by the master." This dough $\Psi_\alpha(\mathbf{R})$ had been sufficiently "matured" before baking, ensuring that "when subjected to QMC calculations, it yields a proper product." Now, how exactly is this dough prepared? The fundamental procedure is as follows:

1. First, prepare the initial framework of the many-body wave function, $\Psi_0(\mathbf{R})$, using density functional theory or molecular orbital methods.
2. Introduce functional degrees of freedom by incorporating adjustable parameters $\boldsymbol{\alpha}$ to refine the wave function into $\Psi_{\boldsymbol{\alpha}}(\mathbf{R})$.
3. Adjust the parameters $\boldsymbol{\alpha}$ so that the VMC energy value given by $\Psi_{\boldsymbol{\alpha}}(\mathbf{R})$ decreases ("maturation of the dough").

In this chapter, we will go through this process through practical exercises, deepening our understanding of VMC.

4.1 Overview: Theoretical Background of Variational Optimization

4.1.1 *Specific Implementation of Trial Functions*

Let's review what we have learned so far. We are tackling the following problem:

$$\left[-\frac{1}{2}\sum_{j=1}^{N}\nabla_j^2 + V(\mathbf{r}_1, \cdots, \mathbf{r}_N)\right] \cdot \Psi_\alpha(\mathbf{r}_1, \cdots, \mathbf{r}_N) = E \cdot \Psi_\alpha(\mathbf{r}_1, \cdots, \mathbf{r}_N) \tag{4.1}$$

R. Maezono, *Ab initio Quantum Monte Carlo Tutorial*,
https://doi.org/10.1007/978-981-95-4134-8_4

where the exact solution minimizes the functional,

$$E(\alpha) = \frac{\int d\mathbf{R} \cdot \Psi_\alpha^* (\mathbf{R}) \left[-\frac{1}{2} \sum_j \nabla_j^2 + V(\mathbf{R}) \right] \Psi_\alpha (\mathbf{R})}{\int d\mathbf{R} \cdot \Psi_\alpha^* (\mathbf{R}) \cdot \Psi_\alpha (\mathbf{R})}. \tag{4.2}$$

We can then minimize to solve the problem numerically (Sect. 1.2.1). Note that $\mathbf{R} = (\mathbf{r}_1, \cdots, \mathbf{r}_N)$, as previously defined.

The solution describing the many-electron system is obtained by solving Eq. (4.1) under the **antisymmetric constraint**:

$$\Psi_\alpha \left(\cdots, \mathbf{r}_j, \cdots, \mathbf{r}_i, \cdots\right) = (-) \cdot \Psi_\alpha \left(\cdots, \mathbf{r}_i, \cdots, \mathbf{r}_j, \cdots\right) \tag{4.3}$$

[1, 2]. Therefore, the trial function used for the variational minimization of Eq. (4.2) must satisfy this antisymmetric constraint.

The most straightforward way to construct a function that satisfies the antisymmetric constraint is to take a linear combination of the antisymmetrized product of single-particle orbital functions $\left\{\psi_j(\mathbf{r})\right\}_{j=1}^N$ [1, 2]:

$$\begin{aligned} \Psi^{(\mathrm{SD})}(\mathbf{r}_1, \mathbf{r}_2, \cdots, \mathbf{r}_N) &= \sum_P (-)^P \cdot \psi_1\left(\mathbf{r}_{P_1}\right) \psi_2\left(\mathbf{r}_{P_2}\right) \cdots \psi_N\left(\mathbf{r}_{P_N}\right) \\ &= \begin{vmatrix} \psi_1(\mathbf{r}_1) & \cdots & \psi_1(\mathbf{r}_N) \\ \vdots & \ddots & \\ \psi_N(\mathbf{r}_1) & \cdots & \psi_N(\mathbf{r}_N) \end{vmatrix} \end{aligned} \tag{4.4}$$

This type of trial function is called the **Slater determinant**.[1]

The single-particle orbital functions $\left\{\psi_j(\mathbf{r})\right\}_{j=1}^N$ can be obtained from conventional ab initio methods such as DFT or molecular orbital methods. In most practical applications of ab initio QMC, Slater determinants are constructed using these precomputed orbital functions. The **bwfn.data** file, which we have been working with in previous exercises, corresponds to this Slater determinant component in the input files.

The Slater determinant is based on the assumption that a many-body wave function can be expressed through separable variables. This assumption is justified in the absence of particle-particle interactions [3]. However, with the introduction of interactions, the many-body wave function is expected to deform beyond the functional forms that can be expressed by separable variables. As a method

---

[1] When the variational minimization problem of Eq. (4.2) is treated using this simplest trial function assumption, one can analytically derive the equation governing the single-particle orbitals $\left\{\psi_j(\mathbf{r})\right\}_{j=1}^N$. This leads to the **Hartree-Fock equation** [1, 2].

to incorporate this **deformation effect of the many-body wave function,**[2] one approach is to extend the Slater determinant-type trial function $\Psi^{(SD)}(\mathbf{R})$ by introducing an additional exponential factor:

$$\Psi_{\alpha}(\mathbf{r}_1, \cdots, \mathbf{r}_N) = \exp\left[J(\mathbf{R}; \boldsymbol{\alpha})\right] \times \Psi^{(SD)}(\mathbf{R}) . \tag{4.5}$$

The exponential factor $\exp\left[J(\mathbf{R}; \boldsymbol{\alpha})\right]$ is called the **Jastrow factor** [4]. This type of trial function is referred to as a **Slater-Jastrow** trial function. In practical-use ab initio variational Monte Carlo methods, including the exercises in this book, the Slater-Jastrow trial function is frequently employed. Regarding this form, we note that

- The Jastrow factor carries the dependence on the variational parameters $\boldsymbol{\alpha}$.
- Since this dependence appears in the exponent, the variational degrees of freedom are set solely to adjust the **amplitude** (a nodal structure fixing approximation will be discussed in Sect. 8.2).
- The antisymmetry constraint is maintained by the Slater determinant part $\Psi^{(SD)}(\mathbf{R})$.

In our exercises, the Jastrow function component corresponds to **correlation.data**. There are various approaches to implementing the Jastrow function [5], but in this book, we adopt a formulation where the function is divided into three distinct terms [6]:

$$J(\mathbf{R}) = \sum_{i>j} u\left(r_{ij}\right) + \sum_{i,I} \chi_I\left(r_{iI}\right) + \sum_{I,\, i>j} f_I\left(r_{iI}, r_{jI}, r_{ij}\right) . \tag{4.6}$$

The indices i and I represent electrons and nuclei, respectively. Here, r_{ij} denotes the distance between electrons, while r_{iI} represents the distance between an electron and an ionic core. The Jastrow function $J(\mathbf{R})$ consists of three terms: the u term, the χ term, and the f term. The u term represents **electron-electron** coalescence, the χ term represents **electron-nuclei** coalescence, and the f term introduces variational degrees of freedom to represent a **three-body** coalescence in which two electrons coalesce near a nucleus.

We use the term "coalescence," but what is meant is the deformation of the wave function as charged particles approach each other. As will be discussed in Sect. 8.1, the asymptotic behavior of this deformation reveals that the wave function exhibits a cusp at the coalescence point. Since this **cusp condition** involves a nonanalytic behavior, it is difficult to express it using only the single-determinant wave function $\Psi^{(SD)}(\mathbf{R})$. By definition, electron correlation effects are those that cannot be described within Hartree-Fock theory. Thus, the cusp behavior that $\Psi^{(SD)}(\mathbf{R})$ fails

[2] This effect is referred to as electronic correlation. Misuse of the term "electronic correlation" is sometimes observed due to its connotation (Sect. 16.2.2) of the word of "correlation," but this interpretation here aligns with its original definition.

to describe is a type of electron correlation known as **dynamical correlation.**[3] Since the Jastrow function is designed with such dynamical correlation in mind, each of the u, χ, and f terms is defined as a function with **explicit dependence on interparticle distances** to describe particle coalescence.

There exist various implementations for variational trial wave functions that describe the asymptotic cusp behavior [5]. In this tutorial, we adopt a **power series expansion** form [6], given by

$$u\left(r_{ij}\right) \sim \left(r_{ij} - L_u\right)^2 \left\{\alpha_0 + \alpha_1 \cdot r_{ij} + \alpha_2 \cdot r_{ij}^2 + \cdots + \alpha_{N_u} \cdot r_{ij}^{N_u}\right\}. \tag{4.7}$$

Here, L_u is the cutoff radius, ensuring that as the interparticle distance r_{ij} increases, $u(r_{ij})$ smoothly approaches zero as a quadratic function.[4] Although not explicitly represented in the equation, $u\left(r_{ij}\right)$ is further suppressed to zero via a step-function-like implementation for $r_{ij} > L_u$. Thus, the effective interaction range of the u-term is determined by L_u. The parameter N_u specifies the number of expansion terms. The χ- and f-terms also follow similar power series expansions, where their respective **cutoff radii** and **expansion order** are specified by the user within **correlation.data**. These parameters collectively constitute the variational parameters $\boldsymbol{\alpha}$ in $\Psi_{\boldsymbol{\alpha}}(\mathbf{R})$.

In principle, the cutoff radius can also be treated as a variational parameter and optimized. However, in most practical applications, it is typically fixed at a recommended value[5] and not optimized. Instead, only the expansion coefficients $\left\{\alpha_j\right\}$ are optimized.[6]

4.1.2 Zero Variance Property and Variance Minimization Method

The variational functional in (2.19),

$$E(\boldsymbol{\alpha}) = \frac{\int d\mathbf{R} \cdot \left|\Psi_{\boldsymbol{\alpha}}(\mathbf{R})\right|^2 \cdot \Psi_{\boldsymbol{\alpha}}^{-1}(\mathbf{R})\, \hat{H} \Psi_{\boldsymbol{\alpha}}(\mathbf{R})}{\int d\mathbf{R} \cdot \left|\Psi_{\boldsymbol{\alpha}}(\mathbf{R})\right|^2} = \left\langle E_L(\mathbf{R}; \boldsymbol{\alpha})\right\rangle_{\left|\Psi_{\boldsymbol{\alpha}}\right|^2} \tag{4.8}$$

[3] Detailed discussion is given in Sect. 10.5.2.

[4] It is also possible to use a different power instead of a quadratic function.

[5] For solid-state periodic systems, a characteristic length scale is given by the unit cell size (approximately the Wigner-Seitz radius). Thus, the cutoff radius is usually set to a certain fraction of this characteristic length.

[6] When solving the optimization problem $\partial F(\alpha)/\partial\alpha = 0$ for $F(\alpha)$, if α appears nonlinearly within $F(\alpha)$, then $\partial F(\alpha)/\partial\alpha$ will still contain α, which generally complicates the solution method. Since the cutoff radius is a nonlinear parameter, its numerical optimization tends to be unstable. Thus, fixing it to a recommended value is a practical approach. On the other hand, the expansion coefficients are linear parameters, making their optimization comparatively straightforward.

suggests that minimizing it with respect to the variational parameters $\boldsymbol{\alpha}$ brings us closer to the exact solution, which is the fundamental idea behind VMC [variational Monte Carlo]. The notation $\langle \cdots \rangle_{|\Psi_\alpha|^2}$ on the rightmost term indicates an expectation value with respect to the probability distribution $|\Psi_\alpha|^2$.

$$E_L(\mathbf{R}; \boldsymbol{\alpha}) = \Psi_\alpha^{-1}(\mathbf{R})\, \hat{H} \Psi_\alpha(\mathbf{R}) \tag{4.9}$$

is called the local energy. The problem of variational minimization is, more generally,

$$I(\boldsymbol{\alpha}) = \frac{\int d\mathbf{R} \cdot |\Psi_\alpha(\mathbf{R})|^2 \cdot F_\alpha(\mathbf{R})}{\int d\mathbf{R} \cdot |\Psi_\alpha(\mathbf{R})|^2} = \langle F_\alpha(\mathbf{R}) \rangle \tag{4.10}$$

which takes the form of minimizing the dependence of $\boldsymbol{\alpha}$ on this quantity.

Given that the expectation value $E(\boldsymbol{\alpha})$ in Eq. (4.8) has been determined, one can consider the expectation of the squared deviation from this mean:

$$\sigma^2(\boldsymbol{\alpha}) = \left\langle [E_L(\mathbf{R}; \boldsymbol{\alpha}) - E(\boldsymbol{\alpha})]^2 \right\rangle \tag{4.11}$$

namely, the variance. If we take $F_\alpha(\mathbf{R})$ in Eq. (4.10) as $F_\alpha(\mathbf{R}) = E_L(\mathbf{R}; \boldsymbol{\alpha})$, then we recover Eq. (4.8), and if we set $F_\alpha(\mathbf{R}) = [E_L(\mathbf{R}; \boldsymbol{\alpha}) - E(\boldsymbol{\alpha})]^2$, it leads to Eq. (4.11). Minimizing $I(\boldsymbol{\alpha})$ for $I(\boldsymbol{\alpha}) = E(\boldsymbol{\alpha})$ is referred to as **energy minimization**, while minimizing $I(\boldsymbol{\alpha}) = \sigma^2(\boldsymbol{\alpha})$ is known as **variance minimization**.

In the case of the local energy in Eq. (4.9), let us assume that the trial function $\Psi_\alpha(\mathbf{R})$ coincides with the exact solution $\Psi_\alpha^{(\mathrm{exact})}(\mathbf{R})$. Since the exact solution satisfies the eigenvalue equation,[7]

$$\hat{H} \cdot \Psi_\alpha^{(\mathrm{exact})}(\mathbf{R}) = E_0 \cdot \Psi_\alpha^{(\mathrm{exact})}(\mathbf{R})$$

the local energy in Eq. (4.9) for this exact solution is given by

$$\begin{aligned} E_L^{(\mathrm{exact})}(\mathbf{R}; \boldsymbol{\alpha}) &= \left[\Psi_\alpha^{(\mathrm{exact})}(\mathbf{R})\right]^{-1} \hat{H}\Psi_\alpha(\mathbf{R}) \\ &= \left[\Psi_\alpha^{(\mathrm{exact})}(\mathbf{R})\right]^{-1} E_0 \cdot \Psi_\alpha^{(\mathrm{exact})}(\mathbf{R}) = E_0 \end{aligned} \tag{4.12}$$

where the numerator and denominator cancel out, leaving only the constant E_0, thus eliminating the dependence of the local energy on $\mathbf{R}$. That is, when the sampling distribution has reached the exact solution, regardless of where the local energy

[7] In general, $\hat{H}\Psi_\alpha(\mathbf{R})$ is not proportional to $\Psi_\alpha(\mathbf{R})$ but becomes a different function. For the exact solution, however, $\hat{H}\Psi_\alpha(\mathbf{R})$ is proportional to $\Psi_\alpha(\mathbf{R})$, with the proportionality factor being the energy value E_0.

Fig. 4.1 When the quality of the trial function $\Psi_{\alpha}^{(\mathrm{exact})}(\mathbf{R})$ is still poor, the error bar (variance) is large, and the statistical estimate of the energy is not very reliable. Even if the mean value decreases (white circle), the sampled value (black circle) may happen to increase within the range of the error bar

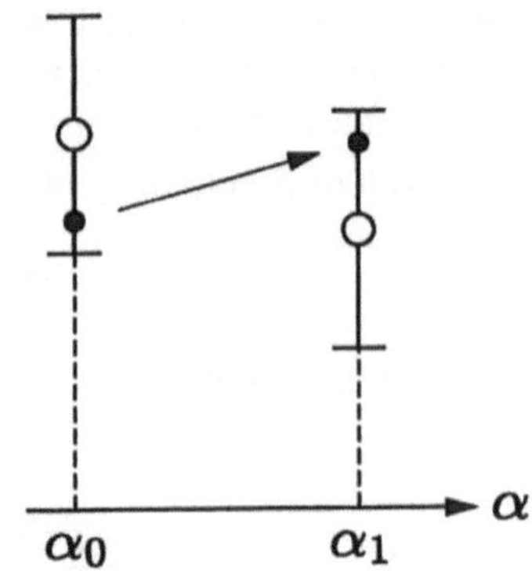

$E_L(\mathbf{R}_j;\boldsymbol{\alpha})$ is sampled, the sampled position $\mathbf{R}$ always yields the constant value E_0, without any fluctuations. This means that for the exact solution, the variance in Eq. (4.11) is zero, a property known as **zero-variance**. Summarizing,

- The exact solution minimizes the expectation of the local energy in Eq. (4.9),
- The exact solution sets the variance of the local energy in Eq. (4.11) to zero.

Originally, VMC aimed to "approach the exact solution by minimizing energy." However, replacing energy minimization with **variance minimization also leads to the exact solution.**

If we reverse the interpretation of zero variance, we can say that "the magnitude of local energy variance indicates the poor quality of the trial function." In the early stages of variational optimization, the quality of the trial function $\Psi_{\alpha}(\mathbf{R})$ is still low, leading to large error bars (variance). This results in situations like the one shown in Fig. 4.1. Specifically, during the parameter update $\boldsymbol{\alpha}_0 \rightarrow \boldsymbol{\alpha}_1$ in variational optimization, the true mean value (white circle in the figure) decreases, indicating that $\boldsymbol{\alpha}_0 \rightarrow \boldsymbol{\alpha}_1$ is an update that should be accepted. However, due to the large error bars, the sampled mean value (black circle in the figure) happens to shift from the lower bound to the upper bound of the error range.[8] As a result, the update is mistakenly judged as "an increase in the mean energy" and rejected. Thus, in wave function optimization via energy minimization, unless the trial function is of sufficiently high quality, updates to $\boldsymbol{\alpha}$ are frequently misjudged. For this reason, variance minimization offers greater "statistical numerical stability" and enables more robust wave function optimization. Therefore, in the tutorials in this book, we adopt variance minimization.[9]

[8] Recall the meaning of error bars. Depending on the sampling method, the estimated mean value can take any value within the range of the error bar. This is the fundamental interpretation of error bars.

[9] As mentioned in Sect. 8.2.4, in advanced applications beyond the introductory level, more sophisticated optimizations combine variance minimization with energy minimization and other statistical minimization methods.

4.1.3 Optimization Using Reweighting Method

The minimization problem in Eq. (4.10)

$$I(\boldsymbol{\alpha}) = \frac{\int d\mathbf{R} \cdot |\Psi_{\boldsymbol{\alpha}}(\mathbf{R})|^2 \cdot F_{\boldsymbol{\alpha}}(\mathbf{R})}{\int d\mathbf{R} \cdot |\Psi_{\boldsymbol{\alpha}}(\mathbf{R})|^2}$$

is formulated to be evaluated using a sampling sequence that follows the distribution of $\Psi_{\boldsymbol{\alpha}}(\mathbf{R})$. If this is implemented straightforwardly, each time $\boldsymbol{\alpha}$ is updated as $\boldsymbol{\alpha}_0 \to \boldsymbol{\alpha}_1 \to \boldsymbol{\alpha}_2 \to \cdots$, millions of new sampling points must be taken, and a comparison is performed as

$$\text{What if} \quad \left\langle F_{\alpha_0}(\mathbf{R}) \right\rangle_{|\Psi_{\boldsymbol{\alpha}_0}|^2} \geq \left\langle F_{\alpha_1}(\mathbf{R}) \right\rangle_{|\Psi_{\boldsymbol{\alpha}_1}|^2} \quad ??$$

to determine whether to update the variational parameter $\boldsymbol{\alpha}$ in the direction that decreases the function $F_{\boldsymbol{\alpha}}(\mathbf{R})$. In such a naive implementation, a large amount of sampling must be generated even for trial updates that will eventually be rejected, resulting in extremely poor efficiency.

To avoid this issue, we adopt the following approach: Equation (4.10) can be rewritten as

$$I(\boldsymbol{\alpha}) = \frac{\int d\mathbf{R} \cdot |\Psi_{\boldsymbol{\alpha}}(\mathbf{R})|^2 \cdot F_{\boldsymbol{\alpha}}(\mathbf{R})}{\int d\mathbf{R} \cdot |\Psi_{\boldsymbol{\alpha}}(\mathbf{R})|^2} = \frac{\int d\mathbf{R} \cdot \left|\Psi_{\alpha_0}(\mathbf{R})\right|^2 \dfrac{|\Psi_{\boldsymbol{\alpha}}(\mathbf{R})|^2}{\left|\Psi_{\alpha_0}(\mathbf{R})\right|^2} \cdot F_{\boldsymbol{\alpha}}(\mathbf{R})}{\int d\mathbf{R} \cdot \left|\Psi_{\alpha_0}(\mathbf{R})\right|^2 \dfrac{|\Psi_{\boldsymbol{\alpha}}(\mathbf{R})|^2}{\left|\Psi_{\alpha_0}(\mathbf{R})\right|^2}}$$

$$= \frac{\int d\mathbf{R} \cdot \left|\Psi_{\alpha_0}(\mathbf{R})\right|^2 w_{\boldsymbol{\alpha}}(\mathbf{R}) \cdot F_{\boldsymbol{\alpha}}(\mathbf{R})}{\int d\mathbf{R} \cdot \left|\Psi_{\alpha_0}(\mathbf{R})\right|^2 w_{\boldsymbol{\alpha}}(\mathbf{R})}, \quad w_{\boldsymbol{\alpha}}(\mathbf{R}) = \frac{|\Psi_{\boldsymbol{\alpha}}(\mathbf{R})|^2}{\left|\Psi_{\alpha_0}(\mathbf{R})\right|^2}$$

which can be expressed as

$$I(\boldsymbol{\alpha}) = \left\langle w_{\boldsymbol{\alpha}}(\mathbf{R}) \cdot F_{\boldsymbol{\alpha}}(\mathbf{R}) \right\rangle_{|\Psi_{\boldsymbol{\alpha}_0}|^2}$$

This formulation allows us to evaluate the function $w_{\boldsymbol{\alpha}}(\mathbf{R}) \cdot F_{\boldsymbol{\alpha}}(\mathbf{R})$ through statistical sampling using just a single distribution $\left|\Psi_{\boldsymbol{\alpha}_0}\right|^2$. More specifically, by inserting the normalization factor $\left[\int d\mathbf{R} \cdot \left|\Psi_{\alpha_0}(\mathbf{R})\right|^2\right]$ into the numerator and denominator, we can use the reasoning outlined in Sect. 2.2.4 to express

$$I(\boldsymbol{\alpha}) = \frac{\left[\int d\mathbf{R} \cdot \left|\Psi_{\boldsymbol{\alpha}_0}(\mathbf{R})\right|^2 w_{\boldsymbol{\alpha}}(\mathbf{R}) \cdot F_{\boldsymbol{\alpha}}(\mathbf{R})\right] / \left[\int d\mathbf{R} \cdot \left|\Psi_{\boldsymbol{\alpha}_0}(\mathbf{R})\right|^2\right]}{\left[\int d\mathbf{R} \cdot \left|\Psi_{\boldsymbol{\alpha}_0}(\mathbf{R})\right|^2 w_{\boldsymbol{\alpha}}(\mathbf{R})\right] / \left[\int d\mathbf{R} \cdot \left|\Psi_{\boldsymbol{\alpha}_0}(\mathbf{R})\right|^2\right]}$$

which allows us to evaluate $I(\boldsymbol{\alpha})$ through statistical sampling as

$$I(\boldsymbol{\alpha}) = \frac{\frac{1}{M}\sum_{j=1}^{M} w_{\boldsymbol{\alpha}}(\mathbf{R}_j) \cdot F_{\boldsymbol{\alpha}}(\mathbf{R}_j)}{\frac{1}{M}\sum_{j=1}^{M} w_{\boldsymbol{\alpha}}(\mathbf{R}_j)} \quad , \quad \{\mathbf{R}_j\} \sim \left|\Psi_{\boldsymbol{\alpha}_0}(\mathbf{R})\right|^2 \tag{4.13}$$

This formulation confirms that $I(\boldsymbol{\alpha})$ can be evaluated through a single statistical sampling. With this formulation, the acceptance or rejection of parameter updates is determined based on the comparison

$$\text{What if} \quad \left\langle w_{\boldsymbol{\alpha}_0} \cdot F_{\boldsymbol{\alpha}_0}(\mathbf{R})\right\rangle_{|\Psi_{\boldsymbol{\alpha}_0}|^2} \geq \left\langle w_{\boldsymbol{\alpha}_1} \cdot F_{\boldsymbol{\alpha}_1}(\mathbf{R})\right\rangle_{|\Psi_{\boldsymbol{\alpha}_0}|^2} \quad ??$$

Thus, a single set of sampled configurations following the distribution $\left|\Psi_{\boldsymbol{\alpha}_0}(\mathbf{R})\right|^2$ can be prepared in advance and reused for multiple parameter updates.[10] This approach is referred to as the **reweighting method**.

The methodology for determining the update direction $\boldsymbol{\alpha}_0 \rightarrow \boldsymbol{\alpha}_1$ in parameter space, so that the evaluation function decreases, has been extensively studied in the field of numerical optimization [7]. Parameter updates in our context leverage generic numerical libraries developed in this field [8].

4.2 Performing Variational Optimization

Up to the previous chapter, we have been performing calculations using the "matured dough $\Psi_{\boldsymbol{\alpha}}(\mathbf{R})$" prepared by the instructor. Now, we will take on the task of preparing this "matured dough" ourselves. The maturation of the dough refers to optimizing the variational parameters $\boldsymbol{\alpha}$ of $\Psi_{\boldsymbol{\alpha}}(\mathbf{R})$. These variational parameter sets are recorded in the file **correlation.data** (Sect. 4.1.1). Therefore, we will first

[10] When updating parameters as $\boldsymbol{\alpha}_0 \rightarrow \boldsymbol{\alpha}_1 \rightarrow \cdots \boldsymbol{\alpha}_k \rightarrow \cdots$, the same set of sampled configurations can be used as long as $\boldsymbol{\alpha}_k$ remains close to $\boldsymbol{\alpha}_0$. However, if the parameter set moves significantly away from $\boldsymbol{\alpha}_0$, a new sampling set following the updated $\left|\Psi_{\boldsymbol{\alpha}_k}(\mathbf{R})\right|^2$ must be regenerated. The timing of such resampling is specified in the input file (Sect. 4.2.2).

prepare a "blank correlation.data" with all initial values set to zero. Then, we will conduct a calculation that updates these numbers ($\boldsymbol{\alpha}$) so that they achieve variance minimization.

4.2.1 Preparing correlation.data

First, prepare the working directory as follows:

```
% pwd
    /Users/maezono/work
% cp -r 04total/ 05opt
% cd 05opt/
% clearup
% rm correlation.data
% ls
    al_pp.data  bwfn.data.bin  input
    bwfn.data   n_pp.data
```

Next, edit the input file as follows:

```
# RUN
runtype              : vmc_opt
newrun               : T
testrun              : T

# VMC
...
```

Setting "**runtype** : vmc_opt" enables the execution mode for variational optimization. Specifying "**testrun** : T" ensures that the program will only check for any input file inconsistencies and terminate without performing the full calculation. This allows the geometry information contained in bwfn.data to be written out to the out file.

Now, we will prepare the framework for the Jastrow function, as described in (4.7), by creating the correlation.data file. The numerical data contained within (such as expansion coefficients) will serve as the variational parameters $\boldsymbol{\alpha}$, which we will optimize using variational Monte Carlo. To construct (4.6), we need to determine the number of electrons, the positions and types of ionic cores, and an appropriate cutoff radius ($\sim$ the Wigner-Seitz radius of the unit cell). To extract this information into the out file, we first perform a test run by setting "testrun : T" in the input file and executing the following commands:

```
% mpirun -np 8 casino &
% ls
% cat out
```

Ensure that the out file has been correctly generated and contains the necessary information.

Next, execute the following commands:

```
% pwd
    /Users/maezono/work/05opt
% cp ~/work/qmcTutorials/utils/editCorr.py .
```

This copies the script **editCorr.py**, which is used to generate the **correlation.data** file, from the downloaded tutorial materials. This script extracts the necessary information from the out file generated by setting "testrun : T" earlier and creates **correlation.data**. After confirming the presence of the out file using the **ls** command, execute the following command:

```
% python3 editCorr.py > correlation.data
```

This will generate the empty framework for **correlation.data**.

Using the **more** command to inspect the contents of **correlation.data**, you will see the following structure:

```
% more correlation.data
 ...
 START JASTROW
 START U TERM
 ...
 START CHI TERM
 ...
 START F TERM
 ...
 END JASTROW
```

This confirms that the file contains the components corresponding to Eq. (4.7): the u term, the χ term, and the f term, each represented in its respective section. For example, in the section corresponding to the u term, you will find entries such as

```
 ...
 Expansion order N_u
   6
 Spin dep (0->uu=dd=ud; 1->uu=dd/=ud; 2->uu/=dd/=ud)
   1
 Cutoff (a.u.)     ;  Optimizable (0=NO; 1=YES)
   0.                 0
 ...
```

Here, the line labeled `Expansion order N_u` specifies 6, which means the power series expansion of the u term is truncated at the sixth order. The entry for `Cutoff` is set to `0`, which means a recommended default value, such as the Wigner-Seitz radius, will be automatically assigned. The parameter `Optimizable` is set to `0`, indicating that the cutoff radius is fixed and will not be optimized. Similar specifications can be found for the χ and f terms. These terms are specified separately for each inequivalent ion core, so they are divided into subgroups labeled as `SET1`, `SET2`, and so on.

4.2.2 Setting Conditions for Optimization

After clearing the previous calculation results with the **clearup** command, open the `input` file and change `testrun : T` to `F` so that the main calculation proceeds. The accumulation settings for the VMC calculation are as follows:

```
vmc_equil_nstep   : 1000
vmc_nstep         : 10000
vmc_nblock        : 1
vmc_nconfig_write : 10000
vmc_decorr_period : 10
```

The setting "`vmc_nconfig_write : 10000`" corresponds to the number of sampling points M in (4.13).[11]

For a single sampling evaluation as in a VMC calculation, it is sufficient to set "`vmc_nstep : 10000`" and "`vmc_nconfig_write : 0`." However, as explained in the previous section, in a **vmc_opt** calculation, the sampling distribution based on

$$p(\mathbf{R}) \sim \left|\Psi_{\alpha_0}(\mathbf{R})\right|^2$$

is stored and reused multiple times. Thus, "`vmc_nconfig_write : 10000`" is set to save the sampling distribution in **config.out** for reuse.

In the input file, pay attention to the following entries:

```
...
opt_method  : varmin_linjas
opt_cycles  : 10
opt_jastrow : T
...
use_jastrow : T
```

"`use_jastrow : T`" specifies that a Slater-Jastrow trial function with a Jastrow factor is used. `opt_jastrow : T` designates the Jastrow function as the optimization target. `opt_method : varmin_linjas` defines variance minimization as the optimization method. The **opt_method** keyword allows for the selection of **emin** (energy minimization), **madmin** (median minimization), **varmin** (variance minimization), and **varmin_linjas** (variance minimization of linear parameters within the Jastrow function).

In this case, we perform "variance minimization of linear parameters within the Jastrow function." For this specific optimization, a "specialized, highly efficient optimization algorithm" is applicable, which is why we set **varmin_linjas**.

[11] The choice of 10,000 points naturally depends on the system. It should be set to a value large enough to ensure stable optimization while minimizing computation time. The number of required samples for stable optimization depends on factors such as whether variance minimization or energy minimization is used.

Although replacing this setting with **emin**, **varmin**, or **madmin** would still allow for similar optimization, in this particular example, **varmin_linjas** provides the most stable and efficient optimization process.

Note that in this tutorial, we omit the optimization of the cutoff length. However, if we were to optimize the cutoff length or handle more advanced topics, such as using trial functions beyond the Slater-Jastrow type to optimize the nodal degrees of freedom, the specialized algorithm of **varmin_linjas** would no longer be applicable. In such general cases, we skillfully select and apply **emin**, **madmin**, or **varmin** to perform the optimization.[12]

In the input file, the setting **opt_cycles : 10** specifies that the parameter update operation will be repeated ten times.

$$
\begin{aligned}
&\alpha_0 \to \alpha_1 \quad \text{using} \quad p(\mathbf{R}) \sim \left|\Psi_{\alpha_0}(\mathbf{R})\right|^2 \\
&\alpha_1 \to \alpha_2 \quad \text{using} \quad p(\mathbf{R}) \sim \left|\Psi_{\alpha_1}(\mathbf{R})\right|^2 \\
&\ldots \\
&\alpha_9 \to \alpha_{10} \quad \text{using} \quad p(\mathbf{R}) \sim \left|\Psi_{\alpha_9}(\mathbf{R})\right|^2
\end{aligned}
$$

Thus, the optimization process is iteratively performed.

4.2.3 Execution of Optimization

With the above input file settings,

```
% pwd
    /Users/maezono/work/05opt
% clearup
% mpirun -np 8 casino &
```

execute CASINO using the mpirun command. While the execution is ongoing, checking the directory contents using the **ls** command reveals that files such as **correlation.out.0**, **correlation.out.1**, **correlation.out.2**, ... are gradually being generated. The results can be analyzed using the utility **ve**, which is included with the CASINO installation, as follows:

```
% ve
    ...
    VMC #1:  E = -23.38(2) ; var = 5.6(3)   (correlation.out.0)
    VMC #2:  E = -23.86(2) ; var = 2.0(2)   (correlation.out.1)
    VMC #3:  E = -23.83(1) ; var = 1.49(5)  (correlation.out.2)
    VMC #4:  E = -23.88(1) ; var = 1.54(7)  (correlation.out.3)
    VMC #5:  E = -23.61(7) ; var = 15(5)    (correlation.out.4)
```

[12] This is where the expertise of a professional group in ab initio QMC comes into play.

```
VMC #6:  E = -23.85(1) ; var = 1.66(7) (correlation.out.5)
VMC #7:  E = -23.84(1) ; var = 1.47(5) (correlation.out.6)
VMC #8:  E = -23.87(1) ; var = 1.55(5) (correlation.out.7)
VMC #9:  E = -23.86(1) ; var = 2.0(5)  (correlation.out.8)
VMC #10: E = -23.83(1) ; var = 1.6(1)  (correlation.out.9)
VMC #11: E = -23.87(2) ; var = 1.7(3)  (correlation.out.10)
...
```

As specified by **opt_cycles : 10**, the optimization process has been iterated up to α_{10} (correlation.out.10). In this example, a significant decrease in energy occurs from $\alpha_0 \to \alpha_1$ (-23.38(2) $\to$ -23.86(2)), after which the energy fluctuates up and down. The values labeled as "var = ..." represent the variance, which should be minimized to obtain a higher-quality trial function. From the perspective of achieving both lower energy and lower variance, the optimal choice appears to be[13]

```
-23.88(1) ; var = 1.54(7) (correlation.out.3)
```

Next, execute the following commands to store the results and clear the working directory:

```
% pwd
    /Users/maezono/work/05opt
% mkdir varmin
% mv input out correlation.* varmin/
% rm config.*
```

To adopt the best result from **correlation.out.3** as the new **correlation.data**, execute

```
% cp varmin/correlation.out.3 .
% mv correlation.out.3 correlation.data
% cp ../04total/input .
```

The **input** file is copied from the **04total** directory, which was used in Sect. 3.4.3, to set up the DMC calculation with the previously established conditions. After clearing the working directory with **clearup**, run the computation using **mpirun**. Once the computation is complete, analyze the results using **graphit** or **reblock**, and confirm that the obtained energy values match those from Sect. 3.4.3 within statistical errors.

Before this chapter, the calculations were performed using trial functions prepared in advance by the instructor. However, in this chapter, the mission was to "prepare the Jastrow function independently," which has now been successfully completed.

[13] Since the results will eventually be used in a later DMC calculation, there is no need to be overly concerned with selecting the absolute best value at this stage. It suffices to choose one that seems reasonably good.

4.3 DMC Without Jastrow Function

For the trial function in DMC, the Slater-Jastrow-type function from (4.5) is used:

$$\Psi_\alpha(\mathbf{r}_1, \cdots, \mathbf{r}_N) = \exp[J(\mathbf{R}; \boldsymbol{\alpha})] \times \Psi^{(SD)}(\mathbf{R}) .$$

The Jastrow function component, $\exp[J(\mathbf{R}; \boldsymbol{\alpha})]$, is pre-optimized in VMC before the DMC calculation, a process metaphorically described as "maturing before baking."

The projection method in DMC, as expressed in (3.7), is a framework that holds for **any initial** trial function. Even if we use a trial wave function without a Jastrow function, $\Psi^{(SD)}(\mathbf{R}) = \Psi(\mathbf{R}, \tau = 0)$, the longtime propagation should still lead to $\Psi(\mathbf{R}, \tau \to \infty) = \Psi_0(\mathbf{R})$, extracting the exact ground-state wave function $\Psi_0(\mathbf{R})$. This raises the following question: can DMC calculations proceed without first optimizing the Jastrow function in VMC? Let us now test this assumption.

First, prepare the working directory as follows:

```
% pwd
    /Users/maezono/work
% cp -r 04total 06dmcWoJ
% cd 06dmcWoJ/
% clearup
% rm correlation.data
```

Since we have removed "correlation.data," the directory contents should now be

```
% ls
    al_pp.data  bwfn.data  bwfn.data.bin  input  n_pp.data
```

Next, open the input file and modify the **use_jastrow** setting as follows:

```
use_jastrow : F
```

Now, proceed with the calculation using the **mpirun** command.

After some time, the calculation completes successfully, and visualization using **graphit** produces a plot as shown in Fig. 4.2. The projection operation successfully lowers the energy value in accordance with the variational principle and achieves convergence. Analysis using **reblock** confirms that the obtained energy value matches the previous results within statistical accuracy. Comparing Figs. 3.5 and 4.2, it is evident that the projection operation starts from a higher energy value. In previous calculations using the Jastrow function, the initial trial function had been sufficiently "refined" through VMC calculations, resulting in a lower starting energy. In contrast, this time, the calculation starts from a trial function without the Jastrow factor, which is of lower quality. As a result, the initial energy value is relatively higher. Nevertheless, the DMC calculation proceeds without issues and converges to the same value as when using the Jastrow function.

Then, why do we refine the Jastrow factor to a certain extent before starting the DMC calculation? In this case, the calculation happened to proceed without

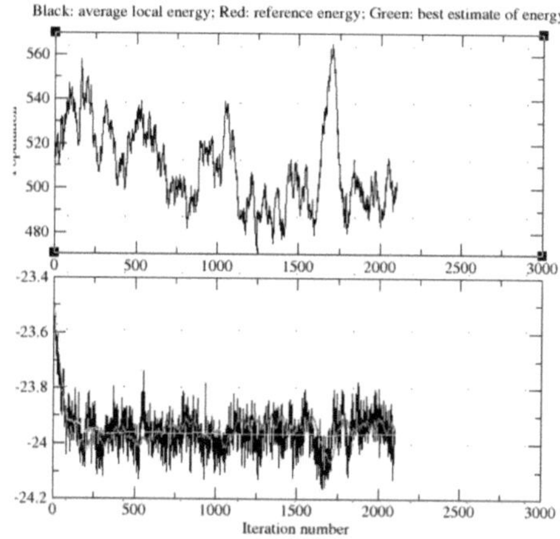

Fig. 4.2 Even with an initial trial function without the Jastrow factor, the DMC calculation proceeds without any issues in this case. The result matches that of the DMC calculation performed with an initial trial function including the Jastrow factor, within the range of statistical error

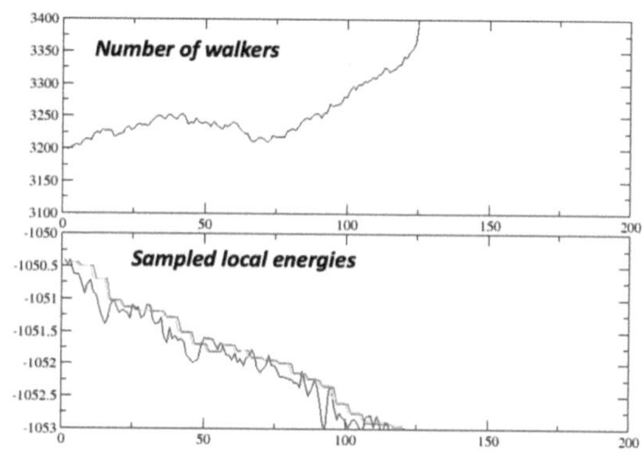

Fig. 4.3 DMC calculation exhibiting population explosion. Beyond 125 steps, the number of configurations diverges (upper plot), leading to a calculation halt

issues, but in general, a trial function without the Jastrow factor often leads to divergent behavior in the DMC calculation, causing it to stop. We will later conduct an exercise on such a case in Sect. 6.2.7. A typical example of such behavior is shown in Fig. 4.3. Why does this happen? This will be discussed later in detail in relation to the **cusp theorem** (Sect. 8.1). For now, a rough explanation is as follows: If, in a randomly sampled snapshot $\mathbf{R}_j$, an electron gets too close to another charge, a **near-contact configuration** occurs, leading to a numerical situation where it "hits the infinity" due to the potential $\sim 1/r_{ij}$. In this case, the number of walkers diverges as shown in Fig. 4.3, causing memory overflow and forcing the calculation to stop. This issue is referred to as the **population explosion** problem.

Population explosion occurs when the trial wave function fails to accurately describe the behavior of the many-body wave function during particle coalescence. The Jastrow factor is introduced to improve this description. As detailed in Sect. 8.1, "the behavior of the many-body wave function during particle coalescence" refers to how $\Psi\left(\cdots, \mathbf{r}_i, \cdots, \mathbf{r}_j, \cdots\right)$ responds when some argument $\mathbf{r}_j$ approaches another, $\mathbf{r}_j \rightarrow \mathbf{r}_i$. In this limit, $\Psi\left(\mathbf{r}_1, \cdots, \mathbf{r}_N\right)$ exhibits cusp-like divergence. The Jastrow factor is designed to ensure that this cusp behavior is properly incorporated.

If the DMC equilibration process proceeds sufficiently, the initial trial wave function gradually relaxes toward the exact solution, ultimately capturing the cusp behavior even without the aid of the Jastrow factor. However, during the initial stages, if the cusp reinforcement from the Jastrow factor is insufficient, the wave function may still be underdeveloped in representing the cusp, leading to numerical instabilities where infinities occur, resulting in a population explosion that halts further relaxation. To ensure that the early relaxation process proceeds smoothly, some level of cusp reinforcement through the Jastrow factor is necessary. This preparation of cusp reinforcement is precisely the role of the variational optimization process described in this chapter.

Why Many Groups that Embarked on DMC Research Disappeared
The role of the Jastrow factor discussed here is highly instructive. No matter how many times one carefully rereads the principles of DMC, one finds that "in principle, it should converge with any trial function," and thus, the necessity of the Jastrow factor is not explicitly stated. However, this assumption holds only in a hypothetical scenario, where even if a population explosion occurs, one has access to **infinite memory** to store the diverging walker count. Recognizing this limitation, only a few research groups realized that "to successfully perform DMC, it is crucial to refine VMC first." Most groups believed that "DMC is based on principles independent of VMC, and its advantage lies in bypassing VMC," failing to recognize this crucial point. As a result, their DMC calculations frequently stalled, and under the guidance of supervisors convinced that "it should work," many graduate students faced "research burnout." Some even speculated that their DMC failures were due to the sign problem (Sect. 8.2), leading them into an unsolvable abyss. On the other hand, groups that realized the importance of VMC in DMC practice had, by the mid-90s, successfully implemented first-principles DMC calculations, even for systems as complex as periodic solids, with relative ease.

References

1. "Molecular Electronic-Structure Theory" (Wiley, 2013), T. Helgaker, P. Jorgensen, J. Olsen, ISBN: 978-11-18531-47-1
2. "Computational Physics" (Cambridge University Press, 2007), J. Thijssen, ISBN: 978-05-21833-46-2
3. "Introduction to Computational Chemistry, Third Edition" (Wiley, 2017), F. Jensen, ISBN: 978-11-18825-99-0
4. "Quantum Monte Carlo Approaches for Correlated Systems" (Cambridge University Press, 2017), F. Becca, S. Sorella, ISBN: 978-11-07129-93-1
5. "Optimization of Many-body Wave function" R. Maezono, J. Comput. Theor. Nanosci., 6, 2474 (2009). https://doi.org/10.1166/jctn.2009.1308

6. “Jastrow correlation factor for atoms, molecules, and solids” N.D. Drummond, M.D. Towler, R.J. Needs, Phys. Rev. B **70**, 235119 (2004). https://doi.org/10.1103/PhysRevB.70.235119
7. “Numerical Recipes with Source Code CD-ROM 3rd Edition” (Cambridge University Press, 2007), W.H. Press, S.A. Teukolsky, W.T. Vetterling, B.P. Flannery, ISBN: 978-05-21884-07-5
8. “Continuum variational and diffusion quantum Monte Carlo calculations”, R.J. Needs, M.D. Towler, N.D. Drummond, P.L. Rios, J. Phys.: Condens. Matter 22, 023201 (2010), https://doi.org/10.1088/0953-8984/22/2/023201

Chapter 5
Generating Trial Nodes with a DFT Package

Continuing from the previous chapter, we now aim to replace the approach where "Chap. 3 relied on preprepared mature dough $\Psi_\alpha(\mathbf{R})$ provided by the instructor" with a method for **self-preparation**. For Slater-Jastrow-type many-body wave functions, the previous chapter assumed "using a preprepared Slater part" while acquiring the technique to optimize the Jastrow part independently. This chapter explains how to generate the Slater component, "bwfn.data," by oneself.

5.1 Constructing a Slater Determinant from One-Body Orbitals

5.1.1 *Generating One-Body Orbitals*

The Slater-Jastrow-type trial function

$$\Psi_{\boldsymbol{\alpha}}(\mathbf{r}_1, \cdots, \mathbf{r}_N) = \exp\left[J(\mathbf{R}; \boldsymbol{\alpha})\right] \times \Psi^{(\mathrm{SD})}(\mathbf{R})$$

has a Slater component $\Psi^{(\mathrm{SD})}(\mathbf{R})$ constructed from the set of one-body orbitals $\left\{\psi_j(\mathbf{r})\right\}_{j=1}^{N}$, forming the Slater determinant

$$\Psi^{(\mathrm{SD})}(\mathbf{r}_1, \mathbf{r}_2, \cdots, \mathbf{r}_N) = \sum_P (-)^P \cdot \psi_1\left(\mathbf{r}_{P_1}\right)\psi_2\left(\mathbf{r}_{P_2}\right)\cdots\psi_N\left(\mathbf{r}_{P_N}\right)$$

$$= \begin{vmatrix} \psi_1(\mathbf{r}_1) & \cdots & \psi_1(\mathbf{r}_N) \\ \vdots & \ddots & \\ \psi_N(\mathbf{r}_1) & \cdots & \psi_N(\mathbf{r}_N) \end{vmatrix} .$$

R. Maezono, *Ab initio Quantum Monte Carlo Tutorial*,
https://doi.org/10.1007/978-981-95-4134-8_5

Thus, the procedure involves computing the same target system using another method that handles the problem in terms of one-body orbitals. The resulting set of orbitals $\{\psi_j(\mathbf{r})\}_{j=1}^{N}$ is then used to construct the Slater determinant. In the examples provided in this book, one-body orbitals are generated using the density functional theory (DFT) package "Quantum Espresso" [1]. These orbitals are subsequently converted into a Slater determinant, which is stored in the "bwfn.data" file.

It is worth noting that ab initio QMC packages are designed for general use, regardless of whether the system is a solid periodic system or an isolated system. For solid periodic systems, one-body orbitals are typically represented using plane-wave basis functions (plane wave). Slater determinants constructed from these plane-wave-expanded orbitals are stored in a file named "pwfn.data," taking the "p" from "plane wave" (see Fig. 5.1).

On the other hand, in the case of isolated molecular systems, implementations commonly employ Gaussian basis functions, such as those in "Gaussian" [2], rather than plane waves. It is also common to use molecular orbital methods instead of density functional theory. When a Slater determinant is constructed from Gaussian-based one-body orbitals, it is stored in a file named "gwfn.data," where "g" is taken from "gaussian."

Additionally, for isolated atoms, high-precision calculations often take advantage of the small-scale, spherically symmetric nature of the system. In such cases, one-body orbitals are represented numerically without basis functions. These numerically represented Slater determinants are stored in a file named "awfn.data," where "a" is taken from "atomic" (see Fig. 5.1).

In this book, we consider a case where one-body orbitals are constructed using the **DFT** with a **plane-wave basis** and generated using **Quantum Espresso**. Alternatively, as shown in Fig. 5.1, "CASTEP" can also be used instead of "Quantum Espresso." The choice of implementation depends on the availability of format converters.

It is also possible to use **molecular orbital methods** instead of DFT. Moreover, instead of a **plane-wave basis**, one can use **Gaussian or numerical bases** to generate orbitals. A **tutorial on isolated molecular systems using Gaussian basis functions** will be provided later in Sect. 6.2.

5.1.2 Plane-Wave Basis Calculations Converted to Localized Basis

In the previous section, we explained that Slater determinants constructed from one-body orbitals expanded in plane-wave basis are stored in **pwfn.data**. However, in this tutorial, we use **bwfn.data**. This indicates that the one-body orbitals are represented using a blip basis expansion. As shown in Fig. 5.1, the workflow follows the following sequence: **QuantumEspresso** first outputs **pwfn.data**, which is then converted into **bwfn.data** before actual use.

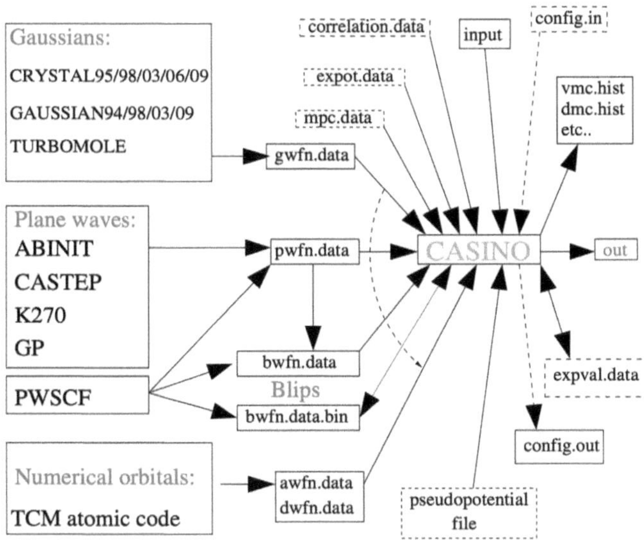

Fig. 5.1 A flowchart for generating many-body wave functions (figure taken from the CASINO manual [3]). One-body orbitals are generated using packages that employ plane-wave or Gaussian basis functions, and they are written in formats such as **pwfn.data** or **gwfn.data** for use as input files in CASINO. The format written in **pwfn.data** is further converted into **bwfn.data** for actual use

The existence of the **pwfn.data** format indicates that QMC calculations using plane-wave basis expansion are also possible. However, it has been found that performing QMC calculations directly in the plane-wave basis results in significantly slow computation. By re-expanding the wave function originally expressed in plane-wave basis into a blip basis, a speedup of approximately **hundreds of times** can be achieved [4]. Here, we briefly explain the reasons behind this significant computational efficiency improvement.

First, recall that the Fourier transform of a delta function is a constant function. A delta function represents a localized event. The fact that its Fourier transform becomes a constant implies that when representing localized events using a plane-wave expansion, nearly all basis functions contribute almost equally. As a result, in Monte Carlo simulations, whenever an electron position is updated, all the coefficients of the plane-wave basis must be updated accordingly. On the other hand, in representations using Gaussian basis functions, a localized update only affects a much smaller number of basis functions, and only those located near the event are updated.[1] In QMC calculations using **pwfn.data**, every electron position update requires **updating thousands or even tens of thousands of plane-wave**

[1] In contrast, plane-wave basis functions extend throughout the entire system, making it impossible to specify their "location."

basis coefficients, leading to a high computational cost. This demonstrates that using **localized basis functions** such as Gaussians allows for significantly faster wave function evaluations.

This concept of locality, handling local events locally without affecting the entire system, is a fundamental principle in numerical processing beyond just Monte Carlo calculations. For example, in PowerPoint or other graphics software, when adjusting the position of pixels forming a curve, only the nearby pixels are slightly adjusted to maintain smooth connections, while distant pixels remain unchanged. This preservation of locality is due to the use of **spline functions**, which serve as highly localized basis functions for curve representation.[2] Gaussian basis functions exhibit good locality, but research into even more localized basis sets has been driven by broader applications, including computer graphics. As a result, spline basis functions have been developed and widely utilized [5]. The **blip basis** used in this book is also a three-dimensional function representation technique based on splines.[3]

QuantumEspresso outputs orbitals expressed in plane-wave expansion. Previously, this required using a CASINO utility to re-expand the orbitals in the blip basis before generating the necessary files. However, QuantumEspresso has since incorporated a built-in feature that directly outputs orbitals in the blip basis. As a result, QuantumEspresso can now generate **bwfn.data** directly, eliminating the need for a separate conversion step.

Handling Periodic Solids with Gaussian Basis Sets
The implementation of ab initio QMC using the blip basis set advanced significantly around 2004, after the author entered this field [6]. Prior to that, the computational cost of plane-wave-based calculations was a major issue. To overcome this, researchers used the package **CRYSTAL** [7] to construct Slater determinants for periodic solid systems using Gaussian basis sets.

5.1.3 Tutorials to Generate bwfn.data

Now, let's use **QuantumEspresso** to generate **bwfn.data** and verify whether it reproduces the same QMC calculation results as before. First, prepare the working directory.

```
% pwd
    /Users/maezono/work
```

[2] The term "spline" originates from flexible strips once used for drawing smooth curves, often made from whale baleen.

[3] The term "blip" refers to a localized bright spot or flash of light.

```
% cp -r qmcTutorials/genwfnPw/ 07genwfn/
% cd 07genwfn/
% ls
    al_ppot.gon   input.in  n_ppot.gon  pw2casino.dat
```

The file **input.in** serves as the standard input for **QuantumEspresso**. Examining its contents reveals that the unit cell of AlN and a $1 \times 1 \times 1$ simulation cell are specified.[4] In particular, note the following section:

```
% cat input.in
...
 &system
    nosym=.true.
...
ATOMIC_SPECIES
 Al  26.9815  al_ppot.gon
 N   14.007    n_ppot.gon
...
```

The line **nosym=.true.** ensures that **Brillouin zone folding is disabled**. In standard density functional theory (DFT) calculations, symmetry reduction techniques are used to compute only the inequivalent regions of the Brillouin zone, reducing computational cost. However, when generating trial wave functions for QMC, this symmetry reduction must be disabled to explicitly compute all k-points in the Brillouin zone.

Within the above **input.in**, it is specified that Al and N use the pseudopotentials **al_ppot.gon** and **n_ppot.gon**, respectively. These correspond to **al_pp.data** and **n_pp.data**, which were used in the QMC calculations but formatted for input in QuantumEspresso. As a general rule, the same pseudopotentials should be used for both DFT and QMC calculations. However, the pseudopotentials employed in QMC differ from those commonly used in DFT, requiring careful attention. This topic will be discussed in detail in later sections (Sects. 5.2, 8.3).

The QuantumEspresso calculation for generating **bwfn.data** is executed with

```
% pwd
    /Users/maezono/work/07genwfn/
% pw.x -pw2casino < input.in | tee out
```

Here, the **-pw2casino** option is used.[5] An auxiliary input file, **pw2casino.dat**, is read during execution:

```
% cat pw2casino.dat
    &inputpp
    blip_convert=.true.
    blip_multiplicity=1.d0
    blip_binary=.false.
```

---

[4] For details on QuantumEspresso input specifications, refer to the author's previous books [8].

[5] As noted in "Version Information," "-pw2casino" seems to be no longer supported in v7.4 and later.

```
        blip_single_prec=.false.
        n_points_for_test=0
        /
```

While detailed descriptions of the options are omitted, note that **blip_multiplicity =1.d0** specifies the spatial resolution of the blip function expansion basis.

When the above **pw.x** calculation is executed, a file named **aln.bwfn.data** is generated in the working directory. This file corresponds to the desired **Slater determinant part expressed in the blip basis**, i.e., **bwfn.data**. Now, let's verify whether we obtain the same computational results using this file as before:

```
% pwd
    /Users/maezono/work
% cp -r 02vmcOnly 08vmcOnlyByOwnBw
% cd 08vmcOnlyByOwnBw
% mv input_org input
% clearup
% rm bwfn* input_* out* vmc.hist*
% ls
    al_pp.data   input
    correlation.data   n_pp.data
```

This prepares a working directory by copying the same input conditions as the previous VMC calculation. Next, set up **bwfn.data** as follows:

```
% pwd
    /Users/maezono/work/08vmcOnlyByOwnBw
% cp ../07genwfn/aln.bwfn.data .
% mv aln.bwfn.data bwfn.data
% ls
    al_pp.data   input  bwfn.data
    correlation.data   n_pp.data
```

Now, execute the CASINO calculation using **mpirun**. As a result, it is confirmed that the same results as the previous calculation conducted in **02vmcOnly** are obtained.

5.2 Preparing Pseudopotentials

In the previous sections, we learned the process of generating initial trial wave functions for ab initio QMC using DFT. Once the target system is determined, the practical workflow involves using a DFT or molecular orbital package to generate single-particle orbitals. At this stage, users attempting to run DFT often encounter the following question: **"Where do I obtain the pseudopotential?"** In practice, this is one of the most challenging aspects, and in most cases, **"It must be prepared by an expert".**

Creating Pseudopotentials
There are not many research groups capable of independently creating or adjusting pseudopotentials. In the early days, the author also relied on international collaborators to prepare them for research. For pseudopotentials in ab initio QMC, as will be discussed in this section, there have been over 20 years of difficult challenges. Thus, obtaining highly reliable pseudopotentials remains a crucial aspect of the process.

As pointed out in the previous section, in principle, the same pseudopotential used in QMC should be employed in DFT. Pseudopotentials are roughly classified into norm-conserving, ultrasoft, and PAW types [8]. The ultrasoft and PAW types were specifically developed for the formalism of DFT and cannot be used in QMC, which belongs to *many-body wave function* theory, not the *density functional* theory. Thus, it is necessary to perform DFT or molecular orbital calculations using norm-conserving pseudopotentials.

Research on desirable pseudopotentials suitable for ab initio QMC began around the 1990s [9]. It has been suggested that pseudopotentials generated from relativistic Dirac-Fock calculations, which do not diverge at the origin, are preferable [10]. In DFT, pseudopotentials are often constructed based on precise atomic calculations within the DFT framework. However, in ab initio QMC, pseudopotentials generated from DFT are rarely used. Since QMC research primarily focuses on the **accurate analysis of electronic correlation**, pseudopotentials derived from Hartree-Fock or Dirac-Fock methods, which do not account for electronic correlation, are generally preferred over those partially incorporating electronic correlation through DFT, for making the baseline with zero correlation clearer.

The requirement for non-divergence at the origin is related to the issue of population explosion discussed in Sect. 4.3 and is explained in connection with the cusp theorem (Sect. 8.3.1). Even after 2000, pseudopotentials for ab initio QMC have continued to face criticisms regarding asymptotic behavior at long distances, local approximation errors, and transferability. As of 2023, improvements are still being proposed, making this an ongoing fundamental challenge [11].

As described above, pseudopotentials for QMC applications are highly specialized for this field. Thus, users must first obtain pseudopotentials provided by the QMC community. These are typically available for download from the portal sites of major ab initio QMC packages. The practical challenge, then, is how to perform DFT calculations using these QMC pseudopotentials.

In the case of CASINO, after downloading QMC pseudopotentials in CASINO format, a utility is used to convert them into a format readable by QuantumEspresso. This utility is provided by CASINO, allowing the conversion of QMC-format "al_pp.data" into QuantumEspresso-compatible "al_ppot.gon." These practical procedures evolve with applications and time, continuously changing even today, so detailed instructions are omitted here.[6] For beginners, the key concept to grasp is "Obtain a QMC pseudopotential, convert it into a DFT-compatible format, run the DFT calculation, and construct the Slater determinant."

5.3 Trial Wave Functions Used in Ab Initio QMC

This book presents an example where initial trial wave functions are generated using "plane-wave DFT with pseudopotentials." In materials science, this approach is considered the most common in practice. Alternatively, trial wave functions can also be generated using "Gaussian basis calculations with pseudopotentials".[7] In this case, not only DFT but also general molecular orbital methods can be applied to generate initial trial wave functions. In Eq. (4.5),

$$\Psi_\alpha(\mathbf{r}_1, \cdots, \mathbf{r}_N) = \exp[J(\mathbf{R}; \boldsymbol{\alpha})] \times \Psi^{(\mathrm{SD})}(\mathbf{R})$$

the antisymmetry of the many-body wave function was represented by a single determinant $\Psi^{(\mathrm{SD})}(\mathbf{R})$. However, calculations are also possible to perform where this part is replaced by a multi-determinant wave function $\Psi^{(\mathrm{MD})}(\mathbf{R})$, obtained from "CASSCF calculations using the Gaussian package" and similar methods [12].

In molecular orbital calculations using Gaussian bases, **all-electron calculations** without pseudopotentials are also common. Such all-electron calculations can also be handled in ab initio QMC (Sect. 6.2). The preparation of pseudopotentials is one consideration that is omitted, making it seemingly simpler. However, from a specialized perspective, there is a sense that **all-electron calculations are more difficult**. One reason is that it becomes necessary to address the cusp associated with the coalescence between electrons and nuclei (Sect. 8.1). Another reason is that the energy range differs significantly between valence and core electrons.[8] This requires sophisticated techniques to adjust the update step size for electron positions. "Using a pseudopotential" is not only a passive approach to reducing computational cost by decreasing the number of electrons considered but also an active strategy to ensure numerical stability and reliability by limiting the calculation to valence electrons, thereby unifying the numerical range.

---

[6] It is likely that generative AI like ChatGPT will handle such tasks in the future.

[7] In molecular orbital methods using Gaussian bases, the term ECP (effective core potential) is sometimes used instead of pseudopotential, but they are essentially the same.

[8] Handling quantities that differ by several orders of magnitude simultaneously often leads to numerical instabilities, such as loss of precision, which is considered a major challenge in numerical computations [13].

DFT without Pseudopotentials
In addition to the standard method using plane-wave basis sets, DFT can also be performed using full-potential methods (such as FLAPW) or Green's function-based approaches (such as KKR). These methods allow for calculations without the use of pseudopotentials. However, because they are formulated differently from methods that explicitly construct many-body wave functions, they are practically difficult to use for generating trial wave functions for ab initio QMC.

When considering **practical applications** in materials science, as assumed in this book, it is generally expected that the same target system can be treated using conventional methods such as DFT or MO. A straightforward approach is to construct the antisymmetric part $\Psi^{(SD)}(\mathbf{R})$ using orbitals obtained from these conventional ab initio methods based on a single-particle picture. However, the initial trial wave function for ab initio QMC **is not necessarily limited** to a many-body wave function expressed in terms of single-particle orbitals.[9] Once an antisymmetric many-body function can be explicitly given, the same method and principles can be applied to analyze any trial wave function.[10] A major advantage of QMC is their flexibility in defining many-body wave functions. This advantage proves highly effective in handling electronic correlations, which are challenging to describe with conventional theories. As a result, the application of various many-body trial wave functions, such as geminal, Pfaffian, and backflow types, has become a key topic at the forefront of research [4, 15].

Breakthroughs Enabled by Many-Body Wave Function Forms
Examples include Cooper's trial wave function, which led to BCS theory, and Laughlin's trial wave function for the quantum Hall effect [16]. During my time as a graduate student, I worked in a field that described condensed matter physics using quantum field theory. I was struck by the approach of explicitly constructing exotic many-body wave functions to achieve breakthroughs. At the time, a senior researcher explained to me, somewhat imprecisely,

(continued)

[9] For DFT, the Hohenberg-Kohn framework was implemented in terms of single-particle orbitals through the Kohn-Sham formulation. Before the development of the Kohn-Sham method, there were studies that did not rely on single-particle orbitals [14], and recently, research in this direction has been pursued as orbital-free DFT.

[10] Conversely, if applicability is restricted to a specific "solution representation," one can sacrifice generality to advance analytical formulations further, leading to "more solvable equations." Molecular orbital methods and Kohn-Sham DFT fall into this category.

"Laughlin had done a lot of work similar to band structure calculations, which led him to that insight."

To envision many-body wave functions that do not rely on single-particle orbitals, a historically famous example is the Hylleraas-type trial function, which serves as a representative case. For the helium atom, instead of constructing the many-body wave function $\Psi_{\mathrm{He}}(\mathbf{r}_1, \mathbf{r}_2)$ as a sum of products of single-electron orbitals, it is explicitly expressed as a function of electron-electron distances

$$s = |\mathbf{r}_1| + |\mathbf{r}_2| \quad , \quad t = |\mathbf{r}_1| - |\mathbf{r}_2| \quad , \quad u = |\mathbf{r}_1 - \mathbf{r}_2|$$

and expanded as

$$\Psi(\mathbf{r}_1, \mathbf{r}_2) = \exp\left(-\frac{1}{2}k \cdot s\right)\left(1 + \beta \cdot u + \gamma \cdot t^2 + \delta \cdot s + \varepsilon \cdot s^2 + \zeta \cdot u^2 \right.$$
$$\left. + \chi_6 \cdot s \cdot u + \cdots + \chi_{(lmn)} \cdot s^l \cdot t^m \cdot u^n + \cdots \right).$$

Variational calculations optimize the coefficients in this expression following the variational principle [17].[11]

Finally, it should be noted that the Slater determinant constructed using Kohn-Sham orbitals from DFT has no intrinsic physical meaning. Hartree-Fock orbitals were introduced to approximate the many-body wave function using a Slater determinant. In contrast, Kohn-Sham orbitals were not introduced with this purpose [8]. Thus, a Slater determinant built from Kohn-Sham orbitals does not have an intrinsic role in approximating physical properties. However, since it satisfies the antisymmetry condition of Eq. (4.3), it can be utilized as a building block for a many-body trial function. This is a formal justification rather than a physical motivation. Regardless of its physical interpretation, if a trial function yields a lower energy under the variational principle, it is guaranteed to be closer to the exact solution. This is the strength of the variational method.

References

1. https://www.quantum-espresso.org (URL confirmed on 2024.10)
2. https://gaussian.com (URL confirmed on 2024.11)

[11] Hylleraas initially proposed a six-parameter variational method in 1929. Later, Pekeris extended this approach to a 1078-dimensional secular equation, further refining the accuracy [17].

3. "Continuum variational and diffusion quantum Monte Carlo calculations", R.J. Needs, M.D. Towler, N.D. Drummond, P.L. Rios, J. Phys.: Condens. Matter 22, 023201 (2010), https://doi.org/10.1088/0953-8984/22/2/023201
4. "Optimization of Many-body Wave function" R. Maezono, J. Comput. Theor. Nanosci., 6, 2474 (2009). https://doi.org/10.1166/jctn.2009.1308
5. "Numerical Recipes with Source Code" (Cambridge University Press, 2007), W.H. Press, S.A. Teukolsky, W.T. Vetterling, B.P. Flannery, ISBN: 978-0521884075
6. "Efficient localized basis set for quantum Monte Carlo calculations on condensed matter" D. Alfe and M.J. Gillan, Phys. Rev. B 70, 161101(R) (2004). https://doi.org/10.1103/PhysRevB.70.161101
7. https://www.crystal.unito.it/index.php (URL confirmed on 2024.11)
8. "Ab initio Calculation Tutorial: For Materials Analysis, Informatics and Design", (Springer Nature, 2023), R. Maezono, ISBN: 978-9819909216
9. "Variational quantum Monte Carlo nonlocal pseudopotential approach to solids: Formulation and application to diamond, graphite, and silicon", S. Fahy, X.W. Wang, S.G. Louie Phys. Rev. B 42, 3503 (1990). https://doi.org/10.1103/PhysRevB.42.3503
10. "Pseudopotentials for correlated-electron calculations", Y. Lee, P.R.C. Kent, M.D. Towler, R.J. Needs, G. Rajagopal, Phys. Rev. B 62, 13347 (2000). https://doi.org/10.1103/PhysRevB.62.13347
11. "Locality Error Free Effective Core Potentials for 3d Transition Metal Elements Developed for the Diffusion Monte Carlo Method", T. Ichibha, Y. Nikaido, C. Bennett, J. Krogel, K. Hongo, R. Maezono, and F. Reboredo. J. Chem. Phys. 159, 164114 (2023). https://doi.org/10.1063/5.0175381
12. "A Benchmark quantum Monte Carlo study of the ground state chromium dimer", K. Hongo and R. Maezono, Int. J. Quant. Chem. 112, 1243 (2012), https://doi.org/10.1002/qua.23113
13. "Computational Physics: Problem Solving with Python" (Wiley-VCH, 2015) R.H. Landau, M.J. Paez, C.C. Bordeianu ISBN: 978-3527413157
14. "Density-functional Theory of Atoms And Molecules" (Oxford University Press, 1989), R.G. Parr, W. Yang ISBN: 978-0195092769
15. "Correlation and Localization" (Springer, 1999), P.R. Surjan, ISBN: 978-3540657545
16. "Quantum Monte Carlo Approaches for Correlated Systems" (Cambridge University Press, 2017), F. Becca, S. Sorella, ISBN: 978-1107129931
17. "Introduction to Computational Chemistry, Third Edition" (Wiley, 2017), F. Jensen, ISBN: 978-1118825990

Chapter 6
Review of Procedures and Molecular System Calculations

Up to this chapter, we have conducted practical exercises while incorporating conceptual explanations related to the computational principles of ab initio QMC. The most efficient order for beginners to grasp concepts does not necessarily match the most efficient workflow for practical applications after mastering the concepts. Based on this idea, the topics have been introduced in a sequence different from the actual workflow of practical computations. In this chapter, we will review our understanding by following the procedural steps in actual workflow.

In the exercises so far, we have focused on ab initio QMC using plane-wave-based DFT orbital functions for solid-state periodic systems, considering the availability of completely free packages. However, ab initio QMC packages are also compatible with software based on Gaussian basis functions, such as **Gaussian13**. This capability makes them highly effective for describing fine electronic correlations targeted in quantum chemistry. In the first half of this chapter, we will review the procedural workflow for solid-state periodic systems. Following that, we will explain how **isolated molecular systems** are handled using Gaussian basis functions.

Communities Focusing on Solid-State Periodic Systems and Isolated Molecular Systems

When I entered this field, the two research communities were quite separate, with little interaction. Young researchers in quantum chemistry were unfamiliar with leading figures in solid-state physics, and vice versa. As a result, the use of Gaussian bases in solid-state physics or plane-wave bases in molecular science was rare, and the differences between the fields often led to misunderstandings. Since my career started directly with ab initio quantum Monte Carlo, I learned both plane-wave and Gaussian basis methods

(continued)

R. Maezono, *Ab initio Quantum Monte Carlo Tutorial*,
https://doi.org/10.1007/978-981-95-4134-8_6

simultaneously, allowing me to view ab initio calculations from a broader perspective before returning to Japan. Looking back at my presentation materials from that time, I can recall how much effort I put into bridging these cultural gaps. As I mentioned in my previous book [1], at that time, almost no textbooks on ab initio methods consciously addressed the contrast between solid-state physics and molecular science. The situation changed with the national supercomputer project in Japan. Previously, the three major fields of solid-state physics, quantum chemistry, and computational materials science operated independently, but the project brought them together, somewhat forcibly.[1] Over the past decade, this "forced integration" of research and training has led to significant improvements. Nowadays, researchers in these fields are at least aware of each other's leading figures, reflecting a positive trend toward interdisciplinary collaboration.

6.1 Practical Tutorials Following the Actual Workflow (Solid-State Periodic Systems)

In the previous sections, we prioritized introducing concepts to beginners, presenting topics in a different order from actual practice. Looking back, we followed this sequence: first, we treated the trial function as a given and ran Monte Carlo simulations; then, we learned the mechanics of Monte Carlo sampling; next, we explored how trial functions relax through diffusion Monte Carlo; and finally, we covered the refinement of initial trial functions via variational optimization, followed by the generation of trial nodes for constructing initial trial functions.

On the other hand, after introducing these concepts, restructuring the workflow for practical execution results in the following sequence: "Trial function generation → Variational optimization of the trial function → Variational Monte Carlo evaluation → Diffusion Monte Carlo computation (configuration generation/equilibration/statistical accumulation)".[2] From this point onward, we proceed with the training following this practical workflow.

```
% pwd
    /Users/maezono/work
% mkdir pwFlow
```

A working directory is prepared at this location to proceed with the calculation.

[1] Funding, which was previously allocated separately for each field, was consolidated under a single budget framework.

[2] When asked to teach ab initio QMC to beginners, one might instinctively follow the exact workflow used in actual practice. However, this approach often forces beginners to perform a series of steps without grasping the overarching picture, which can diminish their motivation to learn.

6.1.1 Generating Trial Function Using Plane-Wave Basis

First, the trial function is generated using the plane-wave basis DFT calculation package "QuantumEspresso" (Sect. 5.1). The working directory is prepared as follows:

```
% pwd
    /Users/maezono/work/pwFlow/
% mkdir 01genPw
% cd 01genPw/
% cp ../../qmcTutorials/genwfnPw/*.gon .
% cp ../../qmcTutorials/genwfnPw/pw2casino.dat .
% cp ../../qmcTutorials/genwfnPw/input.in .
% ls
    al_ppot.gon    input.in
    n_ppot.gon     pw2casino.dat
```

Please review the meaning of each file to be prepared:

- Pseudopotential files (xx.gon)
- QuantumEspresso input file (input.in)
- Additional input file (pw2casino.dat), which is referenced as an option for conversion into the CASINO format.

Refer to Sect. 5.1 for further details.

Prepare the above files and execute **QuantumEspresso**:[3]

```
% pw.x -pw2casino < input.in | tee out
% ls
    input.in        out
    aln.bwfn.data   aln.xml n_ppot.gon
    pw2casino.dat   al_ppot.gon    aln.save/
```

The obtained **aln.bwfn.data** is the Slater determinant constructed by antisymmetrizing the Kohn-Sham orbitals generated via DFT (Eq. (4.4)).

QuantumEspresso generates Kohn-Sham orbitals expanded in a plane-wave basis. The information for these orbitals is stored under the directory **aln.save**, which is created after execution [1]. By specifying the **-pw2casino** option when executing the QuantumEspresso binary **pw.x**, the orbital functions are re-expanded in a localized basis, namely, the blip function basis (Sect. 5.1.2). The computational settings for this re-expansion into blip functions are specified in **pw2casino.dat**.

The number of electrons is determined from the results of the DFT calculation as follows. Extracting information from the standard output file **out**,

```
% pwd
    /Users/maezono/work/pwFlow/01genPw
% grep number out
```

[3] For Mac, QuantumEspresso was installed in Sect. 11.2.1, and the execution binary path was set in Sect. 11.1.3.

```
    ...
    number of atoms/cell       =            4
    ...
    number of electrons        =        16.00
    ...
% more out
    ...
    ...
    atomic species   valence     mass     pseudopotential
    Al                3.00    26.98150      Al( 1.00)
    N                 5.00    14.00700      N ( 1.00)
    ...
    site n.     atom                positions (alat units)
        1        Al       tau(1) = (0.5000000  -0.2886724   0.0000000  )
        2        Al       tau(2) = (0.5000000   0.2886724   0.8005500  )
        3        N        tau(3) = (0.5000000  -0.2886724   0.6114121  )
        4        N        tau(4) = (0.5000000   0.2886724   1.4119621  )
```

From this information, it is determined that

- The valence electron count is three for Al and five for N.
- The unit cell contains two Al and two N atoms.

Furthermore, examining the k-point setting,

```
% more input.in
    ...
    K_POINTS {automatic}
    1 1 1 0 0 0
    ...
```

confirms that the simulation cell is $1 \times 1 \times 1$. Thus, the total number of simulated electrons is 16.

The generated trial wave function, along with the CASINO-format pseudopotentials **xx_pp.data**, will be linked and reused in subsequent steps. These files are stored in the following directory:

```
% pwd
    /Users/maezono/work/pwFlow
% mkdir 02files
% cd 02files/
% cp ../../qmcTutorials/introByPw/*pp.data .
% cp ../01genPw/aln.bwfn.data bwfn.data
% ls
    al_pp.data  bwfn.data  n_pp.data
```

From this point onward, all QMC calculations will access the common file set from the **02files** directory through symbolic links.

6.1.2 VMC Without Jastrow and Preparation of Jastrow Function (Periodic Solid)

Using the prepared materials, we perform a VMC calculation **without a Jastrow factor**. One of the purposes of this step is to **establish a baseline** to measure how much the energy stabilizes with the inclusion of the Jastrow factor.[4] Additionally, executing the **VMC calculation without a Jastrow factor** serves a practical purpose: it generates an output file (**out**) that is used to create the Jastrow function file, **correlation.data** (Sect. 4.2.1). The Jastrow function requires information about the number of electrons and the types of inequivalent ionic cores. Since this information is written to the **out** file, it is extracted using the script **editCorr.py** to generate **correlation.data** (Sect. 4.2.1).

First, prepare the working directory as follows:

```
% pwd
    /Users/maezono/work/pwFlow
% mkdir 03vmcWoJ
% cd 03vmcWoJ/
% ln -s ../02files/* .
% ls
    al_pp.data@  bwfn.data@  n_pp.data@
% cp ../../qmcTutorials/introByPw/input .
```

In the previous section, wave function files (bwfn.data) and pseudopotential files (x_pp.data) were placed in the **02files** directory. Here, instead of copying (cp), they are placed into the working directory using symbolic links (**ln -s**). The input file is copied from the distributed materials directory (qmcTutorials).

Next, confirm the following section in the input file:

```
neu             : 8     #*! Number of up electrons (Integer)
ned             : 8     #*! Number of down electrons (Integer)
periodic        : T     #*! Periodic boundary conditions (Boolean)
atom_basis_type : blip  #*! Basis set type (Text)
%block npcell
1 1 1
```

Ensure that it matches the settings confirmed in the previous section: A total of 16 valence electrons and a $1 \times 1 \times 1$ simulation cell. Since the system is nonmagnetic, the number of spin-up and spin-down electrons is equally set to 8 each.

---

[4] Here, we use a Slater determinant constructed from Kohn-Sham orbitals. If we were to use a Slater determinant based on Hartree-Fock orbitals (Sect. 6.2), the resulting VMC energy would correspond to the Hartree-Fock energy (Sect. 8.4). That energy would serve as a reference for **zero electron correlation**, and the **reduction from this baseline** would represent the correlation energy (the energy stabilization due to the distortion of the wave function caused by interactions). While a VMC calculation without Jastrow using Kohn-Sham orbitals does not strictly follow this interpretation, it still provides a crucial reference value.

Next, carefully edit the following section of the input file (other parts can remain unchanged):

```
...
runtype             : vmc     #*! Type of calculation (Text)
newrun              : T       #*! New run or continue old (Boolean)
...
vmc_equil_nstep     : 1000    #*! Number of equilibration steps (Integer)
vmc_nstep           : 10000   #*! Number of steps (Integer)
vmc_nblock          : 1       #*! Number of checkpoints (Integer)
vmc_nconfig_write : 0         #*! Number of configs to write (Integer)
...
use_jastrow         : F       #*! Use a Jastrow function (Boolean)
...
```

The calculation mode is set to "vmc" for a new run (`newrun : T`), with no configuration output (`vmc_nconfig_write : 0`), and the Jastrow factor is disabled (`use_jastrow : F`). With this, run the calculation:

```
% mpirun -np 8 casino &
    [After the calculation finishes...]
% cat out
% cp bwfn.data.bin ../02files/
```

Once the run finishes, use the `cat` command to check the `out` file and confirm that the calculation terminated successfully. If the calculation has completed properly, you should find an energy value like the following in the final lines of the `out` file, under the heading "Correlation time method":

$$E = -23.365870889585 + / - 0.023059871987$$

Additionally, the binary file `bwfn.data.bin` will be generated as a result of the run. Since it will be used in later stages, copy it into the `02files` directory.

Then, generate the `correlation.data` file:

```
% cp ../../qmcTutorials/utils/editCorr.py .
% python3 editCorr.py > correlation.data
% ls -l
    ...
    ... 2162 ... correlation.data
% cat correlation.data
```

This will create **correlation.data**, which should be around 2162 bytes in size. Use the `cat` command to verify that its contents have been written correctly.

6.1.3 Variational Optimization of the Jastrow Factor (Periodic Systems)

In the previous section, we prepared a skeletal form of `correlation.data`. Now, we determine the parameter set it contains by minimizing the variance, as described in Sect. 4.1.2. First, prepare the working directory:

```
% pwd
   /Users/maezono/work/pwFlow
% mkdir 04opt
% cd 04opt/
% ln -s ../02files/* .
% cp ../03vmcWoJ/correlation.data .
% cp ../../qmcTutorials/introByPw/input_opt input
```

Here, we symlink the shared files from `02files/`, copy over the `correlation.data` created in the previous step, and set up the input file `input` from the provided tutorial set (`input_opt`).

Then, edit the `input` file with attention to the following settings:

```
runtype            : vmc_opt        #*! Type of calculation (Text)
newrun             : T              #*! New run or continue old (Boolean)
testrun            : F              #*! Test run flag (Boolean)
...
vmc_equil_nstep    : 1000           #*! Number of equilibration steps (Integer)
vmc_nstep          : 10000          #*! Number of steps (Integer)
vmc_nblock         : 1              #*! Number of checkpoints (Integer)
vmc_nconfig_write : 10000           #*! Number of configs to write (Integer)
vmc_decorr_period : 10              #*! Decorrelation loop length (Integer)
...
opt_method         : varmin_linjas  #*! Opt method (varmin/madmin/emin/...)
opt_cycles         : 10             #*! Number of optimization cycles (Integer)
opt_jastrow        : T              #*! Optimize Jastrow factor (Boolean)
...
use_jastrow        : T              #*! Use a Jastrow function (Boolean)
```

In this optimization mode (`runtype: vmc_opt`), we choose the method `varmin_linjas`, which minimizes the variance of the linear Jastrow parameters. The Jastrow factor is both used (`use_jastrow: T`) and optimized (`opt_jastrow: T`). This cycle is repeated ten times (`opt_cycles: 10`). To evaluate the variance to be minimized, we use 10,000 sampling points. To generate this sample, 10,000 steps of VMC are performed (`vmc_nstep: 10000`), and all sampled configurations are saved to the `config` file (`vmc_nconfig_write: 10000`).

Then, execute the calculation with the above `input` settings:

```
% !mpi
    mpirun -np 8 casino &
    ...[after the calculation finishes]...
% ls
    al_pp.data@       correlation.out.0  correlation.out.5
    bwfn.data@        correlation.out.1  correlation.out.6
    bwfn.data.bin     correlation.out.10 correlation.out.7
    config.in         correlation.out.2  correlation.out.8
    config.out        correlation.out.3  correlation.out.9
    correlation.data correlation.out.4  input n_pp.data@
    out
```

Once the calculation has finished, the utility **ve** can be executed to analyze the results as follows (Sect. 4.1.2):

```
% ve
    VMC #1: E = -23.37(2) ; var = 5.6(3)  (correlation.out.0)
    ...
    VMC #6: E = -23.87(1) ; var = 1.46(5) (correlation.out.5)
    ...
```

The result indicates that `correlation.out.5` yields the lowest variance and the most reduced energy, so we extract it as the optimized `correlation.data` by

```
% cp correlation.out.5 correlation.opt
```

Here, `correlation.opt` denotes the optimized version of `correlation.data`.[5]

6.1.4 VMC Using the Optimized Jastrow Function (Periodic Systems)

Next, we perform a VMC calculation using the Jastrow function optimized in the previous section and record the resulting energy as a **reference value**. This allows us to evaluate how much correlation energy has been captured by the Jastrow factor alone and later compare it to the correlation energy further captured through DMC-based trial wave function relaxation. First, prepare the working directory as follows:

```
% pwd
    /Users/maezono/work/pwFlow
% mkdir 05vmcWj
% cd 05vmcWj/
% ln -s ../02files/* .
% cp ../04optJ/input .
% cp ../04optJ/correlation.opt .
% mv correlation.opt correlation.data
% ls
    al_pp.data@  correlation.data   n_pp.data@
    bwfn.data@   input        bwfn.data.bin
```

Then, edit the input file with attention to the following settings:

```
runtype           : vmc #*! Type of calculation (Text)
...
vmc_nconfig_write : 0   #*! Number of configs to write (Integer)
...
```

[5] There is no need to be overly meticulous about choosing which output file to adopt as the optimum. Since this serves merely as a preselection before the DMC calculation, selecting, for example, `correlation.out.4` instead would make little difference in practice.

The calculation mode is set back to VMC (`runtype: vmc`), and configuration output is disabled (`vmc_nconfig_write: 0`). Then, execute the calculation by using `% !mpi`. If the calculation completes successfully, you should find an energy value in the final lines of the `out` file under the section labeled "Correlation time method," such as: E = -23.859232837990 ± 0.012723578232.

6.1.5 DMC Calculation (Periodic Solid)

Now that we have prepared an initial trial wave function that has been sufficiently optimized via variational methods (i.e., the dough has been matured before baking), we proceed to the DMC calculation. In DMC calculations, there are three stages: configuration generation (congen), equilibration (equil), and statistical accumulation (stats) (Sect. 3.3.2).

We begin with the **configuration generation** step in the following working directory:

```
% pwd
    /Users/maezono/work/pwFlow
% mkdir 06dmcCongen
% cd 06dmcCongen/
% ln -s ../02files/* .
% cp ../05vmcWj/correlation.data .
% cp ../05vmcWj/input .
% ls
    al_pp.data@  correlation.data  n_pp.data@
    bwfn.data@   input        bwfn.data.bin
```

Edit the input file with attention to the following items:

```
runtype             : vmc     #*! Type of calculation (Text)
...
vmc_equil_nstep     : 1000    #*! Number of equilibration steps (Integer)
vmc_nstep           : 10000   #*! Number of steps (Integer)
vmc_nblock          : 1       #*! Number of checkpoints (Integer)
vmc_nconfig_write : 512     #*! Number of configs to write (Integer)
...
```

Even for a DMC calculation, configuration generation is performed using VMC, so the initial mode is set as `runtype: vmc`. The number of walkers (configurations) N_w is set to 512, a multiple of 8 cores (`vmc_nconfig_write: 512`). When the calculation is executed with "% !mpi," a file named `config.out` will be written in the working directory. This is the initial configuration used in the DMC stage, and it corresponds to the histogram representing the distribution function on the left side of Fig. 3.1.

Next, we perform **equilibration** to relax the "initial histogram shape of the distribution" so that it approaches the "shape of the exact solution." Prepare the working directory as follows:

```
% pwd
    /Users/maezono/work/pwFlow
% mkdir 07dmcEquil
% cd 07dmcEquil/
% ln -s ../02files/* .
% ln -s ../06dmcCongen/correlation.data .
% cp ../06dmcCongen/input .
% cp ../06dmcCongen/config.out .
% mv config.out config.in
```

The "config.out" file generated during the configuration generation process is renamed to "config.in" and used in the calculation. Then, edit the input file with attention to the following parameter settings:

```
runtype            : dmc_equil  #*! Type of calculation (Text)
...
dmc_equil_nstep    : 100        #*! Number of steps (Integer)
dmc_equil_nblock   : 5          #*! Number of checkpoints (Integer)
dmc_stats_nstep    : 500        #*! Number of steps (Integer)
dmc_stats_nblock   : 25         #*! Number of checkpoints (Integer)
dmc_target_weight  : 512.d0     #*! Total target weight in DMC (Real)
```

The calculation mode is set to the equilibration process with **runtype: dmc_equil**. The number of walkers N_w prepared in the configuration generation step should match the **dmc_target_weight: 512.d0** specification. When you execute the calculation with "% !mpi," the directory contents will appear as follows:

```
% ls
    al_pp.data@       config.in           dmc.hist    out
    bwfn.data@        config.out          input
    bwfn.data.bin@    correlation.data@   n_pp.data@
```

Here, **config.out** corresponds to the "3relaxed histogram from the equilibration process," while **dmc.hist** contains the data for the local energy evolution during the equilibration process.

Next, we proceed to the **statistical accumulation stage**. Prepare the working directory as follows, and carry over the "equilibrated distribution (`config.out`)" and the sampling history of local energy values (`dmc.hist`) obtained from the equilibration process.

```
% pwd
    /Users/maezono/work/pwFlow
% mkdir 08dmcStats
% cd 08dmcStats/
% ln -s ../02files/* .
% ln -s ../06dmcCongen/correlation.data .
% cp ../07dmcEquil/input .
% cp ../07dmcEquil/config.out .
% cp ../07dmcEquil/dmc.hist .
% mv config.out config.in
% ls
    al_pp.data@   bwfn.data.bin@    correlation.data@
    bwfn.data@    config.in   input  n_pp.data@ dmc.hist
```

The transferred config file should be renamed to config.in for use as input in the next stage. Edit the input file with attention to the following settings:

```
...
runtype             : dmc_stats  #*! Type of calculation (Text)
newrun              : F          #*! New run or continue old (Boolean)
...
dmc_equil_nstep     : 100        #*! Number of steps (Integer)
dmc_equil_nblock    : 5          #*! Number of checkpoints (Integer)
dmc_stats_nstep     : 500        #*! Number of steps (Integer)
dmc_stats_nblock    : 25         #*! Number of checkpoints (Integer)
dmc_target_weight : 512.d0       #*! Total target weight in DMC (Real)
dtdmc               : 0.01       #*! DMC time step (Real)
```

The calculation mode is the statistical accumulation stage (runtype: dmc_stats). Since this is a continuation from the equilibration stage, the new run flag should be set to false (newrun: F). When executed with % !mpi, the statistical accumulation calculation begins.

6.1.6 Energy Change (Periodic Solid Systems)

At this point, energy values have been obtained for each stage: from "DFT" → "VMC (without Jastrow)" → "VMC (with Jastrow)" → "DMC". We now compile and compare these results.

To obtain accurate values for both VMC and DMC results, the reblocking procedure described in Sect. 2.4 must be applied. However, for simplicity, we will use the values printed in the out files here.[6] For the VMC values, we adopt those extracted using the following:

```
% pwd
    /Users/maezono/work/pwFlow
% grep 'Correlation time meth' */out
    03vmcWoJ/out: -23.365870889585 +/- 0.023059871987  Corr...
    05vmcWj/out:  -23.859232837990 +/- 0.012723578232  Corr...
```

For the DMC value, we adopt the following output found at the end of the out file:

```
mean: -23.956613945510 +/- 0.004300106572
```

As Quantum Espresso is used for the DFT value, the energy can be obtained as follows:

[6] In this example, neither the VMC nor DMC runs were executed as continuation (resume) runs for statistical accumulation (see Sect. 2.3.3). Therefore, the values reported in the out files for each run are effectively identical to the statistical averages over the total sampling. In actual practice, multiple runs are often concatenated to form a complete statistical ensemble. In such cases, the statistical result obtained from the reblocking analysis should be used, as it differs from the value reported in the out file of any single run.

```
% pwd
    /Users/maezono/work/pwFlow/01genPw
% grep ! out
    !  total energy = -47.60714416 Ry
% bc
    >>> -47.60714416*0.5
    -23.80357208
```

The multiplication by 0.5 is to convert from Rydberg units (Ry) to Hartree units.[7] Rounding to a consistent number of significant figures gives

```
DFT(GGA)     -23.8036
VMC(w/o J)   -23.37(2)
VMC(w J)     -23.86(1)
DMC          -23.957(4)
```

Comparing only the QMC results, the inclusion of the Jastrow factor (aging the dough before baking) lowers the energy. Furthermore, imaginary-time evolution by DMC (relaxing the wave function as the dough is baked) lowers the energy even further. These values are comparable along the **same variational track**, where a "lower energy" means "closer to the exact solution." On the other hand, the DFT value may appear unexpectedly low compared to the QMC results. It may also seem puzzling for beginners that "a trial wave function constructed from DFT, which produced $E = -23.8036$, yields a higher energy of $-23.37(2)$ in VMC." Here, it is important to recall that **the DFT energy value does not lie on the same variational track as QMC**. The DFT energy is influenced by the reference of the exchange-correlation potential, and its absolute value cannot be meaningfully compared with QMC energies.[8] This stands in contrast to the next section, where we compare QMC values with HF results in isolated molecular systems (Sect. 6.2).

DFT do not Lie on the Variational Track, but...

A memory comes to mind of a young researcher who had returned to Asia after being a postdoc colleague in the UK. They were struggling with an unreasonable demand to compare DFT and QMC energies as described in this section. Their supervisor, a professor specializing in strongly correlated

(continued)

[7] Concerning the factor $A = \hbar^2/2m$ appearing in the kinetic energy term of the Schrödinger equation, the unit in which one sets $\hbar = m = 1$ (corresponding to measuring energy in multiples of $\hbar$ and mass in multiples of the electron mass m) is called the **natural unit** (atomic unit). The energy unit in the natural unit, where $A = 1/2$, is referred to as the **Hartree unit**. On the other hand, the unit in which $A = 1/4$ is adopted is called the **Rydberg unit**, in which the bounding energy of the hydrogen atom is expressed simply as $E_n = -1/n^2$. In this way, the Hartree and Rydberg unit differ only by a factor of 2.

[8] Comparisons between DFT and QMC are only meaningful when considering energy *differences*, such as bond energies, cohesive energies, or defect formation energies.

systems, was misinformed, as was not uncommon in the 2000s, insisting, "DFT doesn't include electron correlation (*wrong!*), so the energy must be higher than VMC. Your calculation must be wrong!". No matter how clearly the young researcher explained the matter, the supervisor remained unconvinced. As a result, they were unable to proceed with the project, caught in a misunderstanding about the theoretical foundations.

6.2 Ab Initio QMC for Isolated Molecular Systems

Let us now proceed to perform calculations for isolated molecular systems using Gaussian basis sets, following essentially the same workflow as for periodic solid-state systems with plane-wave/blip basis sets described in the previous section.

Trial wave function generation with plane-wave basis sets was carried out using the open-source package *QuantumEspresso* (Sect. 6.1.1). In contrast, for the Gaussian basis set approach, we employ the commercial software **Gaussian13**.[9] Assuming the reader's PC does not have Gaussian03 installed, we will explain the procedure under the assumption of its availability, namely, "how to generate the necessary files if Gaussian03 is available" and "what files are expected as a result." In practice, we will provide such pre-generated files as instructional materials for the exercises that follow.

In this section, we will illustrate the procedure using the ground-state energy calculation of a water molecule, H_2O, as a working example (see Fig. 2.1). First, prepare the working directory *gwFlow* to learn the flow of calculation using Gaussian basis sets (gw) by executing the following:

```
% pwd
    /Users/maezono/work
% mkdir gwFlow
% cd gwFlow/
```

6.2.1 Generating Trial Wave Function Using Gaussian Basis

Next, execute the following commands to copy the necessary files for trial wave function generation using **Gaussian03** from the downloaded tutorial materials:

```
% pwd
    /Users/maezono/work/gwFlow
% mkdir 01genGw
```

[9] At the time of writing, "Gaussian16" (released in 2016) is the latest version, and future updates will likely change this number accordingly.

```
% cd 01genGw
% cp ../../qmcTutorials/genwfnGw/genGw/* .
```

Among the contents

```
% ls
    Test.FChk    h2o.com          h2o.out
```

only **h2o.com** is the essential **input file** for the calculation. The files "Test.FChk" and "h2o.out" are output files generated when **Gaussian03** is executed using the input file "h2o.com."

Looking at the contents of the input file **h2o.com**,

```
% cat h2o.com
    #p hf/sto-3g scf=tight formcheck=(mo,basis)

    H2O HF/STO-3G wfn for QMC

    0 1
    O
    H   1  R1
    H   1  R1  2  A1

    R1      0.96000000
    A1   109.50000006
```

we find that the latter half specifies the geometry of the system shown in Fig. 2.1. Unlike the DFT calculations for periodic solid systems, this is an **all-electron calculation without pseudopotentials**, so only nuclear positions are specified for the geometry. At the top of the file, the line beginning with "#p" contains the keyword `hf/sto-3g`, which indicates that a Hartree-Fock calculation is to be performed using the **STO-3G Gaussian basis set**.

If the reader **owns a licensed copy of "Gaussian03"**, then by deleting `Test.FChk` and `h2o.out` and executing the following command,

```
% g03 < h2o.com > h2o.out
```

both `Test.FChk` and `h2o.out` will be regenerated. For those who **do not own Gaussian03**, please proceed under the assumption that the above command has been executed and that the files `Test.FChk` and `h2o.out` have already been obtained.

`h2o.out` is the standard output file from **Gaussian03** (see Sect. 2.1). On the other hand, `Test.FChk` contains the exported wave function data. This is analogous to `aln.save` in the case of plane-wave basis (see Sect. 6.1.1). The instruction to write the wave function information to `Test.FChk` is specified in the input file `h2o.com`, where the line beginning with #p contains the keyword `formcheck=(mo,basis)`.

Next, the wave function information output by **Gaussian03** in `Test.FChk` must be converted into a format readable by CASINO. The utility `gaussiantoqmc`, which performs this conversion, is provided by the CASINO package. It is installed and made available in the command path when CASINO is compiled.

First, rename the wave function file `Test.FChk` so that the filename begins with `h2o`, for consistency with other related files:

```
% mv  Test.FChk h2o.Fchk
% ls
    h2o.Fchk    h2o.com        h2o.out
```

Next, execute the utility `gaussiantoqmc`:

```
% gaussiantoqmc
```

You will be prompted with

```
Please enter the root of the Gaussian94/98/03 input file name:
```

Enter the shared root name `h2o`. Once the conversion is complete, a new file `h2o.qmc` will be generated:

```
% ls
    h2o.Fchk    h2o.com   h2o.out   h2o.qmc
```

To be safe, let us confirm the file size is as follows:

```
% ls -l
...
-rw-r--r--@ 1 maezono  staff  21474 12 25 21:39 h2o.qmc
```

This file `h2o.qmc` contains the wave function information converted into CASINO-compatible format. As with the plane-wave basis case, we will store this wave function file in a directory named `02files`:

```
% pwd
    /Users/maezono/work/gwFlow
% mkdir 02files
% cd 02files/
% cp ../01genGw/h2o.qmc gwfn.data
% ls
    gwfn.data
```

Here, `h2o.qmc` is renamed as `gwfn.data`. CASINO expects the filename for Gaussian basis trial wave functions (with the *g*aussian label) to be `gwfn.data`. This naming convention corresponds to the case of blip basis wave functions (blip), where the expected filename was `bwfn.data` (see Sect. 6.1.1).

6.2.2 VMC Without a Jastrow and Preparation of the Jastrow Function (Isolated Molecular System)

Using the previously prepared "gwfn.data," we first perform a single VMC without a Jastrow factor to generate the standard output file "out." Based on the information such as the number of electrons and atomic positions written in that output file, we then use the utility **editCorr.py** to create the Jastrow function input file "correlation.data."

First, prepare the working directory as follows:

```
% pwd
    /Users/maezono/work/gwFlow
% mkdir 03vmcWoJ
% cd 03vmcWoJ/
% ln -s ../02files/gwfn.data .
% cp ../../qmcTutorials/introByPw/input .
```

Here, the trial wave function file "gwfn.data" is brought in by symbolic linking. The CASINO input file *input* is copied from the one used in the previous section for periodic solid systems and then edited for use here. By doing so, it becomes easier to understand what settings differ between periodic solid systems and isolated molecular systems.

As for the "input" file, begin by editing the number of electrons. The molecular system handled in this section is a **all-electron calculation without using pseudopotentials.**[10] The target system is H_2O, where oxygen contributes eight electrons and each hydrogen contributes one electron, giving a total of ten electrons in a nonmagnetic configuration. Thus, the input should be edited as

```
% cat input
     ...
     neu : 5  #*! Number of up   electrons (Integer)
     ned : 5  #*! Number of down electrons (Integer)
```

Unlike the periodic system in the previous section, this section deals with an isolated system using a Gaussian basis. Thus, the input should be edited as follows:

```
% cat input
     ...
     periodic        : F        #*! Periodic boundary conditions
                                    (Boolean)
     atom_basis_type : gaussian #*! Basis set type (Text)
```

In addition, remove the block that specifies the supercell for periodic systems:

```
%block npcell
1 1 1
%endblock npcell
```

Next, confirm that the calculation settings are as follows:

```
% cat input
    runtype         : vmc    #*! Type of calculation (Text)
    newrun          : T      #*! New run or continue old (Boolean)
    testrun         : F      #*! Test run flag (Boolean)
    ...
    vmc_equil_nstep : 1000   #*! Number of equilibration steps (Integer)
    vmc_nstep       : 10000  #*! Number of steps (Integer)
    vmc_nblock      : 1      #*! Number of checkpoints (Integer)
```

[10] This does not imply that all isolated systems require all-electron treatment. Even Gaussian basis calculations for isolated systems can sometimes use pseudopotentials [2].

```
vmc_nconfig_write : 0        #*! Number of configs to write (Integer)
...
use_jastrow       : F        #*! Use a Jastrow function (Boolean)
```

Then execute the calculation with

```
% mpirun -np 8 casino &
```

After completing the run,

```
% cat out
```

and check the end of the output. If you find the following message,

```
...
Dumping reblock data for energy:
    mean:     -75.604405245145 +/- 0.147042663828
```

then the calculation has successfully completed.

Then execute the following to copy the utility **editCorr.py** used to generate **correlation.data**:

```
% pwd
    /Users/maezono/work/gwFlow/03vmcWoJ
% cp ../../qmcTutorials/utils/editCorr.py .
```

After copying, run the script to generate the file:

```
% python3 editCorr.py > correlation.data
```

Then verify the content and file size:

```
% cat correlation.data
% ls -l
...
-rw-r--r--@ 1 maezono  staff   2152 12 25 21:48 correlation.data
```

This completes the generation of an empty-form **correlation.data**, which will later be optimized (the parameter entries are currently blank).

6.2.3 *Variational Optimization of Jastrow Factor (Isolated Molecular System)*

Next, we will determine the parameter set to fill in the empty template of **correlation.data** prepared in the previous section, based on the criterion of variance minimization (see Sect. 4.1.2). First, prepare the working directory:

```
% pwd
    /Users/maezono/work/gwFlow
% mkdir 04optJ
% cd 04optJ/
```

```
% ln -s ../02files/gwfn.data .
% cp ../03vmcWoJ/input .
% cp ../03vmcWoJ/correlation.data .
% ls
    correlation.data    gwfn.data@         input
```

Then, edit the input file while paying attention to the following parameter settings:

```
% cat input
...
runtype            : vmc_opt         #*! Type of calculation (Text)
newrun             : T               #*! New run or continue old (Boolean)
...
vmc_equil_nstep    : 1000            #*! Number of equilibration steps (Integer)
vmc_nstep          : 10000           #*! Number of steps (Integer)
vmc_nblock         : 1               #*! Number of checkpoints (Integer)
vmc_nconfig_write  : 10000           #*! Number of configs to write (Integer)
vmc_decorr_period  : 10              #*! Decorrelation loop length (Integer)
...
opt_method         : varmin_linjas   #*! Opt method (varmin/madmin/emin/...)
opt_cycles         : 10              #*! Number of optimization cycles (Integer)
opt_jastrow        : T               #*! Optimize Jastrow factor (Boolean)
...
use_jastrow        : T               #*! Use a Jastrow function (Boolean)
...
```

These settings are the same as those used in Sect. 6.1.3.

Then, run the calculation with

```
% mpirun -np 8 casino &
```

Once the calculation finishes, run the utility **ve** to analyze the results as follows:

```
% ve
...
VMC #1: E     = -75.6(1) ; var  = 69(3) (correlation.out.0)
VMC #2: E     = -75.7(1) ; var  = 59(2) (correlation.out.1)
VMC #3: E     = -75.8(1) ; var  = 62(6) (correlation.out.2)
VMC #4: E     = -75.2(1) ; var  = 52(2) (correlation.out.3)
VMC #5: E     = -75.8(1) ; var  = 57(3) (correlation.out.4)
VMC #6: E     = -75.9(1) ; var  = 54(2) (correlation.out.5)
VMC #7: E     = -75.8(1) ; var  = 56(3) (correlation.out.6)
...
```

Since it appears that "correlation.out.5 gives both the lowest energy and lowest variance," save it as the optimized correlation data:

```
% cp correlation.out.5 correlation.opt
% cp correlation.opt ../02files/correlation.data
```

This optimized file is stored as **correlation.opt** for later use.[11]

[11] You do not need to be overly precise about which one to select as the optimized result. Since this is just a pre-step before the DMC calculation, choosing, for example, correlation.out.4 instead would not significantly affect the final outcome.

6.2.4 VMC Using the Optimized Jastrow (Isolated Molecular System)

In this step, we perform a VMC calculation using the optimized Jastrow function from the previous section and store the resulting energy as a reference value. This reference allows us to evaluate how much of the correlation energy is captured by the Jastrow factor alone. Later, we will compare it to the correlation energy captured through trial wave function relaxation by DMC. First, prepare the working directory as follows:

```
% pwd
    /Users/maezono/work/gwFlow
% mkdir 05vmcWj
% cd 05vmcWj/
% ln -s ../02files/* .
% ls
    correlation.data@  gwfn.data@
% cp ../03vmcWoJ/input .
```

Then, edit the input file with attention to the following settings:

```
runtype           : vmc  #*! Type of calculation (Text)
...
vmc_nconfig_write : 0    #*! Number of configs to write (Integer)
...
use_jastrow       : T    #*! Use a Jastrow function (Boolean)
...
```

These settings are identical to those used in Sect. 6.1.4. Once edited, execute the calculation using

```
% !mpi
    mpirun -np 8 casino &
```

If the run completes successfully, you should find the following line near the end of the **out** file, beginning with "On-the-fly...":

```
E = -75.940520358425 +/- 0.108876172740
```

This value represents the energy obtained using the optimized Jastrow function.

6.2.5 DMC (Isolated Molecular System)

Now that we have prepared an initial trial wave function refined through variational optimization (like dough properly rested before baking), we move on to the DMC calculation. As noted in Sect. 3.3.2, the DMC process consists of three stages: configuration generation (congen), equilibration (equil), and statistical accumulation (stats).

First, perform **configuration generation** in the following working directory:

```
% pwd
   /Users/maezono/work/gwFlow
% mkdir 06dmcCongen
% cd 06dmcCongen/
% ln -s ../02files/* .
% cp ../05vmcWj/input .
```

Be sure to edit the input file with the following settings:

```
runtype            : vmc    #*! Type of calculation (Text)
...
vmc_equil_nstep    : 1000   #*! Number of equilibration steps
vmc_nstep          : 10000  #*! Number of steps (Integer)
vmc_nblock         : 1      #*! Number of checkpoints (Integer)
vmc_nconfig_write : 512    #*! Number of configs to write
...
```

Even in DMC calculations, configuration generation is performed using VMC, so the calculation mode should be set to `runtype: vmc` at this stage. The number of walkers (configurations), N_w, is set to 512, which is a multiple of eight cores used for parallelization (`vmc_nconfig_write: 512`). Run the calculation with `% !mpi`. Upon completion, a `config.out` file will be created in the working directory. This file contains the initial configurations used for the DMC calculation and corresponds to the histogram representation of the distribution function shown on the left side of Fig. 3.1.

Next, we perform **equilibration** to relax the distribution histogram toward the shape of the exact solution. Prepare the working directory as follows:

```
% pwd
   /Users/maezono/work/gwFlow
% mkdir 07dmcEquil
% cd 07dmcEquil/
% ln -s ../02files/* .
% cp ../06dmcCongen/input .
% cp ../06dmcCongen/config.out .
% mv config.out config.in
% ls
    config.in            gwfn.data@
    correlation.data@    input
```

We rename the `config.out` file generated in the configuration generation step to `config.in`, which will serve as the input for this calculation. Edit the input file with attention to the following settings:

```
runtype            : dmc_equil  #*! Type of calculation (Text)
...
dmc_equil_nstep    : 100        #*! Number of steps (Integer)
dmc_equil_nblock   : 5          #*! Number of checkpoints
dmc_stats_nstep    : 500        #*! Number of steps (Integer)
dmc_stats_nblock   : 25         #*! Number of checkpoints
dmc_target_weight : 512.d0     #*! Total target weight in DMC
```

Set the calculation mode to equilibration using `runtype: dmc_equil`. To match the number of walkers N_w prepared in the configuration generation step, set `dmc_target_weight: 512.d0` accordingly. Run the calculation using `% !mpi`. Upon completion, two files will be generated: `config.out` and `dmc.hist`. The former corresponds to the histogram relaxed during equilibration, while the latter records the trajectory of local energy values during the process.

Next, we proceed to the **statistical accumulation** stage. Prepare the working directory as follows, inheriting both the relaxed distribution (`config.out`) and the sampling history of the local energy values (`dmc.hist`) obtained during the equilibration process:

```
% pwd
    /Users/maezono/work/gwFlow
% mkdir 08dmcStats
% cd 08dmcStats/
% ln -s ../02files/* .
% cp ../07dmcEquil/input .
% cp ../07dmcEquil/config.out .
% mv config.out config.in
% cp ../07dmcEquil/dmc.hist .
```

The inherited configuration file is renamed to `config.in` for use as input in this step. Set up the `input` file with the following key settings:

```
...
runtype             : dmc_stats   #*! Type of calculation (Text)
newrun              : F           #*! New run or continue old
...
dmc_equil_nstep     : 100         #*! Number of steps (Integer)
dmc_equil_nblock    : 5           #*! Number of checkpoints
dmc_stats_nstep     : 500         #*! Number of steps (Integer)
dmc_stats_nblock    : 25          #*! Number of checkpoints
dmc_target_weight   : 512.d0      #*! Total target weight in DMC
dtdmc               : 0.01        #*! DMC time step (Real)
```

Set the calculation mode to the statistics accumulation stage using `runtype: dmc_stats`. Since this is a continuation from the equilibration stage, unset the new run flag by setting `newrun: F`. Execute the calculation using `% !mpi`. Once the run finishes, use `graphit` or `reblock` to analyze the results (see Fig. 6.1).

6.2.6 How Energy Changes (Isolated Molecular System)

Up to this point, we have obtained energy values from "Hartree-Fock calculation" → "VMC (without Jastrow)" → "VMC (with Jastrow)" → "DMC." We will now summarize and compare these results.

To obtain accurate values for both the VMC and DMC results, reblocking, as described in Sect. 2.4, should technically be performed. However, for simplicity, we

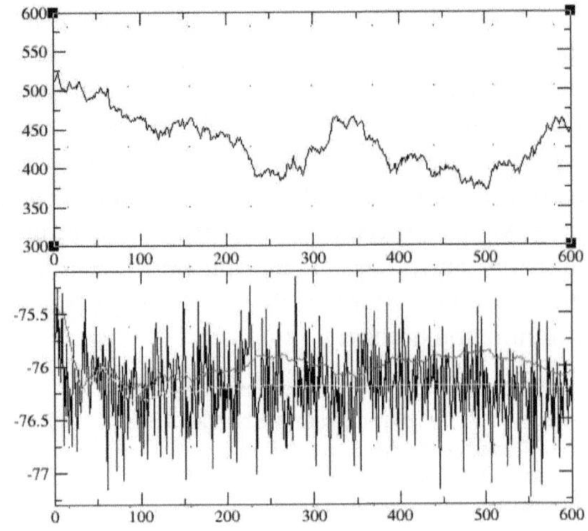

Fig. 6.1 Result of the DMC calculation for the isolated H_2O molecule, visualized using **graphit**

will use the values directly written in the out files here.[12] For the VMC values, we adopt the following extracted results:

```
% pwd
    /Users/maezono/work/gwFlow
% grep 'Correlation time meth' */out
    03vmcWoJ/out: -75.604405245145 +/- 0.139824822996  Corr...
    ...
    05vmcWj/out: -75.940520358425 +/- 0.131630561347
```

For the DMC value, we adopt the result written at the end of the out file:

```
mean: -76.209559044727 +/- 0.021458458266
```

As for the Hartree-Fock calculation, since Gaussian is used, the energy can be retrieved as follows:

```
% pwd
    /Users/maezono/work/gwFlow/01genGw
% grep SCF h2o.out
    ...
    SCF Done: E(RHF) = -74.9607025105 A.U. after 7 cycles
    ...
```

Adjusting the number of significant figures, we obtain

[12] In this case, neither the VMC nor DMC calculations involve continuation runs (resume runs; see Sect. 2.3.3). Thus, the "reported values in the out file for each run" coincide with the "statistical values across the full accumulation," making this simplification acceptable. In practice, statistical accumulation is often constructed by concatenating multiple runs, in which case the reported value in a single out file and the result obtained via reblocking may differ, and the latter should be adopted.

```
HF            -74.96
VMC(w/o J)    -75.6(1)
VMC(w J)      -75.9(1)
DMC           -76.21(2)
```

Comparing the QMC values, we observe that introducing a Jastrow factor (ripening the dough before baking) leads to a reduction in energy, and further relaxation of the wave function via imaginary-time evolution in DMC (ripening during baking) results in even lower energy values. These quantities are comparable on the same variational track, and a "3lower energy value" can be interpreted as being "closer to the exact solution."

What about the Hartree-Fock value? Hartree-Fock theory can be derived analytically as a variational method using a Slater determinant as the trial wave function [3]. When implemented numerically via Monte Carlo sampling, this corresponds to VMC without a Jastrow factor. Therefore, the two represent equivalent approximations implemented by different strategies, and they should yield the same energy value. This is **in contrast** to the case of solid-state systems treated with DFT, as discussed in the previous section. However, when comparing the numerical values, we find that the VMC (w/o J) result here appears slightly lower than the Hartree-Fock value, even accounting for the error bar. This **discrepancy** arises because the VMC (w/o J) calculation includes a **cusp correction for nuclei**. As a result, it becomes variationally slightly better than the SCF-based Hartree-Fock implementation (HF-SCF). Stating the values more precisely,

```
HF-SCF        -74.96
VMC(w/o J)    -75.6(1)...with cusp_correction
VMC(w J)      -75.9(1)
DMC           -76.21(2)
```

In the following section, we provide an overview of the "nuclei cusp correction," clarifying what this adjustment entails. We then verify that, when this correction is disabled, the VMC (w/o J) energy indeed matches the HF-SCF energy, as expected.

6.2.7 Calculation Without Cusp Correction

It is derived that the exact wave function possesses a cusp (a pointed singularity) at the position of nuclei (Eq. (8.6)). As a familiar example, if one looks up the shape of the hydrogen atomic orbital on the Internet, the radial distribution function $R_l(r)$ can be seen to exhibit a sharp peak at $r = 0$ [4]. In the previous section, the Gaussian basis set used was STO-3G. This is known to be **the simplest and least expressive** among Gaussian basis sets [3]. A wide variety of more sophisticated and costly Gaussian basis sets have been developed that offer professional-level expressiveness [3]. The primary difference between inexpensive and costly basis sets lies in the number of adjustable parameters—they consist not of a single Gaussian, but a linear combination of many Gaussian functions, allowing greater flexibility in approximating the exact wave function. Gaussian functions have smooth behavior at the origin and do not describe cusps (see Fig. 10.13). Therefore, in high-level Gaussian basis sets, many sharply rising Gaussian functions with large exponents are superposed to mimic the cusp as closely as possible.

As explained in detail in Sect. 8.1.1, QMC calculations rely on random sampling of electron configurations. If the cusp is not accurately represented, a phenomenon known as population explosion occurs, as shown in Fig. 4.3, making DMC calculations numerically unstable and preventing them from proceeding reliably. This population explosion overwhelms memory capacity due to the sudden increase in walkers, causing the calculation to crash.[13] To address this, CASINO incorporates an *nuclei cusp correction* feature. In all-electron calculations using Gaussian basis sets, this feature estimates the cusp shape from the tail behavior of the wave function and the atomic species and numerically corrects the orbital functions to introduce the appropriate cusp.

The nuclei cusp correction feature is enabled by default (set to "T") and is hidden from the "input" file. By explicitly disabling this nuclei cusp correction, one performs a true "VMC (w/o J)" calculation, which becomes equivalent to the HF-SCF result. Let us now verify this behavior in practice. First, set up the working directory as follows:

```
% pwd
    /Users/maezono/work/gwFlow
% mkdir 10vmcWoCuspCorr
% cd 10vmcWoCuspCorr
% ln -s ../02files/* .
% cp ../03vmcWoJ/input .
% ls
    gwfn.data@  correlation.data@  input
```

Then, open the "input" file and modify the relevant entries as follows:

```
runtype             : vmc_dmc  #*! Type of calculation (Text)
...
vmc_equil_nstep     : 10000    #*! Number of equilibration
vmc_nstep           : 100000   #*! Number of steps (Integer)
vmc_nblock          : 1        #*! Number of checkpoints
vmc_nconfig_write   : 512      #*! Number of configs
...
dmc_equil_nstep     : 100      #*! Number of steps (Integer)
dmc_equil_nblock    : 5        #*! Number of checkpoints (Integer)
dmc_stats_nstep     : 2000     #*! Number of steps (Integer)
dmc_stats_nblock    : 100      #*! Number of checkpoints (Integer)
dmc_target_weight   : 512.d0   #*! Total target weight in DMC (Real)

...
```

[13] When the author was involved in a project as a postdoctoral researcher in the UK, there were no solutions implemented for this issue at the time. Considerable difficulty was experienced when attempting all-electron calculations using Gaussian basis sets.

```
use_jastrow       : F        #*! Use a Jastrow function (Boolean)
cusp_correction   : F
```

Set **runtype** to vmc_dmc so that the run proceeds continuously from VMC through DMC (see Fig. 3.4). Add the line `cusp_correction: F` to explicitly disable the cusp correction applied to orbital functions.[14] In addition, set `use_jastrow: F` to disable Jastrow-based electron-electron cusp corrections. From the perspective of researchers who have been working in this field since the early 2000s, this setup may feel quite risky.[15]

Due to the lack of cusp correction, the sampled values of the local energy $E_L(\mathbf{R})$ are expected to exhibit large fluctuations. Therefore, longer accumulation steps have been specified than in previous runs. For the VMC part, we set 10,000 equilibration steps (**vmc_equil_nstep** = 10,000) and 100,000 steps for statistical accumulation (**vmc_nstep** = 100,000), which is ten times the previous settings. For the DMC part, 2,000 steps of statistical accumulation (**dmc_stats_nstep** = 2,000) are performed in a single run with a block size of 20 (**dmc_stats_nblock** = 100). After completing the above "input" settings, execute the calculation as follows:

```
% pwd
    /Users/maezono/work/gwFlow/10vmcWoCuspCorr
% mpirun -np 8 casino &
```

The results of the DMC calculation can be visualized using **graphit** (see Fig. 6.2). The energy profile exhibits **sharp, spiky behavior** at various points. In calculations without cusp correction, when particle coalescence occurs during sampling, the cusp condition, which compensates divergent potential contributions by a divergent kinetic term of opposite sign, is no longer satisfied, even partially. As a result, the local energy retains divergent contributions, and the presence of unusually large values in the sample becomes the cause of the observed sharp peaks in the plot.

In earlier implementations, this would likely lead to divergent growth in the number of walkers. However, the version used in this book has undergone algorithmic improvements, and the walker population remains stable throughout the simulation.[16] Nevertheless, the total number of walkers shows a gradual dying out over time, indicating that the simulation behavior is not optimal.

Let us now return to the original question posed in this section: "Does the VMC(w/o J) value agree with the HF-SCF value when cusp correction is turned off?" In the previous calculation, VMC was performed with a large number of steps, also serving as the configuration generation phase for the subsequent DMC calculation (**runtype**: vmc_dmc). The resulting VMC energy can be extracted as follows:

---

[14] The default behavior without explicit specification is to apply cusp correction (T).

[15] With the implementations and computing resources available around the year 2000, this kind of setup would often lead to population explosion and calculation failure.

[16] This is because the update of walker weights is now based on local energy values averaged over a number of steps, thereby mitigating sudden surges.

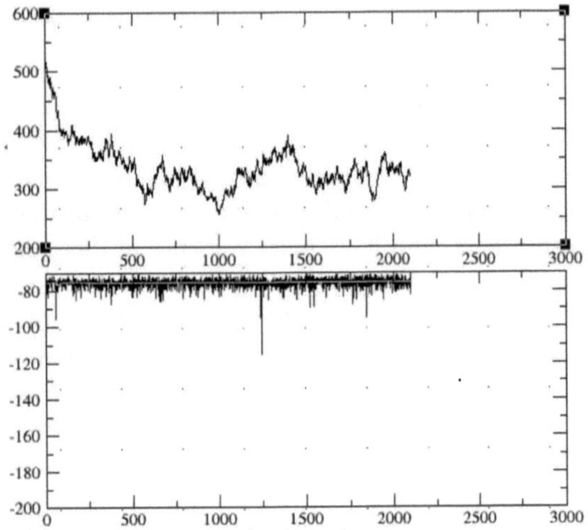

Fig. 6.2 Result of DMC calculation without cusp correction, visualized with **graphit**. In the lower panel, showing the behavior of the local energy, one can observe intermittent sharp peaks indicating instability in the sampling process

```
% grep On-the out
    -75.281048242272 +/- 0.832601962995    On-the-fly ...
```

This value is statistically consistent with the HF-SCF result:

```
HF-SCF          -74.96
VMC(w/o J)  -75.3(8)...w/o  cusp_correction
VMC(w/o J)  -75.6(1)...with cusp_correction
VMC(w J)     -75.9(1)
DMC           -76.21(2)
```

This confirms that, within the margin of statistical error, the VMC(w/o J) energy without cusp correction aligns with the analytical Hartree-Fock SCF result.

6.3 Exercises

In this book, we have used AlN as an example of a solid periodic system and H_2O as an example of an isolated molecular system (see Fig. 2.1). Here, to consolidate what you have learned so far, we provide a few simple exercises that readers can try on their own. No specific computational conditions are given, so please refer to the benchmark results below and challenge yourself to see if you can reach comparable results through your own efforts. For the case of an all-electron calculation on an isolated molecule, we assign methane CH_4 as the target. For a periodic solid system, a pseudopotential calculation on solid silicon using a $2\times2\times2$ simulation cell is assigned. The rough results for each case are shown below.

As an example of an **isolated molecular system with all-electron calculation** without the use of pseudopotentials, we provide a sample case for the methane molecule CH_4. You can find the example files under the following directory:

```
~/work/qmcTutorials/example/mol
```

The directory contains

```
% ls
    Test.FChk ch4.com ch4.out
```

The benchmark results are as follows (note that statistical errors depend on the number of samples accumulated):

```
                              Energy (Ha)
HF                          : -39.726772
VMC Without Jastrow         : -40.087258 ± 0.008173
VMC With Jastrow            : -40.288845 ± 0.002766
DMC                         : -40.479837 ± 0.007420
```

Use these values as a reference and try comparing them with your own results.[17]

For the case of a **periodic solid system** (with pseudopotentials), a sample setup is provided under the following directory:

```
~/work/qmcTutorials/example/solid
```

The directory contains

```
% ls
    pw2casino.dat   si.UPF  si.pwscf.in     si_pp.data
```

The benchmark results (statistical errors depend on the amount of accumulated sampling) are as follows:

```
                    Energy (Ha)
DFT             ... -7.7534
VMC without J   ... -7.521(3)
VMC with optJ   ... -7.770(1)
DMC             ... -7.800(2)
```

Please use these values as a reference and compare them with your own computational results.

References

1. "Ab initio Calculation Tutorial: For Materials Analysis, Informatics and Design", (Springer Nature, 2023), R. Maezono, ISBN: 978-9819909216
2. "Quantum Monte Carlo study of porphyrin transition metal complexes", J. Koseki, R. Maezono, M. Tachikawa, M.D. Towler, and R.J. Needs, J. Chem. Phys. 129, 085103 (2008). https://doi.org/10.1063/1.2966003

[17] Ha (Hartree) is the energy unit in the atomic unit system. For the atomic unit system, see the footnote in Sect. 6.1.6.

3. "Introduction to Computational Chemistry, Third Edition" (Wiley, 2017), F. Jensen, ISBN: 978-1118825990
4. "Introduction to Quantum Mechanics with Applications to Chemistry" (Dover, 1985), L. Pauling, E.B. Wilson, ISBN: 978-0486648712

Chapter 7
Theory of Diffusion Monte Carlo Method

In Chap. 3, we followed the practical workflow of the diffusion Monte Carlo (DMC) method while introducing its conceptual background. The central idea of the method lies in the equivalence between the "time evolution of particle density via random walk governed by the Langevin equation" and the "time evolution of the density function via the Fokker-Planck equation" (Sect. 3.2.2).

However, as noted in Sect. 3.2.4, the explanation provided in Chap. 3 should be regarded as a rough conceptual introduction aimed at beginners. In Chap. 3, it was stated that "what is obtained from the random walk is a histogram distribution of $\Phi(\mathbf{R})$," whereas "in QMC, what we want to sample is the distribution of $|\Phi(\mathbf{R})|^2$.". This discrepancy was mentioned to be resolved by "mapping the simulation target from $\Phi(\mathbf{R})$ to another function $f(\mathbf{R})$," though the details were not elaborated at that point.

In Sect. 3.2.5, the notion that "the number of walkers varies during the simulation" was explained in a highly conceptual manner. There, it was briefly stated that "if we ignore the diffusion term and consider only the reaction term, we can understand why the number of walkers changes," without delving into the mathematical details.

In this chapter, we provide a more in-depth explanation of how the three key points mentioned above are derived and how they fit into the overall theoretical framework. To avoid losing sight of the logical flow amid lengthy mathematical derivations, we will refer to the appendices as needed for detailed steps.

7.1 Random Walk and Diffusion Equation

If we begin with abstract and general notations when discussing the correspondence between random walks and the diffusion equation, the underlying image can be hard to grasp. Therefore, we will start with a simple illustrative example to build an

R. Maezono, *Ab initio Quantum Monte Carlo Tutorial*,
https://doi.org/10.1007/978-981-95-4134-8_7

overall picture of the logic (Sect. 7.1.1). We will then proceed to a discussion using more general notations.

7.1.1 Relationship Between Random Walk and Diffusion Equation

To understand how the diffusion equation can be derived from a random walk, we begin with a simple example to grasp the idea of how an integral representation (random walk) is translated into a differential representation (diffusion equation).

Consider a particle performing a random walk (a drunkard's walk) on a one-dimensional x-axis. Let $\rho(x, t)$ denote the probability of finding the particle at position x at time t. Then, we can write

$$\rho(x, t + \Delta t) = p \cdot \rho(x - \Delta x, t) + q \cdot \rho(x + \Delta x, t) \tag{7.1}$$

This equation describes the arrival of the particle at position x with probability p from the right and probability q from the left. Since they are probabilities, we have $p + q = 1$, and thus we can write

$$p = \frac{1}{2} + r \quad , \quad q = \frac{1}{2} - r \tag{7.2}$$

Here, the imbalance parameter r captures the presence of a **drift flow**, expressing a net bias toward the right (or left) with a single parameter.

By substituting Eq. (7.2) into Eq. (7.1) and evaluating the finite differences, we obtain

$$\begin{aligned}
\frac{\rho(x, t + \Delta t) - \rho(x, t)}{\Delta t} &= \frac{\rho(x - \Delta x, t) + \rho(x + \Delta x, t) - 2\rho(x, t)}{2\Delta t} \\
&\quad - r \cdot \frac{1}{\Delta t} \left[\rho(x + \Delta x, t) - \rho(x - \Delta x, t)\right] \\
&= \frac{\Delta x^2}{2\Delta t} \cdot \frac{\rho(x - \Delta x, t) + \rho(x + \Delta x, t) - 2\rho(x, t)}{\Delta x^2} \\
&\quad - r \cdot \frac{2\Delta x}{\Delta t} \cdot \frac{\rho(x + \Delta x, t) - \rho(x - \Delta x, t)}{2\Delta x}
\end{aligned}$$

Here, if the following limits exist,

$$D = \lim_{\Delta t \to 0} \frac{\Delta x^2}{2\Delta t} \quad , \quad c = \lim_{\Delta t \to 0} \frac{-2r\Delta x}{\Delta t} \tag{7.3}$$

then the above difference relation can be continuously carried over to

$$\frac{\partial}{\partial t}\rho(x,t) = D \cdot \frac{\partial^2}{\partial x^2}\rho(x,t) + c \cdot \frac{\partial}{\partial x}\rho(x,t). \tag{7.4}$$

If the second term is absent, this becomes the well-known diffusion equation [1]. Since the second term originates from $c \sim r$, it is evident that this first-order spatial derivative term describes an overall drift flow.

In this way, the equation describing the position updates due to a random walk, Eq. (7.1), is equivalently expressed as the equation for the time evolution of the density distribution, Eq. (7.4). This correspondence in *form* should be noted.

7.1.2 Formulation of the Markov Process

Equation (7.1) can be written in terms of continuous transitions as

$$\rho(x, t+\Delta t) = \int dx' \cdot P\left(x, t+\Delta t; x', t\right) \cdot \rho\left(x', t\right) \tag{7.5}$$

In this expression, $P(x_2, t_2; x_1, t_1)$ denotes the probability that a particle transitions from (x_1, t_1) to (x_2, t_2), which corresponds to p and q in Eq. (7.1).

In the random walk described by Eq. (7.1), $x(t+\Delta t)$ is determined solely by $x(t)$. In this governing rule description, a particle located at $(x \pm \Delta x)$ does not depend on the history of how it arrived there. If the particle is within that range at time t, then the situations under which it moves to position x with probability p or q are treated equally. A process in which the next step is determined "independently of past history up to the present stage and on equal footing" is called a **Markov process**.

A Markov process can be regarded as a coarse-grained model in which detailed information about past history is absorbed into the "probability of having just transitioned to the current state," thus allowing the process to be formulated. When the transition probability from (x_1, t_1) to (x_2, t_2) is written as $P(x_2, t_2; x_1, t_1)$, a Markov process can therefore be expressed as

$$P(x_3, t_3; x_1, t_1) = \int dx_2 \cdot P(x_3, t_3; x_2, t_2) \cdot P(x_2, t_2; x_1, t_1)$$
$$\cdots (t_1 < t_2 < t_3) \tag{7.6}$$

That is, the process leading to the final state (x_3, t_3) is obtained by multiplying the process $P(x_3, t_3; x_2, t_2)$ on top of the process that led to the intermediate state (x_2, t_2). The "individual histories leading to the intermediate state (x_2, t_2)" are coarse-grained and absorbed into the information $P(x_2, t_2; x_1, t_1)$. The total transition probability $P(x_3, t_3; x_1, t_1)$ is constructed by summing over all possible intermediate states (x_2, t_2). Equation (7.6) is the equation satisfied by transition probabilities in a Markov process and is called the **Chapman-Kolmogorov equation** [2].

By evaluating $P(x_2, t_1 + \Delta t; x_1, t_1)$ in the integral form of the Chapman-Kolmogorov equation and taking the difference to consider the limit $\Delta t \to 0$, one obtains the differential form (see Sect. 13.1.1)

$$\frac{\partial P(x,t;x_1,t_1)}{\partial t} = \sum_{n=1}^{\infty} \frac{1}{n!}\left(-\frac{d}{dx}\right)^n \{\alpha_n(x) \cdot P(x,t;x_1,t_1)\}$$

$$\alpha_n(x) = \lim_{\Delta t \to 0} \frac{\langle \Delta x^n \rangle}{\Delta t} \quad , \quad \langle \xi^n \rangle := \int d\xi \cdot \xi^n \cdot P(x+\xi, t+\Delta t; x, t) \,. \tag{7.7}$$

This differential representation is called the **Moyal expansion**. The quantity $\alpha_n(x)$ defined as a limit corresponds to Eq. (7.3) in the previous subsection.

From Eq. (7.5), which relates the transition probability and the density distribution,

$$\rho(x,t) = \int dx' \cdot P\left(x,t;x',0\right) \cdot \rho\left(x',0\right) \tag{7.8}$$

Taking the time derivative of this expression and substituting Eq. (7.7), the dynamics of the particle density obey the same form as that of the transition probability:

$$\frac{\partial \rho(x,t)}{\partial t} = \sum_{n=1} (-)^n \cdot \frac{1}{n!} \frac{\partial^n}{(\partial x)^n} \cdot [\rho(x,t) \cdot \alpha_n(x)] \tag{7.9}$$

We will now see that a special case of this general expression reduces to the form of Eq. (7.4).

7.1.3 Fokker-Planck Equation

Equation (7.9) is a general equation governing the density of particles whose position fluctuates stochastically according to a Markov process. Once a specific realization of the Markov process is defined, the correlation functions of each order, $\{\alpha_n(x)\}$, can be obtained, and the evolution equation for the density distribution function is determined concretely. Below, we examine the evolution equation that arises from a Markov process in which the random walk is implemented via **Gaussian noise**.

Suppose now that the particle position x evolves according to

$$\frac{dx}{dt} = F(x,t) + \gamma(t) \tag{7.10}$$

Here, $\gamma(t)$ is a random noise term, and $F(x, t)$ is a drift flow with time dependence that is sufficiently slow compared to the random noise. The particle velocity (left-hand side) is thus subject to fine random fluctuations due to $\gamma(t)$ while being overall guided by the drift flow. Such an equation that describes the time evolution of a particle's position under a random force is called a **Langevin equation**.

We model the random noise $\gamma(t)$ as an "ideal white Gaussian noise" (see Sect. 14.1). In the case of white Gaussian noise, its characteristics are defined by the expectation value and correlation function, which are given by Eq. (14.2) as

$$\langle \gamma(t) \rangle = 0 \quad , \quad \langle \gamma(t) \gamma(t') \rangle = \sigma^2 \cdot \delta(t - t') \tag{7.11}$$

Using these, we return to Eq. (7.10) and integrate it to obtain

$$\Delta x = \int_t^{t+\Delta t} F(x, t') + \gamma(t') \cdot dt' \tag{7.12}$$

From this, we find (see Sect. 13.1.2)

$$\langle \Delta x \rangle = F(x, t) \cdot \Delta t \quad , \quad \langle \Delta x^2 \rangle = F^2(x, t) \cdot \Delta t^2 + \sigma^2 \cdot \Delta t \, . \tag{7.13}$$

Therefore, the moments appearing in the Moyal expansion (Eq. (7.7)) are given by

$$\alpha_1(x) = \lim_{\Delta t \to 0} \frac{\langle \Delta x \rangle}{\Delta t} = F(x, t) \quad , \quad \alpha_2(x) = \lim_{\Delta t \to 0} \frac{\langle \Delta x^2 \rangle}{\Delta t} = \sigma^2 \, . \tag{7.14}$$

Due to the nature of the white Gaussian process, all higher-order moments $\alpha_{k \geq 3}(x)$ vanish.[1]

The Moyal expansion for the density distribution, Eq. (7.9), in this case becomes

$$\frac{\partial \rho(x, t)}{\partial t} = \frac{\sigma^2}{2} \frac{\partial}{\partial x} \left[\frac{\partial}{\partial x} - \frac{2}{\sigma^2} F(x, t) \right] \rho(x, t) \tag{7.15}$$

This time-evolution equation is called the **Fokker-Planck equation**. To restate, in a realization of a Markov process driven by random walks induced by Gaussian noise (a **Wiener process**), the evolution equation of the distribution function of the randomly walking particles takes the form of the Fokker-Planck equation.

[1] Evaluating $\langle \Delta x^3 \rangle$ yields $\sim (F^3 \cdot \Delta t^3 + 3F^2 \cdot \Delta t^2 \cdot \int \gamma + 3F \cdot \Delta t \cdot \int \gamma^2 + \int \gamma^3)$, but the third moment term $\int \gamma^3$ can be reduced to lower-order moments for evaluation. By substituting Eq. (7.11) and identifying the order in Δt for each term over the integration interval, it can be shown that all contributions vanish in the limit $\Delta t \to 0$. The same logic applies to moments of order 4 and higher, where reduction formulas that express γ^k in terms of lower-order moments are used [1].

7.1.4 Correspondence of Coefficients

From Eqs. (7.15) and (7.10), if we define

$$D = \frac{\sigma^2}{2} \quad , \quad \tilde{F} = \frac{F}{D} \tag{7.16}$$

then the Fokker-Planck equation

$$\frac{\partial \rho (x,t)}{\partial t} = D \frac{\partial}{\partial x} \left[\frac{\partial}{\partial x} - \tilde{F}(x,t) \right] \rho (x,t) \tag{7.17}$$

corresponds to the Langevin equation

$$\frac{dx}{dt} = D \cdot \tilde{F}(x,t) + \chi_{\mathrm{Gauss}} [0, 2D]$$

Here, the random force previously written as $\gamma(t)$ is now explicitly denoted as white Gaussian noise $\chi_{\mathrm{Gauss}} \left[\mu, \sigma^2 \right]$ (with mean μ and variance σ^2):

$$\gamma(t) = \chi_{\mathrm{Gauss}} \left[0, \sigma^2 \right] = \chi_{\mathrm{Gauss}} [0, 2D] \ .$$

From Eq. (7.12), we obtain (see Sect. 13.1.2)

$$\Delta x = D \cdot \tilde{F}(x,t) \cdot \Delta t + \int_t^{t+\Delta t} \gamma \left(t' \right) \cdot dt' \ .$$

The first term is not a stochastic process. The second term, being a linear combination of a stochastic process, is what renders Δx a stochastic variable. On the other hand, from the discussion near Eq. (7.14), we saw that $\langle \Delta x^k \rangle = 0 \ (k \geq 3)$. Thus, Δx follows a white Gaussian process. Accordingly, we can reexpress the second (stochastic) term of Δx as

$$\Delta x = D \cdot \tilde{F}(x,t) \cdot \Delta t + \chi_{\mathrm{Gauss}} \left[0, \langle \Delta x^2 \rangle \right] .$$

From Eq. (7.13), in the limit $\Delta t \to 0$, we have $\langle \Delta x^2 \rangle = \sigma^2 \cdot \Delta t = 2D\Delta t$,[2] Thus, we obtain

$$\Delta x = D \cdot \tilde{F}(x,t) \cdot \Delta t + \chi_{\mathrm{Gauss}} [0, 2D\Delta t] \tag{7.18}$$

[2] The derivation of Eq. (7.13) is given in the Appendix. However, the result $\langle \Delta x^2 \rangle \propto \Delta t$ is a **key consequence of diffusion processes** and is frequently cited, so it is important to keep it in mind.

From the above, we have established the equivalence between "the time evolution of the distribution $\rho(x, t)$ governed by the differential equation (Eq. (7.17))" and "the time evolution of the distribution histogram generated by the random walk process given in Eq. (7.18)."

7.1.5 Multidimensional and Multiparticle Case

When the particle position is extended to an M-dimensional space, the arguments of the transition probability become

$$P\left(\mathbf{x}', t'; \mathbf{x}, t\right) = P\left(x_1', x_2', \cdots x_M', t'; \mathbf{x}, t\right)$$

Accordingly, the Taylor expansion that underlies the derivation of the Moyal expansion becomes

$$P\left(\mathbf{x}+\Delta\mathbf{x}, t; \mathbf{x}_0, t_0\right) = P\left(\mathbf{x}, t; \mathbf{x}_0, t_0\right) + \sum_i \frac{\partial P}{\partial x_i}\Delta x_i + \frac{1}{2}\sum_{ij}\frac{\partial^2 P}{\partial x_i \partial x_j}\Delta x_i \Delta x_j + \cdots$$

(see Sect. 13.1.3). By following the same procedure based on this, when N particles are present in three-dimensional space, we define

$$\mathbf{R} := \begin{pmatrix} x_{k=1} \\ \vdots \\ x_{k=3N} \end{pmatrix} = \begin{pmatrix} \mathbf{x}_{j=1} \\ \vdots \\ \mathbf{x}_{j=N} \end{pmatrix} \quad , \quad \tilde{\mathbf{F}}^{(3N)} = \begin{pmatrix} \tilde{F}_{k=1} \\ \vdots \\ \tilde{F}_{k=3N} \end{pmatrix} = \begin{pmatrix} \tilde{\mathbf{F}}_{j=1} \\ \vdots \\ \tilde{\mathbf{F}}_{j=N} \end{pmatrix} \tag{7.19}$$

Then, the Fokker-Planck equation in Eq. (7.17) is extended to (see Sect. 13.1.3)

$$\frac{\partial \rho(\mathbf{R}, t)}{\partial t} = D\sum_{j=1}^{N}\nabla_j\left[\nabla_j - \tilde{\mathbf{F}}_j(\mathbf{R}, t)\right]\rho(\mathbf{R}, t) \quad , \quad \mathbf{R}\in\mathcal{R}^{3N} . \tag{7.20}$$

The corresponding random walk process is given by (see Sect. 13.1.3)

$$\Delta\mathbf{R} = D\cdot\tilde{\mathbf{F}}^{(3N)}(\mathbf{R}, t)\cdot\Delta t + \chi^{\mathrm{Gauss}(3N)}[0, 2D\Delta t] . \tag{7.21}$$

Here, the notation $\chi_{\mathrm{Gauss}}[0, 2D\Delta t]$ denotes a multidimensional vector of white Gaussian noise, where each component is independent with mean 0 and variance $2D\Delta t$.

To summarize the conclusions reached in this section Suppose we wish to numerically implement the time evolution governed by the partial differential equation (Eq. (7.20)). To do this, we consider "a set of particles whose distribution forms a histogram representation of the target function $\rho(\mathbf{R}, t)$." If we evolve the positions $\mathbf{R}$ of each of these particles in time using the random walk process described by Eq. (7.21), then the histogram of these particles will evolve in accordance with Eq. (7.20). Through this, we have established a pathway by which the problem of solving a **partial differential equation** can be **equivalently realized** through a **stochastic simulation of particles**.

7.2 Importance Sampling in DMC

In Chap. 3, we introduced the idea underlying diffusion Monte Carlo (DMC): to describe the time evolution of $\Phi(\mathbf{R}, t)$ through its histogram representation. However, in practice, the actual implementation does not work directly with the histogram representation of $\Phi(\mathbf{R}, t)$ itself. Instead, the problem is reformulated as $\Phi(\mathbf{R}, t) \to f(\mathbf{R}, t)$, where

$$f(\mathbf{R}, \tau) = \Phi(\mathbf{R}, 0) \cdot \Phi(\mathbf{R}, \tau)$$

and the histogram representation is taken for $f(\mathbf{R}, \tau)$ instead. The motivation for this reformulation was briefly mentioned in Sect. 3.2.4, but by mapping the problem to the time evolution of $f(\mathbf{R}, \tau)$, the following three advantages are achieved:

1. Improved sampling efficiency (Sect. 7.2.3),
2. The distribution function being evolved in time can be made identical to the distribution function used for sampling (Sect. 7.2.4),
3. Reduced statistical fluctuations in the sampled values (Sect. 7.3.4),

Below, we explain each of these points step-by-step. Some textbooks, perhaps due to space constraints, refer to the reformulation $\Phi(\mathbf{R}, t) \to f(\mathbf{R}, t)$ just as "**importance sampling**" suddenly without much explanation. In what follows, we develop the reasoning step-by-step to clarify why this is referred to as "importance" sampling.

7.2.1 *The Idea of Importance Sampling*

Suppose we wish to evaluate an integral of the form

$$I = \int dx \cdot g(x)$$

If there exists a function $p(x)$ satisfying the conditions

$$\int dx \cdot p(x) = 1 \quad , \quad 0 \le p(x) \le 1 \, ,$$

that is, $p(x)$ is a properly normalized probability distribution function, then we can rewrite the desired integral as

$$I = \int dx \cdot \frac{g(x)}{p(x)} \cdot p(x) = \left\langle \frac{g(x)}{p(x)} \right\rangle_p$$

This leads to the idea of evaluating I as a sampling average:

$$I = \frac{1}{M} \sum_{j=1}^{M} \left[\frac{g(x_j)}{p(x_j)} \right] \quad , \quad \{x_j\} \sim p(x)$$

This is the basic idea behind Monte Carlo integration (see Sect. 2.2.4).

There are no constraints on the choice of $p(x)$ beyond satisfying the definition of a probability distribution. However, consider the case where $g(x)$ is significant only near $x = 0$, but we choose $p(x)$ as a uniform distribution. In such a situation, most of the sampling will avoid the region where $g(x) \approx 0$, resulting in a **low proportion of meaningful contributions**. Consequently, an enormous number of samples would be needed for the estimated value of I to converge to the true value. On the other hand, if we choose a $p(x)$ that concentrates sampling in regions where $g(x)$ has finite weight, we can achieve convergence to the true value of I with far fewer total samples. In this way, we see that the **quality of the choice** of $p(x)$ greatly affects sampling efficiency.

From the above reasoning, it is clear that the choice

$$p(x) \propto g(x)$$

would result in the most efficient sampling. Upon normalization, the ideal sampling distribution becomes

$$p^{\text{ideal}}(x) = \frac{g(x)}{\int dx \cdot g(x)}$$

However, the integral in the denominator is precisely the unknown quantity we are trying to evaluate in the first place. Thus, we realize that this is a "nonviable strategy."

To address this, one introduces a **guiding function** $g_G(x)$ $[\approx g(x)]$, which is estimated to resemble the distribution of $g(x)$ and for which the integral can be computed. Then, the sampling distribution is taken as

$$p(x) = \frac{g_G(x)}{\int dx \cdot g_G(x)}$$

Using this $p(x)$, one can carry out sampling more efficiently. This strategy of "focusing sampling on regions where the integrand takes finite values" is called **importance sampling**.

7.2.2 Importance Sampling Using the Fokker-Planck Equation

In the context of performing importance sampling, we often face the following problem: "given a guiding function $p_G(x)$, we want to generate a sampling distribution $\{x_j\} \sim p_G(x)$." As a method for generating a sampling sequence that follows a given distribution, we introduced in Sect. 2.2.2 the idea of "implementing a Markov process Monte Carlo method using the Metropolis algorithm."

In the previous section, we explained the correspondence that "the time evolution of a distribution function can be described by the Fokker-Planck equation.' Knowing this framework leads to the idea of "evolving a sampling distribution function via the Fokker-Planck equation in such a way that it eventually reaches the desired distribution $p_G(x)$." This provides an alternative strategy. Such a strategy can be realized by appropriately choosing the external force term $\mathbf{F}_j(\mathbf{R}, t)$ in the Fokker-Planck equation:

$$\frac{\partial \rho(\mathbf{R}, t)}{\partial t} = \sum_j D\nabla_j \left[\nabla_j - \mathbf{F}_j(\mathbf{R}, t)\right] \rho(\mathbf{R}, t) \tag{7.22}$$

To state the conclusion first: if the desired distribution is $\rho_G(\mathbf{R})$, then by choosing

$$\mathbf{F}_j(\mathbf{R}, t) = \mathbf{F}_j(\mathbf{R}) = \frac{\nabla_j \rho_G(\mathbf{R})}{\rho_G(\mathbf{R})} \tag{7.23}$$

and letting the system evolve in time, the distribution $\rho(\mathbf{R}, t)$ converges to the target distribution $\rho_G(\mathbf{R})$. In this way, we can "guide the equilibrium solution toward the desired $\rho_G(\mathbf{R})$": For the equilibrium solution $\rho_G(\mathbf{R})$, the time derivative on the left-hand side of Eq. (7.22) vanishes, so we have

$$0 = \sum_j D\nabla_j \left[\nabla_j - \mathbf{F}_j(\mathbf{R})\right] \rho_G(\mathbf{R})$$

It can be shown that the choice given by Eq. (7.23) indeed satisfies this condition (see Sect. 13.2.1). Therefore, if we evolve the distribution function over time using the Fokker-Planck equation with the force term chosen as in Eq. (7.23), we see that the distribution will converge toward the desired target $\rho_G(\mathbf{R})$.

In variational Monte Carlo (VMC), the target distribution we wish to achieve is of the squared-wave function form:

$$\rho_G(\mathbf{R}) = a \cdot \Psi(\mathbf{R})^2 \tag{7.24}$$

(as given in Eq. (3.17)). In this case, we have

$$\nabla_j \cdot \rho_G(\mathbf{R}) = 2a \cdot \Psi(\mathbf{R}) \cdot \nabla_j \Psi(\mathbf{R})$$

Thus, the drift flow term becomes

$$\mathbf{F}_j(\mathbf{R}) = \frac{\nabla_j \cdot \rho_G(\mathbf{R})}{\rho_G(\mathbf{R})} = 2\frac{\nabla_j \Psi(\mathbf{R})}{\Psi(\mathbf{R})} \tag{7.25}$$

By choosing this drift term $\mathbf{F}_j(\mathbf{R})$ in the Fokker-Planck equation, the time-evolved distribution $\rho(\mathbf{R}, t)$ will converge to the distribution $\Psi(\mathbf{R})^2$.

7.2.3 Realization of Importance Sampling via Projection

In the diffusion Monte Carlo method, the time-evolution equation (Eq. (3.15)) is given by

$$\frac{\partial \Phi(\mathbf{R}, \tau)}{\partial \tau} = \left(\frac{1}{2}\sum_{j=1}^{N} \nabla_j^2 + E_T - V(\mathbf{R})\right) \Phi(\mathbf{R}, \tau)$$

This expression does not contain any first-order spatial derivative terms that would correspond to a drift term $\mathbf{F}_j(\mathbf{R})$. Now, let us define a mapping

$$f(\mathbf{R}, \tau) = Y(\mathbf{R}) \cdot \Phi(\mathbf{R}, \tau)$$

where the wave function is reweighted in space by a function $Y(\mathbf{R})$. Under this transformation $\Phi(\mathbf{R}, \tau) \rightarrow f(\mathbf{R}, \tau)$, the time-evolution equation for $f(\mathbf{R}, \tau)$ becomes

$$\begin{aligned}\frac{\partial f(\mathbf{R}, \tau)}{\partial \tau} &= \frac{1}{2}\sum_{j=1}^{N} \nabla_j \left(\nabla_j - 2\frac{\nabla_j Y(\mathbf{R})}{Y(\mathbf{R})}\right) f(\mathbf{R}, \tau) \\ &\quad - \left(V(\mathbf{R}) - E_T - Y^{-1} \cdot \frac{1}{2}\sum_{j=1}^{N} \nabla_j{}^2 \cdot Y\right) f(\mathbf{R}, \tau)\end{aligned} \tag{7.26}$$

as shown in Sect. 13.2.2. In the diffusion term (the first term on the right-hand side), the additional term involving $Y(\mathbf{R})$, namely, $2\nabla_j Y(\mathbf{R}) / Y(\mathbf{R})$, has the same form as the drift term in Eq. (7.25). Thus, in the absence of the reaction term (the second term on the right-hand side), Eq. (7.26) will equilibrate toward a distribution proportional to $Y(\mathbf{R})^2$, which is exactly the conclusion drawn in "importance sampling via the Fokker-Planck equation."

In DMC, the goal is to guide the equilibrium distribution toward $|\Phi(\mathbf{R},\tau)|^2$. To do this, we adopt an estimated guiding function $\Psi_G(\mathbf{R})$ that is expected to be close to the unknown solution $\Phi(\mathbf{R},\tau)$, and set $Y(\mathbf{R}) = \Psi_G(\mathbf{R})$. This choice allows the system to equilibrate toward the desired distribution. As a guiding function that is expected to be close to the unknown solution, it is natural to adopt the trial wave function from VMC, $\Psi_\alpha(\mathbf{R})$, as the guide:[3]

$$Y(\mathbf{R}) = \Psi_G(\mathbf{R}) = \Psi_\alpha(\mathbf{R}) \tag{7.27}$$

Accordingly, the mapping should be chosen as

$$f(\mathbf{R},\tau) = \Psi_\alpha(\mathbf{R}) \cdot \Phi(\mathbf{R},\tau) = \Phi(\mathbf{R},0) \cdot \Phi(\mathbf{R},\tau) \tag{7.28}$$

This form represents a mapping that is well suited to the goals of DMC.[4] Henceforth, we will refer to the mapping $\Phi \to f = \Psi_\alpha \cdot \Phi$ simply as **importance sampling**.

From the above explanation, we now understand that by using the mapping $\Phi \to f = \Psi_\alpha \cdot \Phi$, we can achieve **improved sampling efficiency** via importance sampling based on the Fokker-Planck equation. Since the resulting sampling sequence follows the distribution $|\Psi_\alpha(\mathbf{R})|^2$, this approach can also be applied to VMC. In Sect. 2.2.2, we introduced the method of generating sampling sequences proportional to $|\Psi_\alpha(\mathbf{R})|^2$ using the Metropolis algorithm. In the CASINO program package, the subroutines used for DMC are repurposed to generate the sampling distribution $|\Psi_\alpha(\mathbf{R})|^2$ by evolving the equation for f with the reaction term omitted. This is also applied in VMC calculations, used in conjunction with the Metropolis method. In the input file, the parameter **dtvmc** specifies the "time step in VMC," which corresponds to the time step used when evolving the time-evolution equation for f in the VMC calculation.

Next, we focus on the reaction term (the second term on the right-hand side) in Eq. (7.26):

$$U(\mathbf{R}) := V(\mathbf{R}) - E_T - Y^{-1} \cdot \frac{1}{2}\sum_{j=1}^{N} \nabla_j{}^2 \cdot Y$$

[3] In some specialized literature with limited space, there is a tendency to refer to the VMC trial wave function $\Psi_\alpha(\mathbf{R})$ as the guiding function from the outset, on the assumption that VMC is a preliminary step for DMC. However, for beginners, this naming may feel unintuitive. For that reason, this book has deliberately continued to use the term "trial function" for clarity.

[4] If we temporarily ignore the reaction term (the second term on the right-hand side) in Eq. (7.26), we have understood that the system will equilibrate toward $|\Psi_\alpha(\mathbf{R})|^2$. The treatment of the reaction term will be discussed later.

If the potential $V(\mathbf{R})$ is a c-number,[5] and we may write

$$V(\mathbf{R}) = Y^{-1} \cdot V(\mathbf{R}) \cdot Y$$

and combine the expression as

$$U(\mathbf{R}) = Y^{-1} \cdot \left(-\frac{1}{2} \sum_{j=1}^{N} \nabla_j{}^2 + V(\mathbf{R}) \right) \cdot Y - E_T = Y^{-1} \cdot \hat{H} \cdot Y - E_T$$

Recalling the definition of the local energy, Eq. (2.22),

$$E_L(\mathbf{R}) = \Psi_\alpha^{-1}(\mathbf{R})\, \hat{H}\, \Psi_\alpha(\mathbf{R})$$

and using the choice made in Eq. (7.27), namely, $Y(\mathbf{R}) = \Psi_\alpha(\mathbf{R})$, we find that

$$U(\mathbf{R}) = E_L(\mathbf{R}) - E_T$$

Thus, under the selection of Eq. (7.27), the time-evolution equation satisfied by $f(\mathbf{R}, \tau)$ becomes

$$\begin{aligned} \frac{\partial f(\mathbf{R}, \tau)}{\partial \tau} &= \frac{1}{2} \sum_{j=1}^{N} \nabla_j \left(\nabla_j - 2\frac{\nabla_j \Psi_\alpha(\mathbf{R})}{\Psi_\alpha(\mathbf{R})} \right) f(\mathbf{R}, \tau) \\ &\quad - (E_L(\mathbf{R}) - E_T)\, f(\mathbf{R}, \tau) \end{aligned} \tag{7.29}$$

7.2.4 Energy Estimator

As described in the previous section, DMC program implementations treat the time evolution of the Fokker-Planck equation for $f(\mathbf{R}, \tau)$ rather than directly evolving $\Phi(\mathbf{R}, \tau)$. The corresponding Langevin equation is given by Eq. (7.21), which includes the drift term from Eq. (7.25):

$$\mathbf{F}_j(\mathbf{R}) = 2\frac{\nabla_j \Psi(\mathbf{R})}{\Psi(\mathbf{R})}$$

[5] As will be discussed later, in practical calculations involving nonlocal pseudopotentials, part of $V(\mathbf{R})$ includes operators and thus is not a c-number. In that case, the discussion here requires modification (see Sect. 8.3.4).

and takes the form

$$\mathbf{\Delta R} = D \cdot \tilde{\mathbf{F}}^{(3N)}(\mathbf{R}, t) \cdot \Delta t + \chi^{\text{Gauss}(3N)}[0, 2D\Delta t]$$

A random walk simulation based on this equation is implemented computationally.

In the relation $f(\mathbf{R}, \tau) = \Psi_\alpha(\mathbf{R})\,\Phi(\mathbf{R}, \tau)$, the guiding function $\Psi_\alpha(\mathbf{R})$ is independent of time. Therefore, the time evolution of $f(\mathbf{R}, \tau)$ scales directly with that of $\Phi(\mathbf{R}, \tau)$. Maintaining the norm of $f(\mathbf{R}, \tau)$ constant is thus equivalent to maintaining the norm of $\Phi(\mathbf{R}, \tau)$ constant. Accordingly, the operational principle of DMC, namely, keeping the total number of virtual walkers (representing the histogram of $f(\mathbf{R}, \tau)$) constant, realizes the norm conservation of Φ. This corresponds to the strategy discussed in Sect. 3.2.5.

Since the random walk is performed in accordance with the time evolution of $f(\mathbf{R}, \tau)$, the histogram distribution formed by the walkers $\left\{\mathbf{R}^{(j)}\right\}_{j=1}^{N_w}$ will, through simulation, converge to the distribution $f(\mathbf{R}, \tau \to \infty)$. Using these walkers, we evaluate the local energy as

$$I = \frac{1}{N_{\text{step}}}\sum_{j=1}^{N_{\text{step}}} E_L\left(\mathbf{R}^{(j)}\right) = \langle E_L(\mathbf{R})\rangle_f \quad , \quad \left\{\mathbf{R}^{(j)}\right\}_{j=1}^{N_w} \sim f(\mathbf{R}, \tau)$$

We now consider what kind of value is obtained by sampling the local energy in this way. Substituting $f = \Phi \cdot \Psi_\alpha$ gives

$$\begin{aligned}\langle E_L(\mathbf{R})\rangle_f &= \left\langle \Psi_\alpha^{-1}(\mathbf{R})\,\hat{H}\Psi_\alpha(\mathbf{R})\right\rangle_f = \frac{\int d\mathbf{R} \cdot f(\mathbf{R}, \tau) \cdot \Psi_\alpha^{-1}(\mathbf{R})\,\hat{H}\Psi_\alpha(\mathbf{R})}{\int d\mathbf{R} \cdot f(\mathbf{R}, \tau)} \\ &= \frac{\int d\mathbf{R} \cdot \Phi(\mathbf{R}, \tau) \cdot \hat{H}\Psi_\alpha(\mathbf{R})}{\int d\mathbf{R} \cdot \Phi(\mathbf{R}, \tau)\,\Psi_\alpha(\mathbf{R})} \,.\end{aligned}$$

Since the operator $\hat{H}$ in the numerator is Hermitian, it can be regarded as acting on the left-hand side function Φ rather than on the right-hand side function Ψ_α. Assuming that $\Phi(\mathbf{R}, \tau)$ has sufficiently converged and is close to the exact solution, we may approximate $\Phi(\mathbf{R}, \tau)\,\hat{H} \approx \Phi(\mathbf{R}, \tau)\,E_0$. Then the expression becomes

$$\begin{aligned}\langle E_L(\mathbf{R})\rangle_f &= \frac{\int d\mathbf{R} \cdot \left[\Phi(\mathbf{R}, \tau) \cdot \hat{H}\right]\Psi_\alpha(\mathbf{R})}{\int d\mathbf{R} \cdot \Phi(\mathbf{R}, \tau)\,\Psi_\alpha(\mathbf{R})} \\ &\approx \frac{\int d\mathbf{R} \cdot \left[\Phi(\mathbf{R}, \tau) \cdot E_0\right]\Psi_\alpha(\mathbf{R})}{\int d\mathbf{R} \cdot \Phi(\mathbf{R}, \tau)\,\Psi_\alpha(\mathbf{R})} = E_0\end{aligned}$$

Thus, we obtain $\langle E_L(\mathbf{R})\rangle_f \approx E_0$.

By similar reasoning, for any operator $\hat{O}$ that is Hermitian and commutes with the Hamiltonian, we have

$$\left\langle \hat{O} \right\rangle_{\text{mixed}} = \left\langle \Psi_{\alpha}^{-1}(\mathbf{R})\, \hat{O}\Psi_{\alpha}(\mathbf{R}) \right\rangle_{f} = \frac{\int d\mathbf{R} \cdot \Phi(\mathbf{R}, \tau) \cdot \hat{O}\Psi_{\alpha}(\mathbf{R})}{\int d\mathbf{R} \cdot \Phi(\mathbf{R})\, \Psi(\mathbf{R}, \tau)} \tag{7.30}$$

This expression serves as an estimator for the quantum mechanical expectation value with respect to the exact eigenfunction. The subscript "mixed" reflects the fact that the operator is sandwiched between Φ and Ψ_{α} rather than appearing in a symmetric expectation value. This statistical estimator is referred to as the **mixed estimator**.[6] As discussed in Sect. 3.2.5, the **growth estimator** is defined through the adjustment of E_T to maintain a constant total number of walkers, and its converged value coincides with the ground-state energy E_0. Both the growth estimator and the mixed estimator should converge to the exact ground-state energy E_0 as the DMC simulation proceeds toward convergence. In **graphit**, the mixed estimator $E_L(\mathbf{R}_j)$ $\left[\mathbf{R}_j \sim f(\mathbf{R}, \tau)\right]$ is shown in black, while the growth estimator E_T is shown in red, allowing the user to visually monitor their behavior during the simulation.

7.3 Propagator of the Fokker-Planck Equation

Up to this point, we have explained the principle of DMC based on the idea that the time evolution of a position-distribution histogram generated by a random walk driven by Gaussian noise is governed by the diffusion equation. In other words, by inverting this perspective, we have shown that the time evolution of a distribution described by the diffusion equation (Fokker-Planck equation) can be equivalently realized through a random walk simulation driven by Gaussian noise.

On the other hand, one can also derive the particle updates driven by Gaussian noise by formally integrating the Fokker-Planck equation and rewriting it into the **propagator form**. In the imaginary-time-evolution equation of f used in DMC, the equation contains not only a diffusion term but also a reaction term (Sect. 3.2.3). An intuitive explanation of how to handle the **reaction term** was provided in Sect. 3.2.5, where we discussed simplified scenarios such as "neglecting the reaction term" or "neglecting the diffusion term" to give beginner-friendly interpretations of their individual roles. By contrast, in the propagator formalism, the diffusion and reaction terms can be treated simultaneously, making it possible to derive concrete prescriptions for how to quantitatively incorporate the reaction term. To deepen our understanding of these aspects, this section will focus on the formulation and discussion based on the propagator of the Fokker-Planck equation.

[6] For operators that are not Hermitian, the **extrapolated estimator** is used instead [3].

7.3.1 Rewriting into the Propagator Form and Its Interpretation

We rewrite the Fokker-Planck equation derived in Eq. (7.17) as follows:

$$\frac{\partial \rho (x,t)}{\partial t} = D\frac{\partial}{\partial x}\left[\frac{\partial}{\partial x} - \tilde{F}(x,t)\right]\rho (x,t) = \hat{L}_x \cdot \rho (x,t) \tag{7.31}$$

where we define the operator $\hat{L}_x$ to represent the right-hand side. By formally integrating this equation, we obtain

$$\begin{aligned}\rho (x,t) &= \exp\left[t\hat{L}_x\right]\cdot \rho (x,0)\\ &= \exp\left[t\hat{L}_x\right]\cdot \int dx_0 \cdot \delta (x-x_0)\, \rho (x_0,0)\\ &= \int dx_0 \cdot \exp\left[t\hat{L}_x\right]\cdot \delta (x-x_0)\, \rho (x_0,0)\end{aligned}$$

This expression is written in the form of propagation from $\rho (x_0,0)$ to $\rho (x,t)$. Using the identities

$$\delta (x-x_0) = \langle x_0 \mid x\rangle \quad , \quad \exp\left[t\hat{L}\right]|x\rangle = \exp\left[t\hat{L}_x\right]|x\rangle$$

we can rewrite the expression as follows:[7]

$$\begin{aligned}\rho (x,t) &= \int dx_0 \cdot \exp\left[t\hat{L}_x\right]\cdot \langle x_0 \mid x\rangle\, \rho (x_0,0)\\ &= \int dx_0 \cdot \langle x_0|\exp\left[t\hat{L}\right]|x\rangle\, \rho (x_0,0)\\ &=: \int dx_0 \cdot G(x,t;x_0)\, \rho (x_0,0)\,,\end{aligned}$$

$$G(x,t;x_0) = \langle x_0|\exp\left[t\hat{L}\right]|x\rangle \tag{7.32}$$

This is the form expressed via the propagator kernel function.[8]

By inserting the identity operator associated with the Fourier expansion basis $|p\rangle$ (Sect. 14.2.2) from (Sect. 14.2),

[7] See Eq. (14.10) for the first identity. The second identity expresses that an abstract operator $\hat{L}$ acting on $|x\rangle$ becomes the coordinate representation $\hat{L}_x$. For more on the bra-ket notation used throughout this section, refer to Sect. 14.2.

[8] Also referred to as the Green's function form. See Sect. 17.2.

$$\hat{1} = \int |p\rangle \langle p|$$

we have

$$G(x, t; x_0) = \langle x_0| \exp\left[t\hat{L}\right] |x\rangle$$
$$= \int dpdp' \cdot \langle x_0| \cdot |p\rangle \langle p| \exp\left[t\hat{L}\right] |p'\rangle\langle p'| \cdot |x\rangle$$

Using the relation based on Eq. (14.11),

$$\exp\left[t\hat{L}\right] |p'\rangle = |p'\rangle \exp\left[t\hat{L}_x\left(\frac{\partial}{\partial x} \to ip'\right)\right],$$

we obtain

$$G(x, t; x_0) = \int dpdp' \cdot \langle x_0|p\rangle \langle p|p'\rangle \exp\left[t \cdot \hat{L}_x\left(\partial_x \to ip'\right)\right] \langle p'|x\rangle$$
$$= \int dp \cdot \exp\left[t \cdot \hat{L}_x\left(\partial_x \to ip\right)\right] \langle x_0|p\rangle \langle p|x\rangle$$
$$= \frac{1}{2\pi} \int dp \cdot \exp\left[t \cdot \hat{L}_x\left(\partial_x \to ip\right)\right] \cdot \exp\left[ipx_0 - ipx\right].$$

Here, the following formulas from (Sect. 14.2.2) were used:

$$\langle x|p\rangle = \frac{1}{\sqrt{2\pi}} \cdot e^{ipx} \quad , \quad \langle p|x\rangle = \frac{1}{\sqrt{2\pi}} \cdot e^{-ipx}.$$

From Eq. (7.31), we have

$$\hat{L}_x\left(\partial_x \to ip\right) = ip \cdot D\left[ip - \tilde{F}(x, t)\right]$$

Substituting this into the previous expression, we obtain

$$G(x, t; x_0) = \frac{1}{2\pi} \int dp \cdot \exp\left[ip\left(x_0 - x\right) + tD \cdot ip\left[ip - \tilde{F}(x, t)\right]\right]. \quad (7.33)$$

Since the exponent is a quadratic function of p, we can complete the square and apply the Gaussian integration formula, which also holds for complex variables, to obtain

$$G(x, t; x_0) = \sqrt{\frac{1}{2\pi \cdot 2tD}} \cdot \exp\left[-\frac{\left[x - \left(x_0 - tD\tilde{F}(x, t)\right)\right]^2}{4tD}\right]$$

as the result (Sect. 13.2.3). Comparing this with the standard form of the Gaussian distribution

$$g_G\left(x;\mu,\sigma^2\right)=\frac{1}{\sqrt{2\pi\sigma^2}}\exp\left[-\frac{(x-\mu)^2}{2\sigma^2}\right]$$

we see that this corresponds to a Gaussian distribution with center position $x = x_0 - D\tilde{F}(x,t)\cdot t$ and variance $\sigma = \sqrt{2Dt}$,

$$G(x,t;x_0)=g_G\left(x;\left[x_0-D\tilde{F}(x,t)\cdot t\right],2Dt\right)\ . \tag{7.34}$$

To summarize once again, Eq. (7.32) becomes

$$\begin{aligned}\rho(x,t)&=\int dx_0\cdot\langle x_0|\exp\left[t\hat{L}\right]|x\rangle\,\rho(x_0,0)\\&=\int dx_0\cdot G(x,t;x_0)\,\rho(x_0,0)\\G(x,t;x_0)&=g_G\left(x;\left[x_0-D\tilde{F}(x,t)\cdot t\right],2Dt\right)\ .\end{aligned} \tag{7.35}$$

Let us recall that in formulating the Markov process, we wrote Eq. (7.8) as

$$\rho(x,t)=\int dx'\cdot P\left(x,t;x',0\right)\cdot\rho\left(x',0\right)\ .$$

Here, $P\left(x,t;x',0\right)$ describes the probability that a particle initially at x' transitions to position x after time t. From this perspective, Eq. (7.35) expresses

1. A shift of the particle position to $\left(x_0-D\tilde{F}(x,t)\cdot t\right)$ due to the external force $\tilde{F}$,
2. Followed by random diffusion around that position with Gaussian noise of variance $\sigma=\sqrt{2Dt}$,

which corresponds to the position update described by the "random walk induced by the Langevin equation" in Eq. (7.18).

7.3.2 Formulating DMC in Propagator Form

We rewrite the time-evolution equation in DMC, Eq. (7.29), as

$$-\frac{\partial f(\mathbf{R},\tau)}{\partial\tau}=\hat{L}_R\cdot f(\mathbf{R},\tau)\ ,$$

$$\begin{aligned}
\hat{L}_R &= \hat{\tilde{T}} + U(\mathbf{R}) \\
\hat{\tilde{T}} &= -\frac{1}{2}\sum_{j=1}^{N} \nabla_j \left(\nabla_j - 2\frac{\nabla_j \Psi_\alpha(\mathbf{R})}{\Psi_\alpha(\mathbf{R})} \right) =: -\frac{1}{2}\sum_{j=1}^{N} \nabla_j \left(\nabla_j - \tilde{\mathbf{F}}_j(\mathbf{R}) \right) \\
U(\mathbf{R}) &= E_L(\mathbf{R}) - E_T \ .
\end{aligned} \tag{7.36}$$

As in the previous section, we write

$$\begin{aligned}
f(\mathbf{R}, \tau) &= \langle \mathbf{R} \mid f(\tau) \rangle \\
\hat{L}_R \cdot f(\mathbf{R}, \tau) &= \langle \mathbf{R} | \, \hat{L} \, | f(\tau) \rangle
\end{aligned}$$

so that, using the operator $\hat{L}$ in the abstract space with bra-ket notation, the time-evolution equation for the state vector $\left| f(\tau) \right\rangle$ becomes

$$-\frac{\partial}{\partial \tau} | f(\tau) \rangle = \hat{L} \cdot | f(\tau) \rangle \quad , \quad \therefore \quad | f(\tau) \rangle = e^{-\tau \hat{L}} \, | f(0) \rangle \ .$$

By inserting the identity operator corresponding to the complete set $\left\{ |\mathbf{X}\rangle \right\}$,

$$\hat{1} = \int d\mathbf{X} \cdot |\mathbf{X}\rangle \langle \mathbf{X}| \ ,$$

we obtain the propagator form as

$$\begin{aligned}
| f(\tau) \rangle &= \int d\mathbf{X} \cdot e^{-\tau \hat{L}} \, |\mathbf{X}\rangle \langle \mathbf{X} \mid f(0) \rangle \\
\therefore \quad f(\mathbf{Y}, \tau) &= \int d\mathbf{X} \cdot \langle \mathbf{Y} | \, e^{-\tau \hat{L}} \, |\mathbf{X}\rangle \, f(\mathbf{X}, 0) \ .
\end{aligned} \tag{7.37}$$

We now consider evaluating the propagator function

$$G = \langle \mathbf{Y} | \, e^{-\tau \hat{L}} \, |\mathbf{X}\rangle = \langle \mathbf{Y} | \, e^{-\tau \left(\hat{\tilde{T}} + U \right)} \, |\mathbf{X}\rangle$$

Since $\hat{T}$ is exactly the operator appearing in the Fokker-Planck equation given in Eq. (7.36), in the absence of U, the propagator reduces to the multidimensional version of the Fokker-Planck kernel discussed in the previous section (Eq. (7.34)):

$$G_0 = \langle \mathbf{Y} | \, e^{-\tau \hat{\tilde{T}}} \, |\mathbf{X}\rangle = g_G^{(3N)} \left(\mathbf{Y}; \left[\mathbf{X} - \frac{\tau}{2} \cdot \tilde{\mathbf{F}}^{(3N)}(\mathbf{X}, t) \right], \tau \right) \tag{7.38}$$

Here, the dimensionally extended drift force $\tilde{\mathbf{F}}^{(3N)}$ is defined in Eq. (7.19).

When $U = E_L(\mathbf{R}) - E_T$ is finite,

$$G = \langle \mathbf{Y} | e^{-\tau\left(\hat{\hat{T}}+U\right)} | \mathbf{X} \rangle$$

cannot be evaluated in a straightforward manner because $\hat{\hat{T}}$ and U do not commute. Thus, we cannot naively separate the exponential as $e^{-\tau\left(\hat{\hat{T}}+U\right)} = e^{-\tau\hat{\hat{T}}} \cdot e^{-\tau U}$ and apply Eq. (7.38) only to the $e^{-\tau\hat{\hat{T}}}$ part.

While it is true that $e^{-\delta\tau\left(\hat{T}+U\right)} \neq e^{-\delta\tau\hat{T}} \cdot e^{-\delta\tau U}$, if the time step $\delta\tau$ is sufficiently small, then by considering the power series expansion of the exponential, in the limit $\delta\tau \to 0$, we expect

$$e^{-\delta\tau\hat{L}} = e^{-\delta\tau\left(\hat{T}+U\right)} \approx e^{-\delta\tau\hat{T}} \cdot e^{-\delta\tau U}$$

In fact, the following identity holds [3],

$$e^{-\delta\tau\left(\hat{T}+V\right)} = e^{-\delta\tau V/2} \cdot e^{-\delta\tau\hat{T}} \cdot e^{-\delta\tau V/2} + O\left(\delta\tau^3\right) \tag{7.39}$$

based on the Suzuki-Trotter formula [3], [9] and it justifies the approach of evaluating the $e^{-\delta\tau\hat{\hat{T}}}$ part using Eq. (7.38). To apply this approach, we begin with Eq. (7.37):

$$|f(\tau)\rangle = \int d\mathbf{X} \cdot e^{-\tau\hat{L}} |\mathbf{X}\rangle \langle \mathbf{X} | f(0)\rangle \ .$$

We slice the exponential operator into small time steps

$$e^{-\tau\hat{L}} = e^{-(N\times\delta\tau)\hat{L}} = e^{-\delta\tau\hat{L}} \times e^{-\delta\tau\hat{L}} \times \cdots \times e^{-\delta\tau\hat{L}} \tag{7.40}$$

with $\delta\tau := \tau/N$ and insert the identity operator between each short-time propagator:

$$\hat{1} = \int d\mathbf{X} \cdot |\mathbf{X}\rangle \langle \mathbf{X}| \tag{7.41}$$

This leads to the expression

$$|f(\tau)\rangle = \int d\mathbf{X}_N \cdots d\mathbf{X}_0 \cdot e^{-\delta\tau\hat{L}} |\mathbf{X}_N\rangle$$

[9] This formula is based on the Suzuki-Trotter decomposition and is considered a cornerstone result in quantum Monte Carlo methods, since it enables their practical implementation.

$$\times \langle \mathbf{X}_N | \, e^{-\delta\tau \hat{L}} \, | \mathbf{X}_{N-1} \rangle \cdots \langle \mathbf{X}_2 | \, e^{-\delta\tau \hat{L}} \, | \mathbf{X}_1 \rangle$$
$$\times \langle \mathbf{X}_1 | \, e^{-\delta\tau \hat{L}} \, | \mathbf{X}_0 \rangle \, \langle \mathbf{X}_0 \mid f(0) \rangle \; . \tag{7.42}$$

In this way, Eq. (7.39) can be applied to each $e^{-\delta\tau\hat{L}} = e^{-\delta\tau(\hat{T}+V)}$ term.

By choosing $|\mathbf{X}\rangle$ to be the coordinate-space representation on a continuous space,

$$|\mathbf{X}\rangle = |\mathbf{R}\rangle = |\mathbf{r}_1, \cdots, \mathbf{r}_N\rangle \tag{7.43}$$

we obtain[10]

$$f(\mathbf{R}, \tau) = \int d\mathbf{R}_N \cdots d\mathbf{R}_0 \cdot \langle \mathbf{R} | \, e^{-\delta\tau \hat{L}} \, | \mathbf{R}_N \rangle$$
$$\times \langle \mathbf{R}_N | \, e^{-\delta\tau \hat{L}} \, | \mathbf{R}_{N-1} \rangle \cdots \langle \mathbf{R}_2 | \, e^{-\delta\tau \hat{L}} \, | \mathbf{R}_1 \rangle$$
$$\times \langle \mathbf{R}_1 | \, e^{-\delta\tau \hat{L}} \, | \mathbf{R}_0 \rangle \, f(\mathbf{R}_0, 0) \; . \tag{7.44}$$

This expression represents a superposition of short-time evolutions, such as

$$f(\mathbf{R}_1, \delta\tau) = \int d\mathbf{R}_0 \cdot G(\mathbf{R}_1 \leftarrow \mathbf{R}_0; \delta\tau) \cdot f(\mathbf{R}_0, 0) \; . \tag{7.45}$$

This provides the picture of a many-body system that was at configuration $\mathbf{R}_0 = (\mathbf{r}_1, \cdots, \mathbf{r}_N)$ at time $\tau = 0$, undergoing a sequence of configuration updates via the propagator

$$G(\mathbf{R}_k \leftarrow \mathbf{R}_{k-1}; \delta\tau) = \langle \mathbf{R}_k | \, e^{-\delta\tau \cdot \hat{L}} \, | \mathbf{R}_{k-1} \rangle \tag{7.46}$$

and evolving forward in time until reaching time τ.

The propagation over this sufficiently short time span can be evaluated by applying the result of Eq. (7.39) as

$$\begin{aligned} G &\approx \langle \mathbf{R}_k | \, e^{-\delta\tau U/2} \cdot e^{-\delta\tau \hat{T}} \cdot e^{-\delta\tau U/2} \, | \mathbf{R}_{k-1} \rangle \\ &= e^{-\delta\tau U(\mathbf{R}_k)/2} \, \langle \mathbf{R}_k | \, e^{-\delta\tau \hat{T}} \, | \mathbf{R}_{k-1} \rangle \, e^{-\delta\tau U(\mathbf{R}_{k-1})/2} \\ &= e^{-\delta\tau \bar{U}(\mathbf{R}_k)} \cdot g_G^{(3N)} \left(\mathbf{R}_k ; \left[\mathbf{R}_{k-1} - \frac{\delta\tau}{2} \cdot \tilde{\mathbf{F}}^{(3N)}(\mathbf{R}_{k-1}) \right], \delta\tau \right) \; . \end{aligned} \tag{7.47}$$

[10] Up to this point, the complete set $\{|\mathbf{X}\rangle\}$ has not been specified explicitly. It is also possible to choose a representation other than the coordinate-space basis on continuous space. Starting from this formulation, one can extend the understanding to various forms of quantum Monte Carlo methods, beyond those on continuous spaces, which is the main topic of this book.

Here, the averaged value denoted by $\bar{U}(\mathbf{R}_k)$ is defined as

$$\bar{U}(\mathbf{R}_k) = \frac{U(\mathbf{R}_k) + U(\mathbf{R}_{k-1})}{2}.$$

The drift force $\tilde{\mathbf{F}}^{(3N)}$ in our problem is given by

$$\tilde{\mathbf{F}}^{(3N)}(\mathbf{R}) = \begin{pmatrix} \tilde{\mathbf{F}}_{j=1}(\mathbf{R}) \\ \vdots \\ \tilde{\mathbf{F}}_{j=N}(\mathbf{R}) \end{pmatrix} , \quad \tilde{\mathbf{F}}_j(\mathbf{R}) = 2\frac{\nabla_j \Psi_\alpha(\mathbf{R})}{\Psi_\alpha(\mathbf{R})}$$

which depends only on the spatial coordinates.

Equation (7.47) implies that the many-electron configuration $\mathbf{R}_{k-1}$ at the $(k-1)$-th step evolves over a time span $\delta\tau$ by diffusion due to Gaussian noise, and the corresponding position update is multiplied by a weight $e^{-\delta\tau\bar{U}(\mathbf{R}_{k-1})}$. Thus, if we express $f(\mathbf{R}, \tau)$ as a histogram of particles, the procedure can be implemented as

1. Interpret $f(\mathbf{R}, 0)$ as a histogram formed by N_w particle positions $\left\{\mathbf{R}^{(j)}(\tau)\right\}_{j=1}^{N_w}$ in $3N$-dimensional space.
2. Divide the total time τ from $\tau = 0$ to $\tau = \tau$ into N_{step} sufficiently short time slices $\delta\tau$ (where $\delta\tau = \tau/N_{\text{step}}$).
3. For each configuration $\mathbf{R}_{k-1}^{(j)}$ at the $(k-1)$-th step, shift it in configuration space to $\left(\mathbf{R}_{k-1}^{(j)} - \cdot\tilde{\mathbf{F}}^{3N}\left(\mathbf{R}_{k-1}^{(j)}\right)\cdot\delta\tau/2\right)$ under the external force $\tilde{\mathbf{F}}^{3N}\left(\mathbf{R}_{k-1}^{(j)}\right)$,
4. Then, update the position by applying Gaussian noise centered at that location, with variance $\delta\tau$,
5. Multiply the particle by a weight $e^{-\delta\tau\bar{U}\left(\mathbf{R}_k^{(j)}\right)}$. Repeat this process for the subsequent steps.
6. After time τ, reconstruct the histogram using the virtual particles $\left\{\mathbf{R}^{(j)}\right\}$ not by counting each as one but by assigning each the accumulated weight

$$w^{(j)} = \prod_{k=1}^{N_{\text{step}}} \exp\left[-\delta\tau\cdot\bar{U}\left(\mathbf{R}_k^{(j)}\right)\right] \tag{7.48}$$

and then the resulting histogram gives $f(\mathbf{R}, \tau)$.

In this procedure, the simulation specifications include three integers: N, N_w, and N_{step}. We emphasize once again the importance of not confusing them.

7.3.3 *Weight of Propagation and Branching*

The weight function given in Eq. (7.48) is, in principle, evaluated by accumulating the full history for each walker. However, this approach is computationally inefficient and is rarely used in practical calculations. Ultimately, the weight can be reinterpreted as a way of treating the "unit walker" effectively. For example, if a walker has weight 0.3, it can be regarded as existing with 30% probability and disappearing with 70% probability. Likewise, **if the weight is 2.7, then the walker should be counted as three with 70% probability and two with 30% probability**. Thus, instead of accumulating weights over the entire simulation, we replace this by a scheme where the number of walkers is **probabilistically increased or decreased** at each time step.

For the weight of the j-th walker at step k,

$$w_k^{(j)} = \exp\left[-\delta\tau \cdot \bar{U}\left(\mathbf{R}_k^{(j)}\right)\right] = \exp\left[-\delta\tau \cdot \left[\frac{E_L\left(\mathbf{R}_k^{(j)}\right) + E_L\left(\mathbf{R}_{k-1}^{(j)}\right)}{2} - E_T\right]\right]$$

we can define the fractional part of $w_k^{(j)}$ (i.e., < 1, as in the earlier example of 2.7, where the fractional part is 0.7) using the floor function $[x]$ as $p = w_k^{(j)} - [w_k^{(j)}]$. To determine the survival or death of a particle based on this fractional part, we generate a uniformly distributed random number χ_u in the interval $[0, 1]$. If the drawn number is less than or equal to p, the walker survives; if it is greater than p, the walker is eliminated. This procedure is equivalent to setting the number of copies of walker j in the next time step to $\left[w_k^{(j)} + \chi_u\right]$.[11] This procedure of probabilistically creating or annihilating of walkers $\mathbf{R}_{k-1}^{(j)}$ based on their weights is called **branching**.

Let us now carefully examine the expression for branching:

$$w_k^{(j)} = \exp\left[-\delta\tau \cdot \left[\frac{E_L\left(\mathbf{R}_k^{(j)}\right) + E_L\left(\mathbf{R}_{k-1}^{(j)}\right)}{2} - E_T\right]\right] . \tag{7.49}$$

Although $f \sim \Phi(\mathbf{R}, \tau)$ is mapped from another quantity, it is essentially a many-body wave function, that is, a distribution over values in $3N$-dimensional space. This distribution is represented as a histogram formed by walkers indexed by $j = 1, \cdots, N_w$. Each walker is a "particle" located at a single point in $3N$-dimensional

[11] Let p be the fractional part (remainder) of w. Since χ_u is uniformly distributed in the interval $U[0, 1]$, the sum $p + \chi_u$ is uniformly distributed in $U[p, 1 + p]$. The probability that $(p + \chi_u)$ exceeds 1 is equal to the length of the interval $[1, 1 + p]$, which is p. Therefore, the probability that $[p + \chi_u] = 1$ is p. Thus, the procedure of setting $[w + \chi_u]$ as the number of particles in the next step effectively implements "counting 1 particle with probability p for the fractional part of w."

space, representing a snapshot of the configuration of N electrons. The N_w walkers therefore simultaneously sample multiple possible configurations.

Let us now focus on a specific walker indexed by j. The position of this walker in $3N$-dimensional space evolves over time.[12] Time evolution is discretized as $\tau_k = k \cdot \delta\tau$ and indexed by k. For the configuration $\mathbf{R}_{k-1}^{(j)}$ determined at the $(k-1)$-th step of the simulation, the local energy $E_L\left(\mathbf{R}_{k-1}^{(j)}\right)$ is computed. In regions where the local energy is high, walkers tend to vanish with probability $\sim e^{-E_L}$, whereas in regions of low local energy, walkers tend to split and increase in number. In this way, the simulation implements a mechanism in which **electronic configurations associated with lower many-body potential energy gradually become dominant**.

Due to branching, the total number of walkers fluctuates throughout the simulation. The energy offset E_T contributes uniformly as a baseline shift, regardless of whether the local energy is high or low. If E_T is set higher, walkers tend to increase overall, as $\sim e^{E_T}$; conversely, if E_T is set lower, the growth of walkers is suppressed. Since the histogram formed by the walkers is interpreted as the many-body wave function, the total number of walkers corresponds to the total integral of the wave function, that is, its norm. On the other hand, E_T contributes to the imaginary-time evolution of the many-body wave function as shown in Eq. (3.18):

$$\Phi(\mathbf{R},\tau) = \Psi_0(\mathbf{R}) \cdot \exp\left[-\left(E_0 - E_T\right)\tau\right] + \lambda_1 \cdot \Psi_1(\mathbf{R}) \cdot \exp\left[-\left(E_1 - E_T\right)\tau\right] \\ + \lambda_2 \cdot \Psi_2(\mathbf{R}) \cdot \exp\left[-\left(E_2 - E_T\right)\tau\right] + \cdots$$

When E_T approaches the true ground-state energy E_0, Φ converges to a constant in the limit $\tau \to \infty$, and the norm also becomes constant and invariant.[13] Therefore, by **adjusting E_T as a control parameter so that the total number of walkers remains constant,**[14] E_T converges to E_0. This converged value becomes the exact ground-state eigenvalue of the many-body Schrodinger equation (see the growth estimator described in Sect. 3.2.5).

7.3.4 Fluctuation Reduced by Importance Sampling

Instead of the importance-sampled time evolution given in Eq. (7.36), we consider the time evolution for the raw $\Phi(\mathbf{R},\tau)$:

$$\frac{\partial\Phi(\mathbf{R},\tau)}{\partial\tau} = \left(\frac{1}{2}\sum_{j=1}^{N}\nabla_j^2 + E_T - V(\mathbf{R})\right)\Phi(\mathbf{R},\tau) =: -\hat{L}\cdot\Phi(\mathbf{R},\tau)\ .$$

[12] This corresponds to the time evolution of a configuration pattern formed by N electrons.

[13] If E_T deviates from E_0; the norm either decays or grows exponentially.

[14] There is subtlety in the update schemes for E_T, and various algorithms have been studied and developed for this purpose [3].

Proceeding with the discussion in terms of the propagator formalism, we obtain, analogous to Eqs. (7.45) and (7.47),

$$\Phi(\mathbf{R}_1, \delta\tau) = \int d\mathbf{R}_1 \cdot G_\Phi(\mathbf{R}_1 \leftarrow \mathbf{R}_0; \delta\tau) \cdot \Phi(\mathbf{R}_0, 0)$$

$$G_\Phi(\mathbf{R}_k \leftarrow \mathbf{R}_{k-1}; \delta\tau) \approx e^{-\delta\tau\left[\bar{V}(\mathbf{R}_{k-1}) - E_T\right]} \cdot g_G^{(3N)}(\mathbf{R}_k; \mathbf{R}_{k-1}, \delta\tau) \ .$$

In this case, the time evolution of walker positions is carried out by the following procedure:

1. Diffuse walker positions using a Gaussian distribution centered at $\mathbf{R}_{k-1}$ with variance $\delta\tau$,
2. Then apply the branching weight

$$\tilde{w}_k^{(j)} = \exp\left[-\delta\tau \cdot \left[\frac{V\left(\mathbf{R}_k^{(j)}\right) + V\left(\mathbf{R}_{k-1}^{(j)}\right)}{2} - E_T\right]\right] \tag{7.50}$$

to advance the walker positions in time.

Since the raw many-body potential $V(\mathbf{R})$ fluctuates strongly throughout space, a naive implementation based directly on $\Phi(\mathbf{R}, \tau)$ without using the mapping f leads to large fluctuations in walker weights and, hence, in the number of walkers during the simulation. This causes significant numerical instability, rendering the method practically unusable.[15]

Let us compare the branching expression obtained with importance sampling (Eq. (7.49)) to the branching based directly on the raw many-body wave function Φ without importance sampling (Eq. (7.50)). One notices that importance sampling replaces the raw potential $V(\mathbf{R})$ with the local energy $E_L(\mathbf{R}) = \Psi_T^{-1}(\mathbf{R})\hat{H}\Psi_T(\mathbf{R})$, highlighting its benefit. As seen in Eq. (4.12), the local energy has the property (zero-variance property) that it becomes spatially uniform as the trial wave function Ψ_T approaches the exact one. Therefore, by adopting the mapping $\Phi \to f$ via importance sampling, the fluctuations in branching are replaced by milder variations, greatly improving computational stability.[16] This line of reasoning also explains why it is necessary to first perform VMC to bring the initial trial wave function reasonably close to the exact solution before carrying

[15] The number of walkers can fluctuate so wildly that memory is quickly exhausted, leading to a crash of the simulation.

[16] Many explanations motivate the projection $\Phi \to f$ by stating that it is introduced "to suppress fluctuations in the branching weights by replacing $V(\mathbf{R})$ with $E_L(\mathbf{R})$." In such treatments, the term "importance sampling" is introduced abruptly at this point, often leading beginners to wonder "where exactly is the importance sampling happening?". In this book, we instead introduced the concept by starting from the equilibrium solution of the Fokker-Planck equation.

out DMC. By doing so, the spatial fluctuations of the local energy $E_L(\mathbf{R})$ are suppressed, making the simulation more stable and reliable.

References

1. "Mathematical Methods for Physics and Engineering, Third Edition" (Cambridge University Press, 2006), K.F. Riley, M.P. Hobson, S.J. Bence, ISBN: 978-0521679718
2. "Kinetic Theory: Classical, Quantum, and Relativistic Descriptions, Third Edition" (Springer, 2003), R.L. Liboff, ISBN: 978-0387955513
3. "Quantum Monte Carlo simulations of solids", W.M.C. Foulkes, L. Mitas, R. J. Needs, and G. Rajagopal, Rev. Mod. Phys. 73, 33 (2001), https://doi.org/10.1103/RevModPhys.73.33

Chapter 8
Further Topics on Underlying Theory

In the following chapters, we will address more advanced content. This chapter focuses on the theoretical foundations of ab initio QMC, covering material that may be somewhat advanced for beginners. We will build on the concepts introduced thus far and summarize an overview that serves as a bridge to discussions found in more specialized literature. Topics that are not central to the theory but are practically advanced, such as correction schemes or the use of large-scale supercomputers, will be summarized in Chap. 9. In addition, more advanced topics that aim to provide an understanding of the advantages and positioning of ab initio QMC from the broader perspective of many-body electronic structure theory will be presented in Chap. 10.

In this chapter, we will specifically provide an overview of the cusp condition (Sect. 8.1), the fixed-node approximation (Sect. 8.2), and pseudopotentials (Sect. 8.3). These topics represent important technical challenges for practitioners specializing in ab initio QMC. When moving beyond pre-provided tutorial examples to tackle professional-level real-world problems, it is essential to be aware that these issues must be confronted. By the end of this chapter, you will be able to articulate the core concepts of ab initio QMC precisely using specialized terminology. Therefore, in Sect. 8.4, we will revisit the procedure of ab initio QMC using these technical terms.

8.1 Cusp Condition

The Slater-Jastrow-type trial wave function in Eq. (4.5) is given by

$$\Psi_{\boldsymbol{\alpha}}\left(\mathbf{r}_1, \cdots, \mathbf{r}_N\right) = \exp\left[J\left(\mathbf{R};\boldsymbol{\alpha}\right)\right] \times \Psi^{(\mathrm{SD})}\left(\mathbf{R}\right) .$$

Here, the Jastrow factor is exponential and thus always takes positive values. Therefore, the sign changes of $\Psi_{\boldsymbol{\alpha}}$ are determined solely by the Slater determinant

R. Maezono, *Ab initio Quantum Monte Carlo Tutorial*,
https://doi.org/10.1007/978-981-95-4134-8_8

part $\Psi^{(SD)}(\mathbf{R})$. Issues related to sign changes will be discussed later in Sect. 8.2. Here, we begin by examining the role played by the positive-valued Jastrow factor.

8.1.1 *Behavior of the Wave Function Near Particle Coalescence*

Since the Hamiltonian $\hat{H}$ is the sum of the kinetic energy $\hat{T}$ and the potential V, the local energy in Eq. (2.22) can be written as

$$E_L(\mathbf{R}) = \Psi_\alpha^{-1}(\mathbf{R})\,\hat{H}\Psi_\alpha(\mathbf{R}) = \frac{1}{\Psi_\alpha(\mathbf{R})}\left[\hat{T} + V\right]\cdot\Psi_\alpha(\mathbf{R}) \ .$$

The potential V includes the electron-electron interaction (V_{ee}) and the electron-nucleus interaction (V_{eN}) and can be written as

$$V \sim V_{ee}\left(\left\{\left|\mathbf{r}_j - \mathbf{r}_l\right|\right\}\right) + V_{eN}\left(\left\{\left|\mathbf{r}_j - \mathbf{R}_\alpha\right|\right\}\right)$$

It depends on the electronic positions $\left\{\mathbf{r}_j\right\}$ and the nuclear positions $\{\mathbf{R}_\alpha\}$. Both V_{ee} and V_{eN} take the form of Coulomb potentials:

$$V_{ee}\left(\left\{\left|\mathbf{r}_j - \mathbf{r}_l\right|\right\}\right) \sim \frac{1}{\left|\mathbf{r}_j - \mathbf{r}_l\right|} \quad , \quad V_{eN}\left(\left\{\left|\mathbf{r}_j - \mathbf{R}_\alpha\right|\right\}\right) \sim \frac{1}{\left|\mathbf{r}_j - \mathbf{R}_\alpha\right|} \tag{8.1}$$

Thus, when particle coalescence occurs, such as $\mathbf{r}_j \to \mathbf{r}_l$, the absolute value of the potential diverges.

In the evaluation of the local energy

$$E_L(\mathbf{R}) = \frac{1}{\Psi_\alpha(\mathbf{R})}\left[\hat{T} + V(\mathbf{R})\right]\cdot\Psi_\alpha(\mathbf{R})$$

the sampled configurations $\mathbf{R}$ are chosen randomly. Among these, some configurations may lie close to "two-electron coalescence" $\mathbf{r}_j \to \mathbf{r}_l$, or "coalescence with the nucleus" $\mathbf{r}_j \to \mathbf{R}_\alpha$, which causes the potential in Eq. (8.1) to diverge. Since the local energy $E_L(\mathbf{R})$ includes contributions from these potential terms, it also tends to exhibit divergent behavior. Figure 8.1 shows an actual example of the variation in local energy when a coalescence with a nucleus occurs. As shown by the solid black line, a divergent trend is clearly observed.

However, as discussed in Sect. 4.1.2, if the trial wave function is exact, the local energy $E_L(\mathbf{R})$ exhibits zero-variance behavior, that is, it must remain constant regardless of $\mathbf{R}$. Thus, for the exact solution, the $\mathbf{R}$-dependence seen in the solid line of Fig. 8.1 should not appear, and E_L must be constant. On the other hand, the divergence of $V(\mathbf{R})$ occurs regardless of the quality of the trial wave

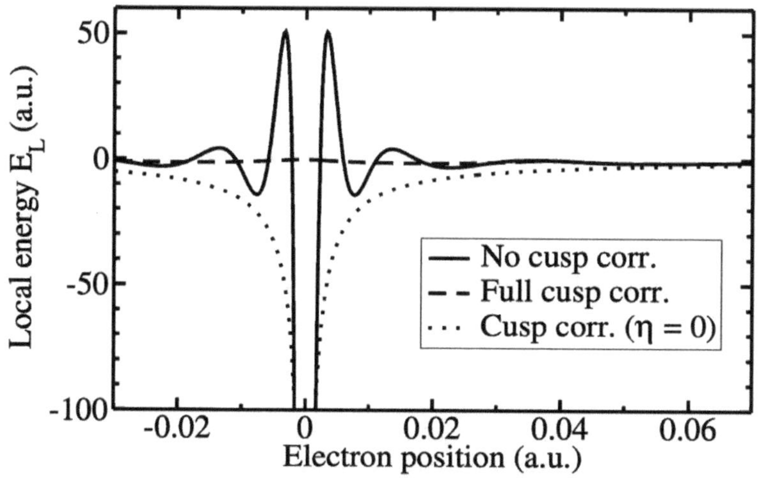

Fig. 8.1 Typical behavior of the local energy E_L near a nucleus. For the exact solution, the local energy E_L should behave as a constant. However, near the nucleus, the theory relies on a precise cancellation between divergent positive and negative terms, so that the net result is constant. Without numerically careful treatment, the behavior can become erratic, as shown by the solid line. By applying a numerical correction scheme known as cusp correction (cusp corr.), this instability can be suppressed, leading to smoother behavior and more stable numerical calculations (Figure adapted from Ref. [1])

function. Therefore, in order for the local energy to exhibit zero-variance behavior, **the divergence of the potential term $V(\mathbf{R})$ must be exactly canceled by an oppositely signed divergence in the kinetic energy term**. For example, in the case of particle coalescence $\mathbf{r}_j \to \mathbf{r}_l$, the kinetic energy term must behave as

$$\hat{T}\Psi_T\left(\left\{\mathbf{r}_j\right\}, \{\mathbf{R}_\alpha\}\right) \sim \nabla^2\Psi_T\left(\left\{\mathbf{r}_j\right\}, \{\mathbf{R}_\alpha\}\right) \sim \frac{-1}{\left|\mathbf{r}_j - \mathbf{r}_l\right|} \quad \left(\text{as} \quad \mathbf{r}_j \to \mathbf{r}_l\right) \tag{8.2}$$

That is, it must exactly cancel the divergence of the potential with opposite sign, so that the local energy $E_L(\mathbf{R})$ remains finite and constant at E_0.[1] The same type of cancellation must also occur for coalescence between an electron and a nucleus, i.e., $\mathbf{r}_j \to \mathbf{R}_\alpha$.

[1] For example, suppose that the exact ground-state energy is -13.6 [eV]. Then, the local energy values sampled using the exact solution must also be -13.6 [eV] at every point. Now imagine that, in a snapshot of a sampled configuration, two particles come very close to each other, and the $1/\left|r_{ij}\right|$ term contributes $+16{,}423{,}221.988$ [eV]. In that case, the kinetic energy term must exactly equal $-16{,}423{,}221.988 - 13.6$ [eV], so that their sum remains precisely -13.6 [eV].

8.1.2 Cusp Condition Equation

We now consider how to rephrase the requirement given in Eq. (8.2), that the kinetic energy, i.e., the Laplacian of the exact wave function, diverges with an inverse power law upon particle coalescence, in terms of the wave function itself, rather than its Laplacian. The conclusion is the emergence of a **cusp condition**, which states that the wave function itself must possess a singular point (cusp) at the particle coalescence point. The procedure to derive this condition is as follows: We begin with the many-body Schrodinger equation for the exact solution Φ:

$$\left(\hat{T}+V-E\right)\cdot\Phi\left(\mathbf{r}\right)=0 \quad , \quad \hat{T}=\frac{1}{2}\nabla^2 .$$

We then extract the dominant divergent contributions associated with the coalescence of two particles $\mathbf{r}_1$ and $\mathbf{r}_2$, expressed in terms of their interparticle distance $r_{12}=|\mathbf{r}_2-\mathbf{r}_1|$, as

$$\hat{T}\sim\hat{K}\left(\nabla_{12}\right) \quad , \quad \Phi\sim\Phi\left(r_{12}\right) \tag{8.3}$$

Then, by requiring that these divergences cancel out, we obtain the cusp condition equation.

We consider a spherically symmetric system centered at the coalescence point and introduce the radial vector from the origin as $\mathbf{r}=(r,\theta,\phi)$. The dominant divergent part of the wave function $\Phi\left(\mathbf{r}\right)$ is then treated using a basis expansion separated in variables via spherical harmonics:

$$\Phi\left(\mathbf{r}\right)=\sum_{l,m}c_{l,m}\cdot R_l\left(r\right)\cdot Y_{lm}\left(\theta,\phi\right)$$

This approach follows standard treatments using spherical harmonics and radial functions [2, 3]. The Laplacian in spherical coordinates [2] is given by

$$\nabla^2\Phi=\frac{1}{r^2}\frac{\partial}{\partial r}\left[r^2\frac{\partial\Phi}{\partial r}\right]+\frac{1}{r^2\cdot\sin\theta}\frac{\partial}{\partial\theta}\left(\sin\theta\frac{\partial\Phi}{\partial\theta}\right)+\frac{1}{r^2\cdot\sin^2\theta}\frac{\partial^2\Phi}{\partial^2\phi} .$$

Applying the above spherical harmonic expansion to this expression, in the case of a Coulomb interaction $V\left(r\right)=Z/r$, the radial function $R_l\left(r\right)$ must satisfy the following radial differential equation:

$$\left[\frac{d^2}{dr^2}+\frac{2}{r}\cdot\frac{d}{dr}+\frac{2Z}{r}-\frac{l\left(l+1\right)}{r^2}+2E\right]\cdot R_l\left(r\right)=0 \tag{8.4}$$

as shown in Ref. [4].

To understand the overall structure, let us begin with the simplest case of $l=0$ (the s-wave). In this case, the $1/r^2$ term in Eq. (8.4) vanishes. Then, as $r\to 0$, the dominant contributions in Eq. (8.4) are those with $1/r$ dependence. The condition that the divergence of the potential energy be canceled by the kinetic energy term leads to the requirement:

$$\left[\left(\frac{2}{r}\cdot\frac{d}{dr}+\frac{2Z}{r}\right)\cdot R_{l=0}\left(r\right)\right]_{r=0}=0$$

From this equation, the cusp condition for $l=0$ (s-wave) is obtained as

$$\left[\frac{1}{R_{l=0}\left(r\right)}\cdot\frac{dR_{l=0}\left(r\right)}{dr}\right]_{r=0}=-Z$$

This leads to

$$R'_{l=0}\left(r\right)=-Z\cdot R_{l=0}\left(r\right)\quad,\quad\therefore\quad R_{l=0}\left(r\right)\sim\exp\left[-Z\cdot r\right]$$

and thus reproduces the well-known result that "$R_{l=0}\left(r\right)$ exhibits a cusp at $r\to 0$," characteristic of s-orbitals [4].

In the general case including $l\neq 0$, we assume the following form to express the leading behavior of $R_l\left(r\right)$ near the origin:

$$R_l\left(r\right)=r^m\cdot\chi\left(r\right)\quad,\quad\chi\left(0\right)\neq 0$$

This expresses the rise of $R_l\left(r\right)$ from the origin as a power law of order m.[2] By analyzing the powers of divergence, we find that the most divergent term is the $1/r^2$ term. The requirement that the divergence arising from the kinetic energy, proportional to $m(m+1)$, cancels the corresponding potential term involving l leads to the condition $m=l$ (see Sect. 15.1). Under this condition, among the remaining terms, we now consider the $1/r$ term. The requirement that the potential energy be canceled by the kinetic energy leads to the cusp condition (see Sect. 15.1):

$$\left[\frac{\chi'\left(r\right)}{\chi\left(r\right)}\right]_{r=0}=-\frac{Z}{(l+1)}\,. \tag{8.5}$$

This is the cusp condition equation. From this equation, we obtain the following asymptotic behavior as $r\to 0$:

$$\chi\left(r\right)\propto\exp\left[-\frac{Z}{(l+1)}\cdot r\right]\quad\therefore\quad R^{(l)}\left(r\right)\propto r^l\cdot\exp\left[-\frac{Z}{(l+1)}\cdot r\right]\,. \tag{8.6}$$

Consider a situation in which a particle coalescence $\mathbf{r}_i\to\mathbf{r}_j$ occurs in the many-body wave function $\Phi\left(\mathbf{R}\right)$. We describe the dependence on the interparticle distance

[2] Setting $m=0$ allows $R_l\left(0\right)\neq 0$ and includes the s-state. For $m\neq 0$, we have $R_l\left(0\right)=0$.

$r_{ij} = \left|\mathbf{r}_i - \mathbf{r}_j\right|$ using the expansion:

$$\begin{aligned}\Phi(\mathbf{R}) &= \Phi\left(\cdots, \mathbf{r}_i, \cdots, \mathbf{r}_i + \mathbf{r}_{ij}, \cdots\right) \\ &= \Phi(\mathbf{R})|_{r_{ij}=0} + r_{ij} \left.\frac{\partial \Phi(\mathbf{R})}{\partial r_{ij}}\right|_{r_{ij}=0} + \cdots . \end{aligned} \tag{8.7}$$

If $\Phi(\mathbf{R})|_{r_{ij}=0} \neq 0$, we treat this term as the leading term and approximate the behavior near $r_{ij} \to 0$ as spherically symmetric to first order. Since the value is finite at $r_{ij} = 0$, the behavior corresponds to an s-wave ($l = 0$). On the other hand, if $\Phi(\mathbf{R})|_{r_{ij}=0} = 0$ is explicitly required, then the second term in Eq. (8.7) becomes the leading term, which we again treat with spherical symmetry in the first approximation. Since this term is linear in r_{ij}, the behavior corresponds to a p-wave ($l = 1$).

In the case of **electron-nucleus coalescence** ($r_{i\alpha} \to 0$), the orbital function of the electron coalescing with a nucleus (of atomic number Z) must, depending on its orbital angular momentum l, take the form

$$\psi_l(r) = r^l \cdot \chi_l(r)$$

and satisfy the following asymptotic condition:

$$\left.\frac{1}{\chi_l}\frac{d\chi_l(r_{i\alpha})}{dr}\right|_{r_{i\alpha}=0} = -\frac{Z}{l+1}$$

In the case of **electron-electron coalescence** ($r_{ij} \to 0$), the condition depends on whether the spins of the two coalescing electrons are parallel or antiparallel. For a **spin-parallel pair**, the Pauli exclusion principle requires that $\Phi\left(r_{ij} = 0\right) = 0$. Therefore, the first term in Eq. (8.7) vanishes, and the leading term becomes linear in r_{ij}, corresponding to $l = 1$. For a **spin-antiparallel pair**, we have $\Phi\left(r_{ij} = 0\right) \neq 0$, and the expansion starts from zeroth order, corresponding to $l = 0$. Unlike the electron-nucleus coalescence case, in electron-electron coalescence, the kinetic energy contribution involves both electrons. To reflect this, a factor of $1/2$ is introduced on the right-hand side of Eq. (8.5). Setting $Z = 1$, we obtain

$$\begin{aligned}\left[\frac{1}{\Phi}\frac{d\Phi}{dr_{ij}}\right]_{r_{ij}=0} &= \frac{1}{4} \quad \text{for} \quad \uparrow\uparrow \\ \left[\frac{1}{\Phi}\frac{d\Phi}{dr_{ij}}\right]_{r_{ij}=0} &= \frac{1}{2} \quad \text{for} \quad \uparrow\downarrow \end{aligned} \tag{8.8}$$

This constitutes the "divergence-cancellation condition that the many-body wave function Φ must satisfy," i.e., the **electron-electron cusp condition**.

Why Gaussian Basis Functions Are Used
In practical all-electron calculations for isolated molecular systems, Gaussian basis functions are almost universally used. A Gaussian basis represents the wave function around a nucleus as a superposition of Gaussian-type functions $\sim\exp\left(-\alpha \cdot x^2\right)$. Gaussian functions are smooth at the origin and do not represent a sharp point (cusp). However, exact orbital functions must exhibit a cusp at the origin, i.e., the electron-ion coalescence point, as required by Eq. (8.6). In Gaussian basis sets, in order to approximate this cusp behavior, many Gaussian functions with large exponents α are superposed to provide a higher-quality all-electron basis. One might wonder: if the exact orbital functions are known to possess a cusp, why not use functions of the form $\sim\exp\left(-\gamma \cdot |x|\right)$ from the beginning, rather than relying on a superposition of Gaussian functions? In molecular orbital methods, including Hartree-Fock calculations, the computation frequently involves evaluating two-electron and four-electron integrals over orbital functions. At the time, such integrals had to be processed using manual or mechanical calculators. For Gaussian-type functions, these integrals can be carried out analytically to a significant extent, allowing the use of precomputed tables. This practicality led to the adoption of the idea that "cusps can be approximated by a superposition of Gaussian functions." In the modern era of digital computing, such detours are no longer necessary. It is now possible to carry out calculations using basis functions of the form $\sim\exp\left(-\gamma \cdot |x|\right)$ (i.e., Slater-type orbitals), and actual implementations of such methods are publicly available [5]. Nevertheless, calculations based on Gaussian basis sets remain mainstream, due to the vast accumulated knowledge regarding their reliability, for example, knowing "how many significant digits can be trusted with a given basis set." Incidentally, the aforementioned precomputed tables were created by Masao Kotani and became known as the "Kotani table" [6], which was widely used worldwide in the era of manual mechanical calculators.

8.1.3 *Role of Jastrow Functions*

The cusp condition is necessary to cancel the divergent contribution from the potential energy at particle coalescence, so that a finite eigenvalue can be obtained. Because of this condition, the sampled local energy values also exhibit internal cancellation of divergent terms, and the variation in value with respect to the sampling point $\mathbf{R}$ is kept small. A large variance in the local energy therefore corresponds to **insufficient enforcement of the cusp condition** in the trial wave function. When using a trial wave function that poorly incorporates the cusp behavior, the divergent contributions are not sufficiently canceled, resulting in large local energy values. Through Eq. (7.49),

$$w_k^{(j)} = \exp\left[-\delta\tau \cdot \left[\frac{E_L\left(\mathbf{R}_k^{(j)}\right) + E_L\left(\mathbf{R}_{k-1}^{(j)}\right)}{2} - E_T\right]\right]$$

this leads to divergent fluctuations in the number of walkers $\sim w_k^{(j)}$. Such fluctuations cause an explosion in the number of configurations, ultimately exhausting memory and halting the calculation, a problem known as **population explosion**.

Among the various particle coalescences, the role of the Jastrow factor in DMC is specifically to suppress the divergent fluctuations arising from electron-electron coalescence. In the Slater-Jastrow-type many-body trial wave function given in Eq. (4.5),

$$\Psi_\alpha(\mathbf{R}) = \Psi_{\text{AS}}(\mathbf{R}) \cdot \exp\left[J_\alpha\left(\left\{\left|\mathbf{r}_j - \mathbf{r}_l\right|\right\}\right)\right]$$

we have designed the Jastrow factor to explicitly depend on the interparticle distances.[3] By identifying Ψ_α with Φ and substituting into Eq. (8.8), the cusp condition can be expressed in terms of J_α as

$$\left[\frac{dJ_\alpha}{dr_{ij}}\right]_{r_{ij}=0} = \frac{1}{4}\left(1 + \delta_{\sigma\sigma'}\right) . \tag{8.9}$$

In the CASINO implementation, the following notation is used:

$$r_{ij} = \left|\mathbf{r}_i - \mathbf{r}_j\right| \quad , \quad r_{iI} = \left|\mathbf{r}_i - \mathbf{R}_I\right|$$

and the Jastrow function is expressed in the form

$$J(\mathbf{X}) = \sum_{i>j} u\left(r_{ij}\right) + \sum_{i,I} \chi_I\left(r_{iI}\right) + \sum_{I,j>j} f_I\left(r_{iI}, r_{jI}, r_{ij}\right)$$

Here, the Jastrow function is divided into components with separate roles: a two-body term $u\left(r_{ij}\right)$, a one-body term $\chi_I\left(r_{iI}\right)$, and a three-body term $f_I\left(r_{iI}, r_{jI}, r_{ij}\right)$ (see Sect. 4.1.1). The electron-electron cusp condition is enforced via the two-body term

$$u\left(r_{ij}\right) \sim \left(r_{ij} - L_u\right)^2 \left\{\alpha_0 + \alpha_1 \cdot r_{ij} + \sum_{l=2}^{N_u} \alpha_l \cdot r_{ij}^l\right\}$$

[3] The subscript "AS" in Ψ_{AS} stands for Anti-Symmetric, indicating the component that enforces antisymmetry. In this text, Ψ_{AS} is constructed using a Slater determinant.

To satisfy the cusp condition of Eq. (8.9),

$$\left[\frac{dJ}{dr_{ij}}\right]_{r_{ij}=0} = \left[\frac{du}{dr_{ij}}\right]_{r_{ij}=0} = \frac{1}{4}\left(1+\delta_{\sigma\sigma'}\right)$$

the coefficients in the expansion with respect to r_{ij}[4] must satisfy a dependency between the first two terms: specifically, the coefficient α_1 is determined by α_0. This constraint ensures that the cusp condition is satisfied.

Role of the One-Body and Three-Body Terms
If the two-body term $u\left(r_{ij}\right)$, which depends on r_{ij}, imposes the electron-electron cusp condition, then it might be natural to expect that the **one-body term** $\chi_I\left(r_{iI}\right)$, which depends on r_{iI}, would impose the electron-ion cusp condition. However, in practice, the electron-ion cusp condition is usually handled not through the Jastrow factor but through the setup of the orbital functions (see the next section). The primary role of the one-body term is to properly describe the charge density in inhomogeneous systems [8]. In realistic systems, unlike the uniform electron gas, the electron density varies due to factors other than electron-electron interactions, such as the presence of nuclei. However, the two-body term in the Jastrow factor has the effect of making it unduly difficult for electrons to approach regions where the electron density is already high. If only the two-body term is included, it is known to interfere with reproducing the correct charge density of realistic systems [8].[5] To address this, a one-body term is introduced, and its variational parameters are adjusted to compensate for the undue influence on the charge distribution. The **three-body term** $f_I\left(r_{iI}, r_{jI}, r_{ij}\right)$ is introduced to avoid unphysical singularities in the local energy that can arise during "three-body coalescence," where two electrons simultaneously approach a nucleus [7].

In this way, by properly incorporating the cusp condition into the two-body term of the Jastrow factor and optimizing it, one can ensure numerically stable simulations without being undermined by divergent contributions. In Sect. 4.3, we deliberately presented a DMC calculation without a Jastrow factor in order to illustrate this very point.

[4] This is known as the Fock expansion [7].

[5] In more advanced terms, this issue is often phrased by stating that the Jastrow factor should "influence only the pair distribution function (Sect. 10.1.3), not the electron distribution itself."

Role of Jastrow in DMC; Often not Accepted by Authorities
In the traditional context of VMC, the Jastrow factor has long been used with the purpose of describing electron correlation. However, in DMC, its role is quite different; it is introduced solely to ensure numerical stability by preventing configuration explosion, a point that is often misunderstood and not readily appreciated. In DMC, it is through the Jastrow-induced "numerically stable imaginary-time evolution" that electron correlation is ultimately captured. This is not a particularly complicated idea; in fact, it's something that beginners can usually grasp without much difficulty. However, for certain senior theorists who have long taken pride in building their "castles," it seems remarkably difficult to let go of the image of the Jastrow factor as it is used in VMC. No matter how logically and clearly I explain the actual role of the Jastrow factor in DMC, there are times when the discussion is abruptly cut off with the statement "I just don't like the Jastrow factor," as if to shut their ears to the argument entirely. To make matters worse, the field was one in which the "entourage of the senior authorities" held considerable influence. At one conference, a certain junior researcher, known to be part of such an entourage, went out of their way to approach my poster presentation and remarked "Mm, this is Mr. Maezono's method, the one the great Professor XX doesn't like, right? It's kind of a *questionable* approach, isn't it?". Later that evening at the conference reception, another senior authority generously introduced me to some young researchers in the field, enthusiastically explaining the merits of the DMC. I was genuinely surprised when that same junior researcher from earlier suddenly changed their tone, approached me, and said "Dear, respectable Dr. Maezono, best regards going forward ..." as if flipping a switch.

8.1.4 Importance of Considering the Electron-Nucleus Cusp

We have explained that the divergent contributions arising from electron-electron coalescence are treated using the two-body term in the Jastrow factor. On the other hand, the electron-nucleus cusp is handled by applying corrections on the orbital function side.

First, it should be noted that the electron-nucleus cusp becomes a practical issue only in *all-electron calculations without pseudopotentials*. The origin of the cusp condition lies in the divergence of the potential at the coalescence point. A pseudopotential effectively averages out both the "divergent influence from the bare Coulomb potential $\sim Z/r$ acting on valence electrons" and "the influence from core electrons." Pseudopotentials can be constructed with functional forms that do not diverge at the origin. Potentials that diverge at the origin give rise to cusp-related issues, so in the field of ab initio QMC, **pseudopotentials that are non-divergent at the origin** are employed (Sect. 8.3). Thus, in calculations using pseudopotentials, there is no divergent potential at a nucleus position, and no cusp problem arises. On the other hand, in all-electron calculations such as those discussed in Sect. 6.2, nuclei are described using the bare Coulomb potential $\sim Z/r$, and it is therefore necessary to account for the electron-nucleus cusp.

In practice, such all-electron calculations are almost always performed using Gaussian basis sets. Because Gaussian basis functions do not exhibit a cusp at the origin, they cannot cancel the potential divergence that arises when an electron approaches an nucleus. As a result, DMC calculations for isolated molecular systems using Gaussian basis sets were historically difficult to realize, as they suffered from the population explosion problem. To overcome this issue, a method was developed in which a **cusp correction** is numerically applied to the orbital function values obtained from Gaussian basis sets [1]. For sampling points located near the nucleus ($r < r_c$), the algorithm returns a corrected orbital value: the value evaluated from the "rounded" orbital function given by the Gaussian basis, plus an additional correction term that restores the proper cusp behavior.

In addition, algorithms have been implemented in which the drift velocity is automatically adjusted when a sampling point approaches a nucleus, so that the walker is updated in a way that moves it away from the nucleus vicinity [9]. These features are now activated by default. Thanks to such careful refinements, DMC calculations for isolated molecular systems using Gaussian basis sets can now be performed with numerical stability.

In principle, DMC is independent of VMC. Indeed, in Sect. 4.3, we saw an example in which the DMC was able to yield values close to the exact solution, even without a Jastrow factor obtained from VMC. However, such cases are rare. In most practical situations, unless the cusp structure is sufficiently incorporated through VMC, the cancellation of divergent potentials fails, leading to population explosion, making it impossible to carry out DMC calculations.[6] Thus, contrary to textbook-level principles, in actual practice, successful application of the DMC often relies heavily on proficiency with VMC, along with group-level tacit knowledge inherited regarding variational optimization.

Jastrow Factors That Break System Symmetry

When using a Jastrow factor, care must be taken regarding its compatibility with the symmetry of the system. In practical materials science applications, this is rarely a serious concern. However, in the context of fundamental science, where highly precise calculations are performed, such symmetry

(continued)

[6] The reason the DMC calculation in Sect. 4.3 succeeded without a Jastrow factor was due to the presence of hidden default algorithms that automatically helped avoid population explosion.

considerations must be kept in mind. Naively, a Jastrow factor can act beyond the boundaries of the periodic simulation cell, thereby breaking the system's translational symmetry. The reason a cutoff length is introduced in the Jastrow factor is to forcibly confine its range of influence within the Wigner-Seitz cell of the simulation cell, so that it does not violate translational symmetry. In high-precision calculations targeting isolated molecular systems, care must be taken to ensure that spin states, such as singlets and triplets, are properly described. This requires that the exact many-body wave function be an eigenfunction of the spin operator. However, once a Jastrow factor is applied, the resulting wave function inevitably ceases to be an eigenfunction of the spin operator, a problem known as **spin contamination** [10]. However, since the cusp cannot be properly represented without using a Jastrow factor, one is faced with a trade-off: either accept spin contamination, or sacrifice the cusp condition. That said, the error introduced by spin contamination is generally considered to be negligible and thus is often ignored in practice [8, 11].

8.2 Fixed-Node Approximation

8.2.1 Premise of a Positive-Definite Distribution Function

The basic idea behind the DMC was introduced in Sect. 3.2.2 and illustrated in Fig. 3.1. At the heart of this idea lies the interpretation of $\Psi(\mathbf{R}, \tau)$, which evolves according to the imaginary-time Schrodinger equation, as a distribution histogram over particle positions. A histogram representing a distribution must be **positive-definite**, so this conceptual framework implicitly assumes that $\Psi(\mathbf{R}, \tau) > 0$. However, $\Psi(\mathbf{R}, \tau)$ is a many-body wave function, which takes both positive and negative values, and in general is a complex function with real and imaginary parts. We must therefore recognize that the naive analogy to a particle distribution function does not strictly hold. In what follows, we examine this issue more closely.

First, regarding the fact that the wave function is a complex-valued function, we can rely on the fact that if the target problem has time-reversal symmetry, the wave function can be taken to be real. In systems with time-reversal symmetry, if $\Psi(\mathbf{R}, \tau)$ is an eigenfunction of the system, then its complex conjugate $\Psi^*(\mathbf{R}, \tau)$ is also an eigenfunction [12]. In such a case, the linearly combined wave function

$$\Psi^{(R)}(\mathbf{R}, \tau) = \Psi(\mathbf{R}, \tau) + \Psi^*(\mathbf{R}, \tau)$$

is a real-valued function and still describes the same eigenstate. By considering this function instead, we are justified in assuming that the wave function is real. However, this assumption holds only when time-reversal symmetry is present. In problems involving magnetic fields, time-reversal symmetry does not hold,[7] so for general systems involving electromagnetic response, where the Hamiltonian includes a **vector potential**, one must keep in mind that a naive application of the DMC becomes problematic.[8]

Allowing spin alignments other than parallel or antiparallel, such as in spiral magnetism or spin canting, is referred to as **noncollinear**.[9] In such cases, spin components other than σ_z contribute to the system, introducing imaginary components. Here as well, naive realification of the wave function is no longer valid, and alternative methodological developments are required.

8.2.2 Nodal Structure of Many-Body Wave Functions

Even when the wave function can be real-valued, it can still take on both positive and negative signs. When the wave function changes sign, it must pass through a point where the wave function value is zero. For a many-body wave function in $3N$-dimensional space, the constraint $\Psi = 0$ defines a $(3N - 1)$-dimensional surface. This is referred to as the **nodal surface**. The electronic systems we consider are fermionic many-body systems. As can be seen from the fact that they change sign under particle exchange (Eq. (4.3)), their ground-state wave functions possess nodes.[10]

The regions of definite sign separated by nodes are called **nodal pockets**. In the case of a fermionic ground state, it can be proven that all nodal pockets are mutually accessible through particle exchange (**tiling theorem** [8]). The amplitude of the many-body wave function over the entire space can thus be reconstructed from the amplitude within a single nodal pocket. Therefore, it becomes possible to implement the above naive assumption—treating the wave function as a **positive-definite distribution function**, by considering only one such region of definite sign. If a trial update of a walker's position causes a sign flip in the wave function, one can reject the move and confine the walker within a single positive-definite region. This procedure appears sufficient to reproduce the form of the many-body wave function describing the ground state.

[7] This is because axial vectors, such as those appearing in the Lorentz force, reverse their direction under time reversal, thus breaking the symmetry.

[8] As an advanced topic in this area, methodological developments such as **fixed-phase approximation** have been actively studied [13].

[9] In antiferromagnetic systems, the two magnetic moments are aligned along the same axis and lie on the same line, so the configuration is called "colinear."

[10] In contrast, the ground-state many-body wave function of a bosonic system has no nodes. Similarly, the ground state of a single-particle orbital is also node-free and strictly positive. Such systems are free from the sign problem discussed in this section and provide ideal settings for applying DMC.

Won't Understand Nodes of Many-Body Wave Function!?
There are several "languages" for describing many-body electronic systems: the field-theoretical formalism, the density functional formalism, and the many-body wave function formalism (Sect. 10.2). There was a time when **many-body perturbation theory** was regarded as the royal road in many-body physics.[11] In particular, when engaging in discussions with researchers from the *field-theoretical formalism*, there were occasions where no effort was made to understand the many-body wave function formalism at all. Because they had no experience in considering many-body wave functions, their dismissive arguments often stemmed from conflating the nodal structures of single-particle orbitals with those of the many-body wave function, resulting in preconceived conclusions. When I pointed out, "Actually, this is the node of a many-body wave function," they once retorted with confidence, "The node of a many-body wave function?! Such a thing doesn't exist!!". As time has passed, one would hope that the current generation of younger researchers no longer operates within just a single framework when engaging with well-established theories. The ability to survey and navigate across parallel formalisms remains an essential skill.

8.2.3 *Fixed-Node Approximation*

For the ground-state description, the tiling theorem provides a guarantee, so the correspondence between the probability density function and the histogram of particle positions can be safely constructed within a single positive-definite nodal pocket. By rejecting walker position updates that would cause them to exit the nodal pocket under consideration, one can perform a random walk simulation confined entirely within a single nodal pocket. Nonetheless, in a simulation involving relaxation toward an *unknown wave function shape*, the associated **nodal surface is also unknown**, raising the question of how one can determine whether a *sign change* (i.e., nodal crossing) has occurred. To address this, an approximate scheme is adopted: the nodal surface provided by the initial trial wave function, $\Phi(\mathbf{R}, 0) = \Psi_\alpha(\mathbf{R})$, is used as the reference to judge whether a sign change has occurred. Under this implementation, the amplitude is relaxed only within the positive-definite region delimited by the nodal surface of the initial trial wave function.

When relaxing the shape of the many-body wave function toward the exact solution using DMC, the relaxation process can be conceptually separated into

[11] The preface of a Japanese monograph from that era [14] contains such critical remarks.

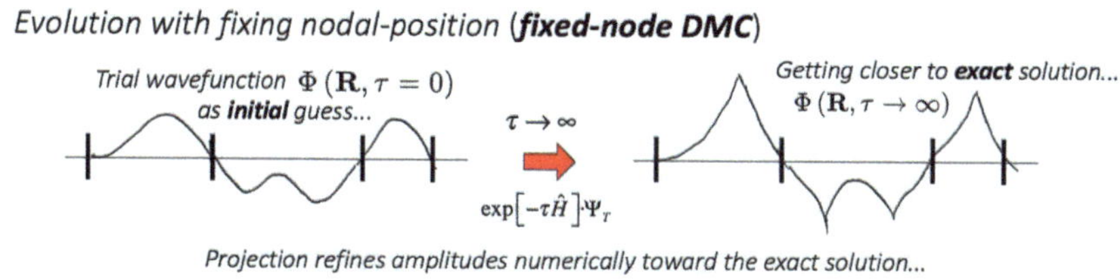

Fig. 8.2 In DMC with the fixed-node approximation, the amplitude degrees of freedom are adjusted in such a way that they approach the exact solution as closely as possible, under the constraint of the given fixed nodes

a nodal shape relaxation and an amplitude relaxation within the nodal pocket. The approximate implementation described above corresponds to fixing the nodal surface to that of the initial trial wave function, without updating it, and relaxing only the amplitude. This is called the **fixed-node approximation** (see Fig. 8.2). If we describe the imaginary-time evolution of the wave function in DMC using the propagator formalism introduced in Sect. 7.3, then it can be written as

$$\Phi\left(\mathbf{R}', \delta\tau\right) = \int d\mathbf{R} \cdot \left\langle \mathbf{R}' \right| \exp\left[-\delta\tau \hat{H}\right] \left|\mathbf{R}\right\rangle \Phi\left(\mathbf{R}, 0\right).$$

In this expression, particle transitions that cross a node and flip the sign correspond to negative-valued matrix elements $\left\langle \mathbf{R}' \right| \exp\left[-\delta\tau \hat{H}\right] \left|\mathbf{R}\right\rangle$. The fixed-node approximation amounts to simply discarding these negative elements. From the perspective of how negative elements are handled, the issue associated with this approximation is referred to as the **sign problem.**

As a potential alternative to the very motivation for introducing the fixed-node approximation, one might consider an extension where the positive and negative regions of the many-body wave function are each handled separately by distinct sets of walkers. Since each of these is a positive-definite region, it may seem that a naive correspondence to a positive-definite distribution function could be restored—for example, by letting red walkers represent the positive region and blue walkers the negative region. However, it is known that such a naive implementation leads to a gradual decay of each amplitude, ultimately drowning in noise [15].[12] To avoid this, it is necessary to devise a time evolution in which walkers in the positive region and those in the negative region are correlated with each other [15]. Such approaches are referred to as "fermion Monte Carlo methods". [13] However, these methods are computationally expensive and are impractical for applications to large-scale systems.

[12] This is often described as "falling into the bosonic ground state."

[13] The name "fermion" is used because, without constraints that correlate positive and negative walkers, the system tends to fall into the bosonic ground state, as noted above.

An alternative approach is to restore, in some manner, the negative elements of $\langle \mathbf{R}' | \exp\left[-\delta\tau\hat{H}\right] | \mathbf{R}\rangle$ that were discarded as an approximation [16]. Since the nodal structure is fundamentally unknown, it is, of course, impossible to know precisely where sign changes occur and therefore to exactly estimate the discarded contributions. However, one may attempt to approximate them using a trial wave function and restore these contributions accordingly. This approach can be regarded as a quasi-fixed-node approximation.

The fixed-node approximation is sometimes misunderstood as introducing a further approximation that is absent in conventional methods. However, this is not true. Whereas conventional approaches are both "fixed-node and fixed-amplitude," the fixed-node DMC is better viewed as a method that, while not overcoming the node fixing, *does* overcome the amplitude fixing, thus representing a method that incorporates an *additional* level of effort. The nodal surface of the trial function is assumed, but its amplitude is treated only as an initial guess and is automatically relaxed toward a form closer to the exact solution (see Fig. 3.1).

The Cause Lay in the Cusp Rather Than the Fixed-Node Approximation
It has been noted that many groups attempting DMC implementation failed to carry out stable calculations and eventually disbanded. The root cause of this was the population explosion stemming from the cusp, which, in retrospect, is rather straightforward to understand.

Why, then, was this not recognized at the time? It is likely that the impact of the fixed-node approximation and the sign problem drew too much attention. One would faithfully implement the method as described in textbooks, only to be met with severe population explosion, leading to overwhelming statistical noise and unusable results. This often led to the assumption that "the sign problem must be the reason this isn't working." Many researchers tended to seek the cause in fundamental directions, such as the sign problem or potential conflicts with the Markov process.

As a result, effort was heavily concentrated on methodological validation studies using model systems. What ultimately overcame the challenges facing practical DMC calculations for real materials was, somewhat ironically, not the sign problem, but rather the amplitude problem, specifically, ensuring the cusp condition in advance using VMC. This carries a rather instructive lesson. Ambitious challenges such as the sign problem, relativistic extensions, or going beyond the Born-Oppenheimer approximation are common themes in proposals by early-career theorists. But to put it bluntly, they can also appear as "tempting buzzwords," topics that even beginners are readily drawn to. Questions like "What are you going to do about the fixed-node approximation?" or "What about relativistic effects?" are common, somewhat overused queries from nonexperts at conferences. It was striking how certain seasoned

(continued)

authorities responded to such questions with "Fixed-node approximation? Well, electrons hardly go near the nodes anyway" or "Relativistic effects? Roughly speaking, it's just atomic contraction, isn't it?".

8.2.4 Improving the Nodal Structure via VMC

Based on accumulated experience in the field, the fixed-node approximation works well for most systems. If one could explicitly present a case in which the fixed-node approximation clearly yields poor results, that alone could serve as the basis for a high-impact publication. Perhaps it is indeed because "electrons hardly go near the nodes in the first place" that this approximation performs so well. Nonetheless, in systems with strong quasi-degeneracy (i.e., strong geometrical (static) correlation), the nodal structure provided by a single Slater determinant is clearly insufficient in representability (Sect. 10.5.2). For such systems, research has been conducted on how to further relax the nodal structure to enhance its representability. Below, we describe several such approaches.

A **variational principle also holds** with respect to the **fixed nodes assumed** in the fixed-node approximation. That is, the closer the fixed nodal structure is to that of the exact solution, the lower the evaluated energy becomes. Therefore, if one embeds variational parameters that modulate the nodal structure into the trial function and tunes them such that the energy evaluated via VMC decreases, one can effectively relax the nodes. This constitutes a **VMC-based node-relaxation scheme**. On the other hand, methods such as the aforementioned fermionic Monte Carlo aim to "automatically approach the exact solution via imaginary-time evolution, including nodal relaxation," representing a **DMC-based node-relaxation scheme**. In practical applications targeting real materials, VMC-based nodal relaxation is the more realistic approach.

As the most straightforward node-relaxation approach, as mentioned in Sect. 5.3, one may replace the Slater part of the Slater-Jastrow trial function with a **multi-determinant** expansion, and variationally optimize its coefficients. This allows the nodal structure given by a single Slater determinant to be relaxed toward a more accurate form (Sect. 10.2.1). At this point, one encounters the issue of how many multi-determinant terms to retain in the expansion of the many-body wave function. Modern implementations of ab initio QMC can handle several thousand to tens of thousands of multi-determinant terms. However, even for small molecules, the number of terms that should be considered with comparable weights can immediately reach hundreds of thousands [17], making this approach impractical for general use. Consequently, efforts have shifted toward implementing schemes that relax the nodal structure by imposing variational parameters, *without increasing the number of expansion terms* in the many-body wave function.

One can generate fixed nodes for the fixed-node approximation using a variety of different methods and perform DMC calculations for each case to compare the resulting energies. Based on the variational principle with respect to the nodes, one can then determine which nodal generation method yields better performance. For example, studies have been conducted where fixed nodes were generated using different values of U in DFT+U methods, and the results were analyzed to determine which U produces the most variationally favorable fixed node [18]. In DFT used to generate nodal structures, employing different exchange-correlation functionals is generally expected to produce different fixed nodes. At the bottom of Fig. 1.6 in Sect. 1.3.2, results from DMC calculations using fixed nodes generated by LDA and GGA are compared. The figure presents binding curves as a function of inter-base distance in DNA. While the shapes of the binding curves differ significantly between LDA and GGA when using DFT alone, the final DMC results using either trial node converge to essentially the same binding curve. This suggests that, for this system, the many-body nodal structure is largely insensitive to whether LDA or GGA is used.

Keeping the form of the orbital functions in the Slater determinant fixed, but modifying their arguments via the transformation

$$\mathbf{r}_j \quad \to \quad \tilde{\mathbf{r}}_j = \mathbf{r}_j + \sum_{j \neq i} \eta_{ij} \cdot \left(\mathbf{r}_i - \mathbf{r}_j\right)$$

by introducing variational parameters $\left\{\eta_{ij}\right\}$ that shift the coordinates, and optimizing them to lower the energy, is referred to as the **backflow-type** variational trial wave function method. These variational degrees of freedom modulate the nodal structure given by the original Slater determinant and thus correspond to a method of "improving the conventional fixed-node approximation via variational Monte Carlo." As for the term **backflow**, it originates from a historical approach in which the electronic correlation beyond Slater determinants was interpreted as an increase in effective mass. This was physically conceptualized as a "backflow," a trailing current generated when an electron moves through an interacting medium, and was subsequently cast into mathematical form as a trial wave function [19]. The argument of the backflow shift reflects how the direction of motion $\mathbf{r}_j$ of a given electron is influenced by interaction forces pulling it toward each of the other electrons through terms like $\left(\mathbf{r}_i - \mathbf{r}_j\right)$.

Another variational strategy for nodal structure relaxation that has seen relatively broad implementation is the **geminal-type** trial wave function:

$$\Phi\left(\mathbf{r}_1, \cdots, \mathbf{r}_N\right) = \hat{A}\left[\chi\left(\mathbf{r}_1, \mathbf{r}_2\right) \cdots \chi\left(\mathbf{r}_{N-1}, \mathbf{r}_N\right)\right]$$

The term "geminal" might be unfamiliar to beginners, but it comes from the same root as "Gemini" (twins) and is essentially an adjective describing "pairing." Here, $\chi\left(\mathbf{r}_i, \mathbf{r}_j\right)$ denotes a pairing orbital, and the many-body wave function is constructed by antisymmetrizing the product of such pairs via the antisymmetrizer $\hat{A}$ [20]. For readers with a physics background, the necessity of introducing pairing orbitals may immediately recall the BCS theory of superconductivity. However, the motivation behind geminal wave functions originates from the **cusp condition** discussed in the previous section [21]. To improve the ability to describe cusps, one must incorporate not only the individual degrees of freedom such as $\mathbf{r}_j$ but also explicit pairwise dependencies like r_{ij}. This motivation led to the historical development of pairing orbital wave functions [21].

The one-particle orbital functions $\left\{\psi_\mu(i)\right\}$ used to construct Slater determinants can be readily obtained using standard implementations of DFT or molecular orbital methods. But how can one construct pairing orbitals in practical calculations? In practice, the approach is to define the pairing orbital as

$$\chi(i,j) = \sum_{\alpha\geq\beta} a_{\alpha\beta}\cdot\left[\psi_\alpha(i)\,\psi_\beta(j) + \psi_\beta(i)\,\psi_\alpha(j)\right] \tag{8.10}$$

by combining products of one-particle orbitals.[14] The expansion coefficients $\left\{a_{\alpha\beta}\right\}$ are optimized variationally [23, 24].

In papers dealing with the topic of this section, you may encounter the term **Pfaffian**-type trial wave function. It is sufficient to understand this as a geminal-type trial function constructed such that it reduces to a Slater determinant in special cases. Although the term Pfaffian may be unfamiliar to beginners, some readers may have come across the notion of Pfaff forms (Pfaff) in mathematics. The Pfaffian is a polynomial defined for even-order skew-symmetric matrices. Its general definition is provided in Appendix Sect. 14.3, but, for example, in the case of a 4×4 matrix, the Pfaffian is given by

$$\mathrm{Pf}(A) = a_{12}a_{34} - a_{13}a_{24} - a_{14}a_{23}$$

In essence, one selects products of elements like a_{ij} such that the index pairs (i, j) are disjoint, e.g., $a_{12}a_{34}$, and sums over all such combinations with appropriate signs according to the parity of the index permutations. This signed sum defines the Pfaffian. In the case of a 6×6 matrix, the Pfaffian is constructed as

$$\mathrm{Pf}(A) = a_{12}a_{34}a_{56} - \cdots$$

and so on, by summing over all possible pairwise products with the appropriate sign conventions. The Pfaffian satisfies the property

$$\mathrm{Pf}^2(A) = \det(A) \ .$$

Therefore, it is often heuristically described as "the square root of a determinant."

[14] More precisely, the set $\left\{\psi_\mu(i)\right\}$ is not composed of raw orbitals, but rather of *natural orbitals* [22] derived from them. Routines to generate these natural orbitals are typically available in molecular orbital packages.

A Pfaffian-type trial wave function is constructed by using the pairing function $\chi(i, j)$ from Eq. (8.10) as the matrix element in place of a_{ij} in the Pfaffian [23–27]. If the indices α, β in the pairing function Eq. (8.10) are restricted to occupied orbitals only, then, from the identity $\mathrm{Pf}^2(A) = \det(A)$, the Pfaffian-type trial function reduces to the Slater determinant. By allowing nonzero $\{a_{\alpha\beta}\}$ for unoccupied orbitals as well, and optimizing these variationally, the nodal structure assumed in the Slater determinant can be relaxed toward a more accurate form.

We have introduced variational trial wave functions such as backflow, geminal, and Pfaffian types, which can relax the nodal degrees of freedom. The variational optimization of these functions is performed using the optimization methods explained in Chap. 4. Compared to the optimization of Jastrow functions (i.e., optimization of amplitude degrees of freedom) as done in the exercises so far, variational optimization of the nodal degrees of freedom is significantly more challenging. The nodal surface refers to the region where the wave function takes the value $\Psi = 0$. At such points, the local energy behaves as $E_L \sim 1/\Psi$ and diverges, causing both energy and its variance to fluctuate wildly. When applying standard energy or variance minimization methods, one often encounters severe numerical instabilities. As a result, the idea of "variationally optimizing the nodal surface" can easily become a mere theoretical ideal. To overcome this, more robust optimization methods have been developed, including median minimization, linear combinations of energy and variance minimization, techniques that cleverly utilize the covariance matrix, and natural gradient methods [28].

8.3 Pseudopotentials

Pseudopotentials are concisely explained in § 5.4 of the author's previous book [45], where the key aspects are highlighted. That section outlines three types of pseudopotentials that users should be aware of when specifying input parameters: norm-conserving, ultrasoft, and PAW (projector augmented wave). Due to historical reasons, pseudopotentials are also commonly referred to as ECPs (effective core potentials) in the field of molecular science. In the context of ab initio quantum Monte Carlo as well, the term ECP may appear depending on the usage, but it refers to the same concept.

As will be discussed in the following sections, the pseudopotentials used in ab initio QMC are significantly more restricted in choice compared to the wide variety available in DFT. Moreover, they present certain fundamental methodological challenges that are unique to this framework.

8.3.1 *Pseudopotentials Used in Ab Initio QMC*

It is crucial to first understand that **only norm-conserving pseudopotentials can be used** in ab initio QMC. One of the motivations for extending pseudopotentials beyond norm-conserving to ultrasoft and PAW types was the desire to reduce the number of plane-wave basis functions, that is, to enable the use of smaller cutoff energy values. In DFT, the underlying idea is that "as long as the charge density is reproduced accurately, the calculation is acceptable." Based on this principle, the concept of compensation charge was introduced to allow a reduction in the basis size, leading to the development of ultrasoft pseudopotentials [29]. Since ab initio QMC is not a **density functional** theory but a **many-body wave function** theory, the conceptual framework of ultrasoft pseudopotentials is not applicable. At the time of writing, at least, only norm-conserving pseudopotentials can be employed. As a result, it is generally necessary to use a **higher plane-wave cutoff energy**. Due to computational cost constraints, it is often the case that calculations must be performed using a cutoff energy value that would be insufficient for density functional theory. However, in DMC, the shape of the wave function can undergo self-healing, and in some cases, the impact of using a slightly reduced cutoff energy is negligible.

Pseudopotentials can be categorized as "those constructed using density functional theory" or "those constructed using quantum chemical calculations." Because pseudopotentials are typically generated by performing all-electron calculations on isolated ions, they are usually constructed using highly accurate methods that incorporate relativistic effects. In ab initio QMC, it is common to use **pseudopotentials generated from quantum chemical calculations**. The primary objective of using ab initio QMC is to explicitly account for electronic correlation, which cannot be accurately captured by other methods. If pseudopotentials derived from density functional theory, which already incorporate some degree of electronic correlation, are used, it becomes difficult to disentangle which parts of the results stem from genuine correlation effects. As a result, pseudopotentials generated from Hartree-Fock theory, which contains no electron correlation and thus serves as a clear baseline, or from Dirac-Fock theory, which also includes relativistic effects, have traditionally been employed. More recently, however, pseudopotentials that incorporate correlation effects at the quantum chemical level during their construction have also become available [30].

Some pseudopotentials exhibit divergence at the origin. Originally, in SCF-based methods such as DFT or quantum chemistry calculations, only the pseudoorbital region outside the core radius is treated explicitly. Thus, whether the potential diverges at the origin was not a practical concern. However, in ab initio QMC, sampling points may, regardless of whether pseudopotentials are used, randomly fall near the origin. As discussed in the section on cusp conditions (Sect. 8.1), such divergences in the potential can lead to numerical instabilities and large statistical errors. Therefore, in ab initio QMC, **pseudopotentials that are non-divergent at the origin** are employed. In this context, even a "pseudopotential for hydrogen" may be employed. When pseudopotentials are explained simply as "effective potentials that replace the core electron region for valence electrons," one might reasonably wonder "How can hydrogen, which has no core electrons, have a pseudopotential?". However, in ab initio QMC, a pseudopotential for hydrogen is understood as a pair of a potential and a pseudoorbital that does not diverge at the origin and effectively reproduces the results of all-electron calculations.

8.3.2 Practical Treatment of Pseudopotentials

Pseudopotentials represent the effective influence of the nucleus as felt by the valence electrons. For negatively charged electrons, the positively charged nucleus exerts a strong attractive interaction. At the same time, however, the inner-shell electrons, which occupy orbitals orthogonal to those of the valence electrons, repel the valence electrons due to the Pauli exclusion principle. This repulsion acts effectively as a repulsive interaction. As a result, the attractive potential from the nucleus is weakened, and the effective potential becomes a shallower attraction.

The effective repulsive part depends on "which core orbitals the valence electron must be orthogonal to." Thus, the potential felt by a valence electron differs depending on whether it has s, p, or d character:

$$V_{\text{ps}} \left|\psi\right\rangle = \begin{cases} V_s \left|\psi\right\rangle \cdots (\left|\psi\right\rangle = \left|s\right\rangle) \\ V_p \left|\psi\right\rangle \cdots (\left|\psi\right\rangle = \left|p\right\rangle) \\ V_d \left|\psi\right\rangle \cdots (\left|\psi\right\rangle = \left|d\right\rangle) \end{cases} .$$

This introduces **orbital-component dependence** into the pseudopotential. Such "case-by-case responses depending on the orbital character" can be expressed using the formalism of **operator representation**:

$$\hat{V}_{\text{ps}} = V_s \left|s\right\rangle \left\langle s\right| + V_p \left|p\right\rangle \left\langle p\right| + V_d \left|d\right\rangle \left\langle d\right| = \sum_{l=0}^{L} V_l \left|l\right\rangle \left\langle l\right| . \tag{8.11}$$

In this case, we have

$$\hat{V}_{\text{ps}} \left|\psi\right\rangle = \sum_{l=0}^{L} V_l \left|l\right\rangle \left\langle l|\psi\right\rangle .$$

Here, the term $\left\langle l|\psi\right\rangle$ involves global evaluation via spatial integration. Since "a value at a particular point (a local value) cannot be determined without integrating over the full domain," the term **nonlocal** operator-type pseudopotential is used. [15]

In practice, instead of explicitly specifying the three components V_s, V_p, and V_d, the pseudopotential is typically expressed as

$$\hat{V}_{\text{ps}} = V_{\text{loc}} + \delta V_{l_1} \left|l_1\right\rangle \left\langle l_1\right| + \delta V_{l_2} \left|l_2\right\rangle \left\langle l_2\right| .$$

This separates the potential into a local part V_{loc} and nonlocal correction terms δV_l. Among the components V_s, V_p, and V_d, the one that is most similar to the others is

[15] More specifically, the terms nonlocal and semi-local are distinguished in technical contexts. In Eq. (8.11), the projection operator is classified by l, based on the assumption that the radial part is local. This treatment is referred to as semi-local. In this case, a $3d$ and a $3p$ orbital feel different potentials, whereas a $3d$ and $4d$ orbital feel the same potential.

typically chosen as the local part V_{loc}. The local component refers to "a c-number component that acts identically on all l-channels".[16] For example, by choosing $V_{\mathrm{loc}} = V_s$, we can write

$$\begin{aligned}\hat{V}_{\mathrm{ps}} &= V_s + \left(V_p - V_s\right) |p\rangle \langle p| + (V_d - V_s) |d\rangle \langle d| \\ &= V_s + \delta V_p |p\rangle \langle p| + \delta V_d |d\rangle \langle d| \end{aligned} \tag{8.12}$$

In this way, Eq. (8.11) can be expressed with fewer projection operators. In practical implementations, it is not possible to realize perfect projection operations.[17] Therefore, using a local component and reducing the number of projection operators tend to minimize the overall error. However, *careful selection of which channel to use as the local part* is important. If a deeply attractive potential channel from the set $\{l\}$ is chosen as the local component, a **ghost state**, a spurious bound state, may often arise. For example, for the p state, the potential effectively felt in the calculation becomes

$$V_p^{(\mathrm{cal})} = V_{\mathrm{loc}} + \delta V_p \cdot (|p\rangle \langle p|)$$

However, as discussed earlier, the projection operator $|p\rangle \langle p|$ cannot be implemented perfectly in practice. Symbolically, we may write its effective action as

$$(|p\rangle \langle p|) \sim \gamma$$

Given that $\delta V_p = V_p - V_{\mathrm{loc}}$, the resulting effective potential becomes

$$V_p^{(\mathrm{cal})} = V_{\mathrm{loc}} + \left(V_p - V_{\mathrm{loc}}\right) \cdot \gamma = (1-\gamma)\, V_{\mathrm{loc}} + \gamma V_p \,.$$

If the projection operator were evaluated perfectly, we would have $\gamma = 1$, yielding

$$V_p^{(\mathrm{cal})} = V_p$$

and no issues would arise. In practice, however, the projection operation is evaluated with some degree of imperfection, leading to $\gamma < 1$. As a result, the effective potential becomes

$$V_p^{(\mathrm{cal})} = (1-\gamma)\, V_{\mathrm{loc}} + \gamma V_p$$

and thus includes a residual contribution from V_{loc}. If V_{loc} is a deeply attractive potential, the term $(1-\gamma)\, V_{\mathrm{loc}}$ can introduce a fictitiously deep potential. This may

[16] Since V_{loc} does not require nonlocal evaluation, it is referred to as the *local* part.

[17] A perfect projection would require that the projected components are mutually orthogonal and that, together, they span the full space. However, this represents an idealized limit that is difficult to achieve in practice.

give rise to bound states that do not physically exist, leading to unphysical results, a phenomenon known as the pathology of **ghost states**.

Pseudopotentials are provided to the simulation in the form of a text file that contains the radial grid values of $V_s(r)$, $V_p(r)$, and $V_d(r)$ from Eq. (8.11). In the case of CASINO, this corresponds to the file named "xx_pp.data." At the beginning of this file, the specification of which channel is to be treated as the local component is given. Inside the program, this information is used to convert the potentials into the form of Eq. (8.12) for further processing.

As also mentioned in Sect. 5.2, pseudopotentials are typically obtained from public libraries. In practical ab initio QMC calculations, an important point is that "the format must also be compatible with density functional theory or molecular orbital calculations for generating orbitals." Such compatibility is usually provided either by the QMC package itself or by the pseudopotential distribution site. As of the time of writing, representative pseudopotentials used in QMC include the Trail-Needs type [31], the BFD type [32, 33], the SBKJC type [34, 35], the RRKJ type [36], and the ccECP type [37].

If a desired pseudopotential, such as one with a small-core or large-core specification, is not available in libraries for ab initio QMC applications, it is possible to obtain a norm-conserving DFT pseudopotential from existing libraries and manually convert it into the required format [38]. However, when discussing precise methodological benchmarks, reviewers may strongly criticize the use of such converted pseudopotentials, claiming that they have not undergone sufficient reliability validation.

8.3.3 Other Topics Related to Pseudopotentials

One method for generating pseudopotentials involves the use of Dirac-Fock calculations, as mentioned in Sect. 8.3.1. In this case, **relativistic effects** are, to some extent, taken into account within the pseudopotential itself. When fully treating relativistic effects using a four-component formalism, there is a difficulty: the antiparticle states lie at lower energies than the particle states of interest. As a result, the variational principle, which forms the basis of ab initio QMC, no longer holds in a straightforward implementation. This issue is known as the problem of **variational collapse** [39].

To avoid dealing with the full four-component treatment, relativistic effects are incorporated into the pseudopotential through the use of the averaged relativistic effective potential (AREP) approach [31]. In this method, isolated ion calculations are first performed using fully relativistic four-component molecular orbital methods. Then, a pseudopotential is constructed such that these four-component orbitals are replaced with pseudoorbitals, and the resulting potential is averaged for practical use. In relativistic treatments, quantum states are classified not by orbital angular momentum l but by the total angular momentum j, which arises from spin-orbit coupling [39]. To convert pseudopotentials defined for each relativistically split j

state back into a form usable for nonrelativistic l components, the contributions from each j are averaged to construct an effective V_l, which is then used in simulations.

Some readers may wonder to what extent **spin-orbit interactions** can be treated, especially in relation to topological materials. Within the framework of pseudopotentials discussed above, there exists an ab initio approach to estimate the coefficient of spin-orbit coupling in the form $V_{SO} \cdot (\mathbf{l} \cdot \mathbf{s})$ [40]. However, these effects are not handled entirely from first principles without parametrization. The approach is based on assuming a form $V_{SO} \cdot (\mathbf{l} \cdot \mathbf{s})$, and then treating it accordingly. That said, at the time of writing, there are almost no practical implementations of ab initio QMC that include spin-orbit coupling to the extent discussed in model calculations.

For an ion with atomic number Z, a pseudopotential constructed with N_{core} core electrons should appear, from the perspective of distant valence electrons, as an effective positive charge of $Z_{\text{eff}} = (Z - N_{\text{core}})$. Therefore, the pseudopotential must exhibit the correct asymptotic form $\sim Z_{\text{eff}}/r$ at large distances. However, in the case of pseudopotentials used in ab initio quantum Monte Carlo methods that include Fock terms, it has been pointed out that this asymptotic behavior may not be correctly reproduced [31]. In the treatment of pseudopotentials, the true orbitals within the core radius are replaced by pseudoorbitals. The difference in charge density between the true and pseudoorbitals propagates beyond the core radius through the nonlocality of the Fock term,[18] and affects the numerical construction of the pseudopotential itself [31]. This issue is intrinsic to any method that employs pseudopotentials constructed using Fock exchange, whether within SCF-based frameworks or QMC approaches. However, in SCF-based methods, the locality of the basis functions tends to suppress the long-range propagation of such effects in a somewhat discontinuous manner, and this issue is generally not regarded as serious [31]. On the other hand, when used in ab initio QMC, especially in DMC, the long-range propagation, initially truncated in the trial wave function, can be revived through imaginary-time evolution. Thus, the problem of failing to reproduce the correct asymptotic behavior must be treated with great care, and pseudopotentials designed with this consideration in mind have been developed and made available.

8.3.4 Local Approximation of Pseudopotentials

The simulation target in DMC is Eq. (7.29), to which importance sampling has been applied:

$$\frac{\partial f(\mathbf{R}, \tau)}{\partial \tau} = \frac{1}{2} \sum_{j=1}^{N} \nabla_j \left(\nabla_j - 2 \frac{\nabla_j \Psi_\alpha(\mathbf{R})}{\Psi_\alpha(\mathbf{R})} \right) f(\mathbf{R}, \tau)$$

[18] In the context of Eq. (8.11), the nonlocal nature of the pseudopotential was explained as arising from the fact that the value at one point cannot be determined without a global integral. The Fock term in Eq. (10.1) shares this characteristic: even information outside the core radius requires a global integration including the region inside the core.

$$- (E_L(\mathbf{R}) - E_T) f(\mathbf{R}, \tau)$$

as explained earlier, where $f(\mathbf{R}, \tau) = \Psi(\mathbf{R}, \tau) \cdot \Phi(\mathbf{R})$. The appearance of the local energy E_L in the second term of Eq. (7.29) was due to the assumption that the potential V is a c-number. Under this assumption, one can write $V = \Psi^{-1} V \Psi$, and thus,

$$\begin{aligned}\left(V + \Psi^{-1}\hat{T}\Psi\right) \cdot f &= \left(\Psi^{-1}V\Psi + \Psi^{-1}\hat{T}\Psi\right) \cdot f \\ &= \Psi^{-1}\left(V + \hat{T}\right)\Psi \cdot f = E_L \cdot f\end{aligned}$$

so that the kinetic term $\hat{T}$ and the potential V can be absorbed into the local energy E_L.

However, as noted in Sect. 8.3.2, the pseudopotentials we use are nonlocal operators, so that $\hat{V} \neq \Psi^{-1}\hat{V}\Psi$. In such cases, the evolution equation becomes

$$\begin{aligned}-\frac{\partial}{\partial\tau} f(\mathbf{R}, \tau) = &-\frac{1}{2}\sum_{j=1}^{N} \nabla_j \left[\nabla_j - 2\left(\Psi_T^{-1}\nabla_j\Psi_T\right)\right] \cdot f \\ &+ \left[E_L(\mathbf{R}) - E_T\right] f + \delta_{\mathrm{NL}}\end{aligned}$$

and an additional term appears:

$$\delta_{\mathrm{NL}} = \left(\frac{\hat{V}_{\mathrm{NL}}\Phi}{\Phi} - \frac{\hat{V}_{\mathrm{NL}}\Psi_T}{\Psi_T}\right) \cdot f \tag{8.13}$$

as derived in Sect. 13.2.4. This residual term is expected to be negligible if the trial wave function Ψ_T is sufficiently close to the DMC converged solution Φ, but it constitutes an error that is difficult to estimate. In practical computations, this residual is often neglected altogether as an approximation. This corresponds to treating an originally nonlocal pseudopotential as a localized one and, thus, is referred to as the **localization error of the pseudopotential**.

The localization of pseudopotentials corresponds to modifying the Hamiltonian being described, and hence, there have been concerns that it may break the variational property, which is one of the greatest strengths of ab initio QMC.[19] To address this, correction schemes that are considered to at least preserve the variational property have been developed [41].

Moreover, an idea is being explored and implemented in which nonlocal pseudopotentials are not absorbed into the potential term but rather **incorporated into the kinetic energy term** so as to avoid the localization error. Recalling that the

[19] This would mean the loss of a reliable guiding principle: the lower the computed energy, the closer to the exact solution.

operator $\hat{L}^2$ acts as

$$\hat{L}^2 = \sum_{l}^{\infty} l\,(l+1)\,|lm\rangle\,\langle lm|\,,$$

one notices that the projection operators over l as in Eq. (8.11) can be constructed by manipulating the operator $\hat{L}^2$. On the other hand, since $\hat{L}^2 = (\mathbf{r} \times i\nabla)^2$, it appears feasible to represent the effect of the nonlocal operator in Eq. (8.11) by replacing the mass in the kinetic energy term with an effective tensor that depends on the direction, in association with the ∇^2 operator. Localization error arose because the nonlocal potential was approximated as a "localized, no longer operator-like" form and incorporated into the potential term E_L. By instead incorporating it via the $\hat{L}^2$ operator as a correction to the kinetic energy term, not the potential term, localization error can be fundamentally avoided. This is referred to as the **L^2-type pseudopotential** [42].

8.4 Overview of the Computational Procedure

By this point, we have introduced the majority of technical concepts that form the foundation of ab initio QMC. Let us now revisit the computational procedure of ab initio QMC, using the terminology we have developed to describe precisely how each step corresponds to these theoretical components. Figure 8.3 shows a typical computational flow, illustrated using a silicon cluster calculation as an example [43].

In this example, the molecular orbital software "Gaussian" is used to perform a GGA-DFT calculation,[20] from which Kohn-Sham orbitals are generated (energy: − 385.1445 hartree). These orbital functions are then assembled into a Slater determinant, which is used as the trial wave function for an ab initio QMC calculation (**NJVMC**/energy: − 372.19(2) hartree). The term "NJVMC" in the figure refers to a calculation that uses only the Slater determinant without a Jastrow factor (No Jastrow), which was referred to as "VMC (w/o J)" in Chap. 6.

It is also possible to generate orbital functions not by DFT, but by SCF calculations within the Hartree-Fock method. By definition, Hartree-Fock SCF yields the variationally optimal energy among trial wave functions of Slater determinant form. An NJVMC calculation using such orbital functions likewise achieves the variationally optimal energy via Monte Carlo sampling. Therefore, the energy values obtained from the two methods must agree within statistical uncertainty.[21] On the other hand, for cases like that shown in Fig. 8.3, where NJVMC is performed

[20] Using the PBE exchange-correlation functional.

[21] If they do not agree, it implies an error in some part of the numerical implementation, and the discrepancy can be used for calibration purposes.

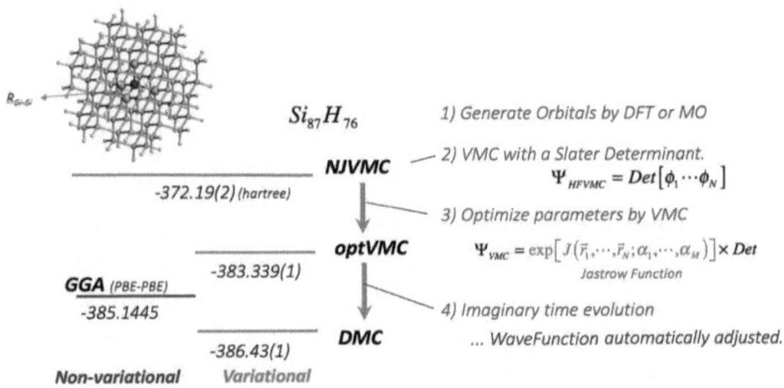

Fig. 8.3 Although the electronic state calculation by QMC is represented as a single box in Fig. 1.4, in practice, it constitutes a workflow composed of multiple kernel calls, as indicated by (1)–(4) in the figure (see Sect. 9.2.2). The numerical values shown in the diagram represent total energies, serving as indicators of proximity to the exact solution, with lower values indicating closer agreement (variational principle (Sect. 1.2.1)). Note, however, that the energy obtained by GGA is not subject to the variational principle. (Figure adapted from Ref. [43])

using Kohn-Sham orbitals, no such equivalence holds. This setup merely provides a mathematically convenient way to impose antisymmetry, without any deeper significance.

From the definition above that "HFSCF gives the variationally best energy among Slater-determinant-type trial functions," it follows that the NJVMC energy using Kohn-Sham orbitals should be higher than that of HFSCF. However, it is important to note that this does not imply that "the nodal structure of the Slater determinant formed from Kohn-Sham orbitals" is inferior to that formed from Hartree-Fock orbitals. Indeed, many examples are known in which, once the amplitude is relaxed toward the exact solution via DMC, the nodal structure based on Kohn-Sham orbitals yields a better variational result [44]. At the NJVMC level, the higher variational energy of the Kohn-Sham Slater determinant is due to the poor description of its amplitude.

The next step in the procedure is to append a Jastrow factor to the Slater determinant and optimize the variational parameters contained therein (**optVMC**, meaning optimized VMC). For the optimization method, one may choose appropriately between energy minimization or variance minimization, depending on the case (Sect. 4.1.2). The Jastrow factor allows for the effective inclusion of dynamic electron correlation, primarily associated with cusp behavior (Sect. 10.5.2), leading to a significant stabilization of the calculated energy value (− 383.339(1) hartree). If the NJVMC were based on a Slater determinant constructed from Hartree-Fock orbitals, then the energy drop from NJVMC to optVMC would correspond precisely to the electron correlation energy captured by the Jastrow factor. In the case shown in Fig. 8.3, however, the NJVMC uses a Slater determinant constructed from Kohn-

Sham orbitals, so the energy drop becomes larger than the electron correlation energy captured by the Jastrow factor.[22]

With optVMC, a trial wave function with a well-optimized variational description of the cusp has been obtained. Using this as the initial trial function, one performs imaginary-time evolution as in Eq. (3.7),

$$\Phi(\mathbf{R}, \tau) = \Psi_0(\mathbf{R}) \cdot \exp[-(E_0 - E_T)\tau] + \lambda_1 \cdot \Psi_1(\mathbf{R}) \cdot \exp[-(E_1 - E_T)\tau] \\ + \lambda_2 \cdot \Psi_2(\mathbf{R}) \cdot \exp[-(E_2 - E_T)\tau] + \cdots$$

and thereby suppresses all components other than the exact ground state $\Psi_0(\mathbf{R})$ in the first term, relaxing the many-body wave function $\Phi(\mathbf{R}, \tau)$ toward the exact solution $\Psi_0(\mathbf{R})$. This procedure yields the result of diffusion Monte Carlo (DMC). The energy is further stabilized in a variational sense, reaching the value of " − 386.43(1) Hartree." As discussed in Sect. 8.2 on the fixed-node approximation, this relaxation updates only the amplitude, while the nodal surfaces are fixed to those of the Slater determinant used in NJVMC (see Fig. 3.1).

As discussed in Sect. 4.3, in principle, it is possible to describe electron correlation and relax toward the exact solution even when using a trial wave function without a Jastrow factor, by performing DMC in the later stage. That is, from a formal perspective, one might proceed directly from NJVMC to DMC, omitting optVMC altogether. In this sense, employing a high-quality trial wave function with an appropriately optimized Jastrow factor in DMC via optVMC could appear to be of secondary importance, merely accelerating convergence. However, as emphasized in Sect. 8.1.3, imposing the cusp condition on the trial wave function via the Jastrow factor plays an essential role in ensuring the numerical feasibility of the calculation. Only with the cusp condition properly implemented is the finiteness of the sampled local energy values guaranteed. Without the cusp condition, local energy fluctuates severely, not only worsening statistical errors but also inducing unreasonably large population fluctuations in the subsequent DMC process,[23] which can exceed memory capacity and terminate the simulation (see Fig. 4.3).

Note that the arrows NJVMC →optVMC →DMC in Fig. 8.3 lie on the same variational footing, that is, they fall under the principle that "the closer one approaches the exact solution, the lower the energy." In contrast, DFT calculation used for orbital generation lies outside the track of many-body variational methods. The fact that the DFT energy, − 385.1445 hartree, is lower than that of NJVMC or optVMC does not imply that "DFT is more accurate as an approximation".[24]

---

[22] As mentioned in the previous paragraph, the NJVMC(KS) energy is higher than that of NJVMC(HF).

[23] The total number of walkers is referred to as the population.

[24] This can be understood by noting that the overall energy level can shift downward depending on the reference of the exchange-correlation potential V_{XC}. What is meaningful in DFT is the energy difference, typically discussed as ΔSCF [45].

With the above understanding in place, the bread-baking analogy used throughout the first part of this book may now resonate more fully with the reader:

- **DFT/MO**
  Orbital functions used to construct the Slater determinant part of the trial wave function are generated by density functional theory (DFT) or molecular orbital (MO) methods. This corresponds to "preparing dough from flour."
- **NJVMC**
  A VMC calculation using the Slater determinant composed of these orbitals directly as the trial wave function. This corresponds to "baking the dough without fermentation."
- **OptVMC**
  A VMC calculation in which a Jastrow factor is added to the Slater determinant, and its variational parameters are optimized. This corresponds to "baking the dough after proper fermentation."
- **DMC**
  A DMC calculation that takes the trial wave function optimized in optVMC and evolves it in imaginary time to relax toward the exact ground state. This corresponds to "further fermentation and maturation of the dough during baking, resulting in a fully risen loaf."

References

1. "Scheme for adding electron-nucleus cusps to Gaussian orbitals", A. Ma, M.D. Towler, N.D. Drummond, R.J. Needs, J. Chem. Phys. 122, 224322 (2005) https://doi.org/10.1063/1.1940588
2. "Mathematical Methods for Physics and Engineering, Third Edition" (Cambridge University Press, 2006), K.F. Riley, M.P. Hobson, S.J. Bence, ISBN: 978-0521679718
3. "Lectures on Quantum Mechanics, Second Edition" (Cambridge University Press, 2015), S. Weinberg, ISBN: 978-1107111660
4. "Schaum's Outline of Quantum Mechanics, Second Edition" (McGraw-Hill, 2010/5/4), Y. Peleg, R. Pnini, E. Zaarur, E. Hecht, ISBN: 978-0071623582
5. https://www.scm.com/amsterdam-modeling-suite/adf/ (URL confirmed on 2024.11)
6. "Table of Molecular Integrals" (Maruzen, 1955), M. Kotani, A. Amemiya, E. Ishiguro, T. Kimura, ISBN: 978-0598976918
7. "Optimized trial wave functions for quantum Monte Carlo calculations", C.J. Umrigar, K.G. Wilson, and J.W. Wilkins Phys. Rev. Lett. 60, 1719 (1988). https://doi.org/10.1103/PhysRevLett.60.1719
8. "Quantum Monte Carlo simulations of solids", W.M.C. Foulkes, L. Mitas, R. J. Needs, and G. Rajagopal, Rev. Mod. Phys. 73, 33 (2001), https://doi.org/10.1103/RevModPhys.73.33
9. "A diffusion Monte Carlo algorithm with very small time-step errors", C.J. Umrigar, M.P. Nightingale, K.J. Runge, J. Chem. Phys. 99, 2865 (1993) https://doi.org/10.1063/1.465195
10. "Multiconfiguration wave functions for quantum Monte Carlo calculations of first-row diatomic molecules", C. Filippi, C.J. Umrigar, J. Chem. Phys. 105, 213 (1996), https://doi.org/10.1063/1.471865
11. "Spin contamination in quantum Monte Carlo wave functions", C-J. Huang, C. Filippi, C.J. Umrigar, J. Chem. Phys. 108, 8838 (1998), https://doi.org/10.1063/1.476330

12. “Quantum Mechanics” (Pergamon Press, 1965) L.D. Landau, E.M. Lifshits, ISBN: 978-0080091013
13. “New stochastic method for systems with broken time-reversal symmetry: 2D fermions in a magnetic field”, G. Ortiz, D.M. Ceperley, and R.M. Martin, Phys. Rev. Lett. 71, 2777 (1993), https://doi.org/10.1103/PhysRevLett.71.2777
14. “Tatai Mondai” (in Japanese, Baifukan Co., Ltd, 1975) F. Takano, ISBN: 978-4563024185
15. “Monte Carlo Methods In Ab Initio Quantum Chemistry (World Scientific, 1994), B.L. Hammond, W.A. Lester Jr., P.J. Reynolds, ISBN: 978-9810203221
16. “Fixed-Node Quantum Monte Carlo Method for Lattice Fermions”, H.J.M. van Bemmel, D.F.B. ten Haaf, W. van Saarloos, J.M.J. van Leeuwen, G. An, Phys. Rev. Lett. 72, 2442 (1994), https://doi.org/10.1103/PhysRevLett.72.2442
17. “A Benchmark quantum Monte Carlo study of the ground state chromium dimer”, K. Hongo and R. Maezono, Int. J. Quant. Chem. 112, 1243 (2012), https://doi.org/10.1002/qua.23113
18. “Ab initio Quantum Monte Carlo Calculations of Spin Superexchange in Cuprates: The Benchmarking Case of Ca_2CuO_3”, K. Foyevtsova, J.T. Krogel, J. Kim, P.R.C. Kent, E. Dagotto, F.A. Reboredo, Phys. Rev. X 4, 031003 (2014), https://doi.org/10.1103/PhysRevX.4.031003
19. “Energy Spectrum of the Excitations in Liquid Helium”, R.P. Feynman and M. Cohen Phys. Rev. 102, 1189 (1956), https://doi.org/10.1103/PhysRev.102.1189
20. “Natural Spin Orbital Analysis of Hydrogen Molecule Wave Functions”, H. Shull, J. Chem. Phys. 30, 1405 (1959). https://doi.org/10.1063/1.1730212
21. “Correlation and Localization” (Springer, 1999), P.R. Surjan, ISBN: 978-3540657545
22. “Introduction to Computational Chemistry, Third Edition” (Wiley, 2017), F. Jensen, ISBN: 978-1118825990
23. “Pfaffian Pairing Wave Functions in Electronic-Structure Quantum Monte Carlo Simulations”, M. Bajdich, L. Mitas, G. Drobny, L.K. Wagner, K.E. Schmidt, Phys. Rev. Lett. 96, 130201 (2006), https://doi.org/10.1103/PhysRevLett.96.130201
24. “Pfaffian pairing and backflow wavefunctions for electronic structure quantum Monte Carlo methods”, M. Bajdich, L. Mitas, L.K. Wagner, K.E. Schmidt Phys. Rev. B 77, 115112 (2008), https://doi.org/10.1103/PhysRevB.77.115112
25. “Superfluid Fermi Gases with Large Scattering Length”, J. Carlson, S.-Y. Chang, V.R. Pandharipande, K.E. Schmidt, Phys. Rev. Lett. 91, 050401 (2003). https://doi.org/10.1103/PhysRevLett.91.050401
26. “Geminal wave functions with Jastrow correlation: A first application to atoms”, M. Casula, S. Sorella, J. Chem. Phys. 119, 6500 (2003), https://doi.org/10.1063/1.1604379
27. “Correlated geminal wave function for molecules: an efficient resonating valence bond approach” M. Casula, C. Attaccalite, S. Sorella J. Chem. Phys. 121, 7110 (2004), https://doi.org/10.1063/1.1794632
28. “Optimization of Many-body Wave function” R. Maezono, J. Comput. Theor. Nanosci., 6, 2474 (2009). https://doi.org/10.1166/jctn.2009.1308
29. “Electronic Structure: Basic Theory and Practical Methods” (Cambridge University Press, 2004), R.M. Martin, ISBN: 978-0521782852
30. “Correlated electron pseudopotentials for 3d-transition metals”, J.R. Trail, R.J. Needs, J. Chem. Phys. 142, 064110 (2015) https://doi.org/10.1063/1.4907589
31. “Smooth relativistic Hartree-Fock pseudopotentials for H to Ba and Lu to Hg”, J.R. Trail, R.J. Needs, J. Chem. Phys. 122, 174109 (2005), https://doi.org/10.1063/1.1888569
32. “Energy-consistent pseudopotentials for quantum Monte Carlo calculations”, M. Burkatzki, C. Filippi, and M. Dolg, J. Chem. Phys. 126, 234105 (2007), https://doi.org/10.1063/1.2741534
33. “Energy-consistent small-core pseudopotentials for 3d-transition metals adapted to quantum Monte Carlo calculations”, M. Burkatzki, C. Filippi, and M. Dolg, J. Chem. Phys. 129, 164115 (2008). https://doi.org/10.1063/1.2987872
34. “Compact effective potentials and efficient shared-exponent basis sets for the first- and second-row atoms”, W.J. Stevens, H. Basch, M. Krauss, J. Chem. Phys. 81, 6026 (1984). https://doi.org/10.1063/1.447604

35. "Relativistic compact effective potentials and efficient, shared-exponent basis sets for the third-fourth- and fifth-row atoms", W.J. Stevens, M. Krauss, H. Basch, and P.G. Jasien, Can. J. Chem. 70, 612 (1992). https://doi.org/10.1139/v92-085
36. "Pseudopotentials for quantum Monte Carlo studies of transition metal oxides", J.T. Krogel, J.A. Santana, F.A. Reboredo Phys. Rev. B 93, 075143 (2016), https://doi.org/10.1103/PhysRevB.93.075143
37. "A new generation of effective core potentials for correlated calculations", M.C. Bennett, C.A. Melton, A. Annaberdiyev, G. Wang, L. Shulenburger, L. Mitas J. Chem. Phys. 147, 224106 (2017) https://doi.org/10.1063/1.4995643
38. "GaN band-gap bias caused by semi-core treatment in pseudopotentials analyzed by the diffusion Monte Carlo method", Y Nikaido, T. Ichibha, K. Nakano, K. Hongo, and R. Maezono, AIP Adv. 11, 025225 (2021). https://doi.org/10.1063/5.0035047
39. "Relativistic Quantum Mechanics: With Applications in Condensed Matter and Atomic Physics" (Cambridge University Press, 2008), P. Strange, ISBN: 978-0521565837
40. "Many-body core-valence partitioning", E.L. Shirley, R.M. Martin Phys. Rev. B 47, 15413 (1993). https://doi.org/10.1103/PhysRevB.47.15413
41. "Beyond the locality approximation in the standard diffusion Monte Carlo method" M. Casula, Phys. Rev. B 74, 161102 (2006), https://doi.org/10.1103/PhysRevB.74.161102
42. "Locality Error Free Effective Core Potentials for 3d Transition Metal Elements Developed for the Diffusion Monte Carlo Method", T. Ichibha, Y. Nikaido, C. Bennett, J. Krogel, K. Hongo, R. Maezono, and F.A. Reboredo. J. Chem. Phys. 159, 164114 (2023), https://doi.org/10.1063/5.0175381
43. "Size dependence of the bulk modulus of semiconductors nanocrystals", R. Cherian, C. Gerard, P. Mahadevan, N.T. Cuong and R. Maezono, Phys. Rev. B 82, 235321 (2010). https://doi.org/10.1103/PhysRevB.82.235321
44. "Wave functions for quantum Monte Carlo calculations in solids: Orbitals from density functional theory with hybrid exchange-correlation functionals", J. Kolorenc, S. Hu, L. Mitas, Phys. Rev. B 82, 115108 (2010), https://doi.org/10.1103/PhysRevB.82.115108
45. "Ab initio Calculation Tutorial: For Materials Analysis, Informatics and Design", (Springer Nature, 2023), R. Maezono, ISBN: 978-9819909216

Chapter 9
Practical Topics

This chapter covers advanced topics that, while not forming the core theoretical foundation of DMC, are still likely to arise in practice, even for beginners. Specifically, we discuss several correction schemes in Sect. 9.1 and the operation of large-scale supercomputers in Sect. 9.2.

9.1 Correction Schemes

9.1.1 Time-Step Error

The implementation of DMC is based on the short-time approximation described in Eq. (7.39). Recalling this, the time evolution of the histogram representation $\langle \mathbf{R} | f(\tau) \rangle = f(\mathbf{R}, \tau) = \Phi(\mathbf{R}, \tau) \cdot \Psi_T(\mathbf{R})$ is written as

$$|f(\tau)\rangle = \int d\mathbf{X} \cdot e^{-\tau\left(\hat{T}+U\right)} |\mathbf{X}\rangle \langle \mathbf{X} | f(0) \rangle \,. .$$

In this expression, the inability to separate $e^{-\delta\tau\left(\hat{T}+U\right)} \neq e^{-\delta\tau\hat{T}} \cdot e^{-\delta\tau U}$ was an obstacle to further evaluation. To overcome this, the exponential operator $e^{-\tau\left(\hat{T}+U\right)} = e^{-\tau\hat{L}}$ is sliced into short-time evolutions as

$$e^{-\tau\hat{L}} = e^{-(N\times\delta\tau)\hat{L}} = e^{-\delta\tau\hat{L}} \times e^{-\delta\tau\hat{L}} \times \cdots \times e^{-\delta\tau\hat{L}},$$

and for sufficiently small time step $\delta\tau$, the approximation

$$e^{-\delta\tau\hat{L}} = e^{-\delta\tau\left(\hat{T}+U\right)} \approx e^{-\delta\tau\hat{T}} \cdot e^{-\delta\tau U}$$

R. Maezono, *Ab initio Quantum Monte Carlo Tutorial*,
https://doi.org/10.1007/978-981-95-4134-8_9

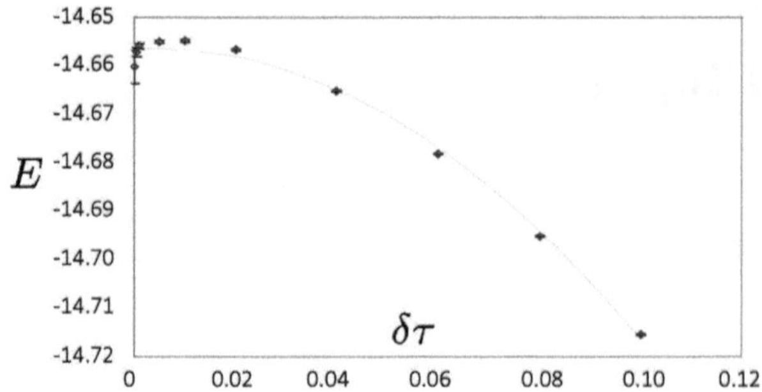

Fig. 9.1 Time step error in the ground-state energy calculation of the Be atom. The ground-state energy (vertical axis, in Hartree) varies depending on the time step $\delta\tau$ (horizontal axis, in 1/Hartree). A reliable extrapolated estimate is obtained by evaluating the bias due to time step selection through an extrapolation procedure as $\delta\tau \to 0$

is employed.

The choice of time step $\delta\tau$ was set to "dtdmc: 0.01" in the tutorials presented in the first half of this book. Regarding the selection of the time step, there is accumulated experience within the development community of each software package, and it is generally safe to follow the "recommended value of the package" at the given time. However, it is important to remain aware that there are various considerations behind this. Details are provided in Appendix Sect. 13.2.5, but the choices of the number of equilibration steps N_{eq} and the time step $d\tau$ in DMC are interrelated. It is important to set these parameters while envisioning: "Will the walkers have fully spread over the system after advancing N_{eq} steps with this time step?"

In practice, as shown in Fig. 9.1, one typically plots results for several selected values of $d\tau$, and extrapolates to $d\tau \to 0$ to obtain a reliable estimate. In doing so, statistical inference techniques are required, such as how to determine the error bar of the extrapolated value from the plot points with error bars. Recently, such extrapolation procedures have been provided as utility tools within packages and can be utilized accordingly. Research has also been conducted on questions like "how many $d\tau$ points are optimal in terms of computational efficiency for such extrapolation?" [1]. Since best practices for this continue to evolve, it is practically advisable to use utilities that reflect the most up-to-date schemes.

9.1.2 *Finite-Size Error*

When performing calculations for periodic solid-state systems, we employ an approximate modeling that treats a system that is "in reality infinitely extended"

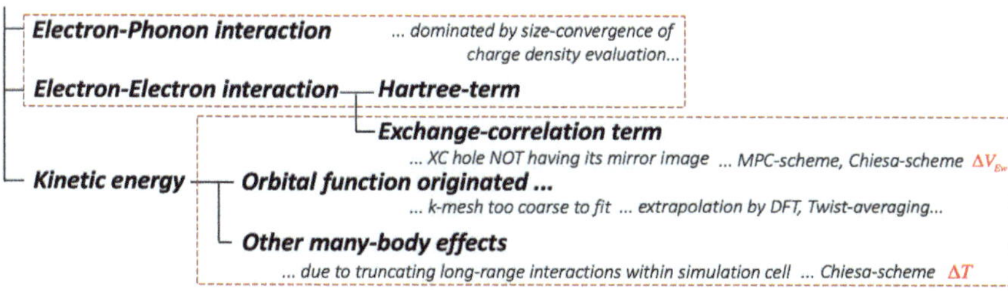

Fig. 9.2 In ab initio QMC, it is not straightforward to perform calculations using large simulation cells. Therefore, it is important to carefully analyze the sources of finite-size errors and disentangle them. Correction schemes have been developed based on the error generation mechanisms of each component

as "an infinite repetition of a finite-size simulation cell".[1] To obtain a reliable extrapolated value, it is necessary to apply corrections for finite-size errors arising from this approximation. In the case of DFT, a practical understanding such as "how finely the Brillouin zone is divided" may often suffice. However, when performing ab initio QMC calculations, which are heavily based on real-space images, it is crucial to swiftly switch between real-space and reciprocal-space perspectives, as emphasized in the author's previous works [2].[2]

The computational cost of ab initio QMC is substantially higher compared to DFT, making calculations with large cells, such as $24 \times 24 \times 24$, easily achievable by DFT, virtually impossible in practice. At best, a cell size of about $4 \times 4 \times 4$ is manageable.[3] Therefore, it is crucial to decompose the sources of finite-size errors and to develop careful correction methods tailored to each specific mechanism. Various analyses and correction schemes have been extensively studied [3, 4].

The total energy to be evaluated can be divided into **(1)** kinetic energy, **(2)** electron–electron interaction, and **(3)** electron–lattice interaction. Furthermore, the electron–electron interaction can be subdivided into **(2a)** Hartree energy and **(2b)** exchange-correlation energy (see Fig. 9.2). The Hartree energy is the classical

[1] Care is needed regarding terminology here. If one loosely says "an infinite system is treated with a finite size," some beginners may misunderstand this as "using a single isolated finite cell." A $1 \times 1 \times 1$ finite-size calculation is referred to as a "Γ-point calculation", , but it also employs "an infinite repetition of cells." In the case of $1 \times 1 \times 1$, the wavefunction phase is identical across all cells, which means that wavefunctions with wavelengths longer than the unit cell cannot be taken into account. See the author's previous book [2] for details.

[2] What is referred to in DFT as an "$M \times M \times M$ k-mesh calculation" corresponds in real-space language to a calculation using a simulation cell consisting of an $M \times M \times M$ multiple of the structural unit cell.

[3] In addition to computational costs, memory capacity also becomes a critical issue. As discussed in Sect. 5.1.2, periodic solid systems initially handled with a plane-wave basis are converted into a localized basis set composed of blip functions. During this process, representing the same orbital functions demands significantly larger file sizes compared to the original plane-wave basis, thus requiring substantial memory capacity to store this information during the calculation.

Coulomb energy arising from the charge density, while the exchange correlation energy refers to the "remaining quantum effects," namely, corrections due to statistics and those due to distortions of the wavefunction (see Sect. 10.1).

The finite-size effects in **(3)** electron–ion interaction and **(2a)** Hartree energy arise only through the "size dependence of the charge density," and their errors converge rapidly with system size, rendering them negligible [3, 5]. In contrast, the size dependence of the exchange correlation part and kinetic energy is significant, and these are the "main subjects of investigation." Evaluation schemes for finite-size error that remove such effects themselves constitute advanced research topics in quantum many-body problems [3, 4].

For **(1)** the kinetic energy, there are two sources of size dependence. The first is the effect that the k-point mesh does not fully resolve the shape of the Fermi surface. This is referred to as the **(1a)** one-body finite-size error. The second arises from the contribution of the **(1b)** Jastrow factor. Including the Jastrow factor in the wave function modifies the amplitude profile, such as cusp behavior, and consequently alters the spatial curvature of the wave function; this contributes to the kinetic energy. Although the Jastrow factor is originally meant to capture long-range behavior, when treated within a finite simulation cell, its long-range tail must be truncated at the cell boundary. This truncation becomes a source of finite-size error.

The finite-size error in the **(2b)** exchange-correlation energy can be broadly understood through the concept of image effects. In periodic solid systems, a single electron interacts over long ranges with all other electrons in the system. Instead of truncating this interaction within a finite-size simulation cell, the effect is accounted for by including interactions with periodic images of the reference electron. This is the essential idea behind the treatment based on Ewald interactions. [4] The charge distribution evaluated through simulation, including quantum mechanical effects, reflects modulation due to quantum contributions such as exchange and correlation effects. These modulations can be visualized as an exchange-correlation hole, when focusing on a specific electron, other electrons tend to avoid it, resulting in a deficit in charge distribution around the reference electron (Sect. 10.1.3). It should be noted that the exchange-correlation hole is an **excess** arising from quantum mechanical effects that have no counterpart in classical interactions.

If one naively evaluates electron–electron interactions using the Ewald summation on charge distributions obtained from quantum mechanical simulations, one ends up replicating the images of the exchange, correlation hole as well. Classical interactions act over long ranges, whereas the quantum mechanical corrections, expressed as the exchange-correlation hole, are short-range effects occurring only in the vicinity of the reference electron. Therefore, contributions from the exchange-correlation hole must be corrected by evaluating them without mirror images. Based on the mechanism of such bias, methodologies like the **MPC interaction**

---

[4] Mathematical-formally, the Ewald interaction is understood as the solution to the Poisson equation for a periodic system [5].

evaluation scheme, which replace the Ewald summation, have been developed and implemented.

The above discussion was derived from the real-space perspective. On the other hand, from the viewpoint of reciprocal space, the origin of finite-size errors is understood as the contribution from integration near the Γ point of the Brillouin zone. When the Brillouin zone is divided into cubes of width ΔG, the integration contribution over the interval $[0, \Delta G]$ is omitted. If we let $\Delta G \sim 1/\lambda$, this corresponds to neglecting the contribution for $G < \Delta G$, that is, ignoring the contribution from wavelengths $l > \lambda$. In other words, the finite-size error stems from the inability to capture phenomena extending beyond the supercell size λ.

With regard to **(1a)** "Fermi surface shape capture via k-point mesh," we perform **twist averaging**. Details are summarized in Appendix Sect. 13.2.5, but from a procedural viewpoint, this method mitigates errors by performing calculations with the origin of the k-point mesh shifted in various ways within the range $G < \Delta G$ and taking the average. As mentioned in the author's previous work [2], this requires attention in the case of metallic systems. On the other hand, for insulators, the Fermi surface does not become complex, so this issue requires much less concern.

The **(1b)** finite-size error from the Jastrow factor and **(2)** finite-size error of the exchange-correlation energy can be separated and expressed in terms of the structure factor $S(\mathbf{k})$ (Sect. 10.2.3). Their contributions in the region $[0, \Delta G]$ can be estimated analytically by applying asymptotic analysis to the k-space integrals. These values are then added back to apply the correction (Chiesa scheme [6]).

9.1.3 Practical Procedure for Finite-Size Error Correction

In current general-purpose packages, various finite-size error corrections can be applied by specifying instructions in the input file. The corrected results are then displayed in the output. In the case of CASINO, the correction based on the MPC scheme can be applied by changing the line **interaction = ewald** in the input file to **mpc**, which triggers the calculation of correction values using the MPC interaction scheme. Additionally, by setting **finite_size_corr** to **T**, the program will estimate correction values for **(1b)** the finite-size error due to the Jastrow factor and **(2)** the finite-size error in the exchange-correlation energy, based on the structure factor $S(\mathbf{k})$.

For twist averaging, one must run multiple ab initio QMC calculations with different twisted boundary conditions and aggregate the results. Therefore, this procedure is not handled within CASINO itself, but rather via wrapper scripts, and utility scripts for this purpose are available. When dealing with real materials, the corresponding set of density functional theory calculations under twisted boundary conditions must be prepared in accordance with the subsequent ab initio QMC simulations. The selection of how many and which specific twist angles to include must be carefully designed on a per-project basis [4]. In addition, once the phase factor $\exp[i \cdot \varphi_T]$ is introduced into the wavefunction via

$$\psi_k (r + L \cdot a) \stackrel{!}{=} \exp[i \cdot \varphi_T] \, \psi_k (r) \, ,$$

as shown in Eq. (15.1).[5] the wavefunction can no longer be treated as real-valued (Sect. 8.2.1) and must instead be handled as a complex function. Such **ab initio QMC with complex wavefunctions** has already been implemented, [6] and in such cases, one should set **complex_wf** = T in the input file.

9.2 Topics on Computational Environments

9.2.1 Guidelines for Choosing a Software Package

For readers newly interested in this field, a natural initial question is often: "Which package should I use?" The historical development of general-purpose ab initio QMC packages has been outlined in Sect. 1.3.4. When making a decision, three main considerations, highlighted in bold below, should be taken into account.

The first factor is: **(1) Is it for educational or practical use?** This is the primary branching point that users should consider. If the goal is "first to understand the implementation principles of the program," as is the case at the level of this book, then trying to operate a "large passenger aircraft" designed for practical use may obscure the core logic, such as "which variables are passed to which algorithms for processing," as discussed in the column at the end of Sect. 7.3.3 titled "thick-layered implementation." In such cases, it is better to start with code tailored for educational use, i.e., a "trainer aircraft," to gain familiarity and conceptual understanding, before transitioning to full-scale practical packages. The **CASINO** package used in this book is implemented in FORTRAN, but there is also **PyQMC**, a Python-based implementation that is more aligned with educational use [8], and it is considered particularly well-suited for understanding algorithms due to its high readability.[7]

The second factor to consider is: **(2) Is your focus on general materials science or methodological development?** Once you have acquired sufficient understanding

[5] The notation "$\stackrel{!}{=}$" represents the equality used in the sense of "... is set as...".

[6] In materials with ferromagnetic or antiferromagnetic order, one can often align the spin polarization along the z-axis, allowing the wavefunction to be expressed as a pair of real-valued functions. This is possible because the corresponding Pauli matrix σ_z is real and diagonal. However, in materials with spiral or canted magnetic structures [7], where spin directions vary across space, the wavefunction includes x and y spin components. In such *noncollinear magnetism*, the relevant Pauli matrices σ_x and σ_y are non-diagonal and include imaginary entries, and the wavefunction cannot be represented as real. These considerations are among the reasons for implementing ab initio QMC calculations using complex wavefunctions.

[7] PyQMC relies entirely on "PySCF," a Python-based SCF calculation package [9]. However, PySCF has limited practicality when dealing with periodic systems.

to use practical-level codes, you will face the decision of whether to specialize in general-purpose materials science applications or pursue cutting-edge methodological development. Each direction calls for a different optimal choice of code. For general materials science applications, tackling realistic materials requires large-scale massively parallel and high-performance computing. As of the time of writing, the most prominent code for this purpose is **QMCPack** from the United States [10], which can be likened to a "large passenger aircraft." On the other hand, codes more focused on methodological development tend to come from Europe. In the author's personal view (at the time of composing): the Eulopean-led **CHAMP** [11] excels in excited-state calculations; the Italian-developed **TurboRVB** [12] specializes in electron correlation using exotic wavefunctions like geminals and Pfaffians; and the French-developed **QMC=Chem** [13] is strong in geometric correlation descriptions for multi-determinant systems. Each of these packages exhibits distinct strengths aligned with its developmental direction.

The third factor to consider is: **(3) Ease of installation**. Packages like QMCPack, which can be likened to a "large passenger aircraft," have undergone extensive optimization efforts to fully exploit the capabilities of large-scale parallel computing and cutting-edge GPU architectures in pursuit of extreme high-performance computation.[8] As a result, such packages are typically compiled to make full use of platform-specific optimized libraries for particular flagship systems. Attempting to port and use them on a personal PC or a department-level server often leads to significant difficulties, even for experts. Trying to compile the code might result in errors like "missing library" or "unavailable function," and attempts to resolve one issue can lead to cascading dependencies, creating a situation that is hard to manage. In contrast, educational packages implemented in interpreted languages like Python, which require no compilation, are designed to completely eliminate such struggles.

Middleware as a Solution

When performing computations on a computer, aspects such as "from which register to transfer data to the processing unit," "where to hold numerical data temporarily," and 'whether dedicated units like floating-point processors are available" can differ significantly between general-purpose CPUs and GPUs designed for numerical computation. At the supercomputing level, where extreme high performance is pursued, these architectural distinctions are fully exploited. There is intense focus on devising efficient strategies to transfer as much data as possible, as quickly as possible, through optimal data paths, ideally into well-matched dedicated processors. However, users implementing

(continued)

[8] This is often necessary to secure access to flagship supercomputers, as resource allocations usually require benchmark evidence demonstrating a certain percentage of performance improvement through the use of the machine's latest features.

programs cannot be expected to account for all such machine-dependent details. It is unrealistic to rewrite numerical computation packages, many of which are maintained for over 50 years, every time a hardware upgrade occurs. Therefore, a strategy is adopted in which frequently used generic algorithms, such as "matrix multiplication" or "eigenvalue calculation," are not implemented from scratch by each user, but rather accessed via designated libraries. These libraries, or **middleware**, are provided and utilized explicitly for such purposes. The responsibility for the inner workings below the middleware layer is left to "experts in computational hardware performance," who optimize the implementation to exploit the full potential of the memory and arithmetic unit architectures mentioned earlier. This establishes a division of labor that enables high performance without requiring every user to be a low-level hardware expert.

From the perspectives discussed above, the CASINO package used in this book offers a well-balanced implementation. It is relatively easy to install, the code is not excessively "fatty," and it is sufficiently robust to be used in practical research on real materials.

9.2.2 Workflow Automation

As reviewed in Chap. 6, the computational procedure follows a series of steps: "01/orbital generation" → "02/construction of the trial wave function" → "03/variational optimization of the trial wave function" → "04/imaginary-time propagation via DMC". Each of these steps, from 01 to 04, can be further subdivided into more granular tasks. There is an aspect of these procedures that requires mastery of certain expert techniques. Without automation, the spread of users will remain limited. The process of scripting and automating this sequence of operations, rather than relying on manual execution, is referred to as **workflow automation**. For well-established and mature procedures, it is ideal that workflow automation is implemented and made accessible to all users. This allows researchers to dedicate their cognitive resources to more advanced aspects of scientific development.

For each ab initio QMC package, implementations of Python scripts to realize workflow automation have been developed and made available. At each step of the process, these scripts handle tasks such as: "Which files are used as input?," "Which executable binary should be run?," "On which supercomputer?," "With how many parallel processes?," "Under which job class?," and "Which output files should be renamed and passed to the next step?." By encoding such procedures in Python scripts in advance, these workflows enable automated execution of the entire process. In the case of the US-based QMCPack project, a workflow system called

NEXUS has been developed and maintained [14]. In Europe, for the TurboRVB package, the associated workflow system **TurboWorkflows** has been developed and maintained as well [15].

> **Developed and Maintained by Surprisingly Small Teams...**
> General-purpose implementations of ab initio QMC methods are often developed, maintained, and managed by a core group of around five people at most. The author was part of the development team for **CASINO** and can confirm this firsthand. However, when this fact was relayed to a well-known professor in Japan, he insisted, "That can't possibly be true. I'm sure they outsource the programming to professional contractors." As discussed in the author's previous book [16], there is a deeply rooted belief in the "conservation law of competence," which makes it hard for some to accept the reality that there are people who "can do it all." A European professor who leads the development of **TurboRVB** is also renowned in the field of abstract model theory. During a research visit, the author witnessed this professor, after receiving an error report, sit down beside them with a MacBook, pinpoint the exact source code issue, and provide detailed guidance on compiler options, demonstrating firsthand their ability to carry out numerical ab initio computations themselves. Even in this case, some researcher in my country confidently insisted, "There's no way! That professor is a theoretical model person. Someone else in his group, likely a younger member, must be doing the ab initio work." Perhaps what the author witnessed was a ghost.

9.2.3 Using Supercomputers

As discussed in Sect. 2.3.2, ab initio QMC calculations are, by their computational principles, highly parallelizable. General-purpose packages have been developed from the outset with parallel processing in mind. Around the 2010s, implementations typically employed MPI-based parallelization, which is conceptually intuitive: dividing the large number of required Monte Carlo samples across multiple parallel cores.

In VMC, which does not involve any concept of "temporal progression," for example, a total sample size of 10,000 can be divided among 500 computational cores, assigning 20 samples per core. This allows the computation time to be reduced to approximately 1/500. In contrast, in DMC where "time evolution along imaginary time" is involved, samples accumulated along the imaginary time axis cannot be distributed across computational cores.[9] In DMC, it is the total number of walkers that is distributed among the computational cores. For instance, 2000 walkers can be divided such that each core handles four walkers, and the computation is distributed across 500 cores to reduce execution time. If there were no creation or annihilation of walkers, this approach would reduce the computation time by nearly 1/500 compared to processing all 2000 walkers on a single core. However, in DMC, due to walker creation-annihilation events, as the simulation progresses, load imbalance arises: e.g., one core might end up with nine walkers, while another has only one. In such cases, the overall computation must wait for the heavily loaded core with nine walkers to finish, so the speed-up is less ideal than in VMC. Therefore, in MPI-parallelized DMC simulations, special care is taken to maintain **load balance** across cores, with implementation-level techniques designed for this purpose.

When using supercomputers, it becomes necessary to plan "how many computational cores will be used to parallelize the total number of samples or total number of walkers," and accordingly select an appropriate job class. The block count should be adjusted such that intermediate results are output within the time limit of the selected job class, so that even if a job is interrupted, a resume calculation can continue from the saved data. During the era when commonly accessible queue classes allowed only a few thousand-core parallel runs, this kind of workflow posed no significant problem. However, from the mid-2010s, with the advent of tens-of-thousands-core parallel capabilities, the traditional workflow, consisting of a short equilibration process followed by a long statistical accumulation process, has had to be reconsidered.

The statistics accumulation process can be shortened inversely with the number of parallel cores by distributing the workload across many cores. On the other hand, the equilibration process requires a certain amount of warm-up for **each parallel core**, so its computational time remains constant regardless of the degree of parallelization. If tens of thousands of cores are used for parallel execution, the time for the statistics accumulation can be drastically reduced, eventually approaching the time required for equilibration. In such cases, there is **no further room for speed-up** unless the equilibration itself can be shortened. Thus, the **equilibration process becomes the bottle-neck**.

With tens of thousands of cores, the situation described above arises. Attempting to scale further to hundreds of thousands or even millions of cores, **flat-MPI parallelization** becomes insufficient, as it cannot contribute to further accelerating the bottleneck equilibration process. To address this, the surplus parallel computing resources must be redirected from *distributing the number of samples* to *accelerating each individual sample calculation itself*. This shift requires advanced implementations that employ **hybrid parallelization**, combining **process-level**

[9] Later samples depend on earlier ones, so if the imaginary time axis is divided, sectors from the future and past cannot run concurrently.

parallelization such as inter-node MPI with **thread-level parallelization** such as intra-node OpenMP.

At the time of writing, flagship-class supercomputers have adopted **high-density core architectures**, with configurations such as "512 cores per node." However, the memory capacity per node does not scale proportionally with the number of cores, resulting in *smaller memory capacity available per core*. Meanwhile, with the use of blip functions to re-expand orbital functions (Sect. 5.1.2), the *amount of memory each core must access* has increased significantly. It is becoming increasingly difficult under high-density core architectures to provide such large memory allocations as **private memory dedicated per core**.[10] Consequently, efficient access schemes to **shared memory** among multiple cores within a node have become necessary. Rather than the naive approach of storing a full copy of wavefunction information *per core*, implementations must now adopt designs where multiple cores efficiently *refer to shared wavefunction data* residing in the node's shared memory.

As time progressed, the emergence of GPU-accelerated supercomputers has further increased the demands on implementation techniques for **speeding up the sampling calculation at a single point**, as described above. By the 2000s, general-purpose packages for ab initio QMC had already been established with codes such as **CASINO**, **CHAMP**, and **QWalk**, leaving little room for new entries. However, the development of **QMCPack** emerged in the mid-2000s, and one reason for this may lie in the shift in implementation trends: from the former focus on *readability and portability*, toward a new paradigm that *de-emphasizes portability in favor of architecture-specific optimizations*, aiming to extract maximum performance by leveraging the features of a particular system configuration.

In this modern trend of **architecture-specialized programming**, as discussed in the column of Sect. 9.2.1, a buffer is established between methodological development and HPC engineering by actively leveraging *middleware maintained by domain experts*. In the United States, practical packages have been consolidated into **QMCPack**, allowing for the focused allocation of personnel and funding for its maintenance and development. In contrast, Europe, as also mentioned in Sect. 9.2.1, appears to pursue a direction that values the *diversity of packages*, each specialized in advanced methodological approaches. Rather than forcing integration, a cooperative European project called **QMCkl** has been launched to develop middleware at the level of electronic structure subroutines (kernel libraries) and promote its adoption [17]. Since most electronic structure packages share subroutines such as *SCF loops, pseudopotential evaluations, charge density calculations, and wavefunction updates per sampling*, this initiative aims to modularize these common layers into middleware to ensure maintainability under architecture-specific programming. In addition, to facilitate interoperability among the remaining

[10] For example, suppose a computational resource offers "512 cores per node." To secure the same per-core memory as before using traditional implementations, one might be forced to use only 32 cores per node while leaving the rest idle.

diverse packages, a standardized I/O format called **TREXIO** is being developed, enabling easier implementation of converters [18].

References

1. "Strategies for improving the efficiency of quantum Monte Carlo calculations", R.M. Lee, G.J. Conduit, N. Nemec, P. Lopez Rios, N. D. Drummond, Phys. Rev. E 83, 066706 (2011), https://doi.org/10.1103/PhysRevE.83.066706
2. "Ab initio Calculation Tutorial: For Materials Analysis, Informatics and Design", (Springer Nature, 2023), R. Maezono, ISBN: 978-9819909216
3. "Finite-size errors in continuum quantum Monte Carlo calculations", N.D. Drummond, R.J. Needs, A. Sorouri, and W.M.C. Foulkes, Phys. Rev. B 78, 125106 (2008). https://doi.org/10.1103/PhysRevB.78.125106
4. "Diamond to beta-tin phase transition in Si within quantum Monte Carlo", R. Maezono, N.D. Drummond, A. Ma, and R.J. Needs, Phys. Rev. B 82, 184108 (2010). https://doi.org/10.1103/PhysRevB.82.184108
5. "Quantum Monte Carlo simulations of solids", W.M.C. Foulkes, L. Mitas, R. J. Needs, and G. Rajagopal, Rev. Mod. Phys. 73, 33 (2001), https://doi.org/10.1103/RevModPhys.73.33
6. "Finite-Size Error in Many-Body Simulations with Long-Range Interactions", S. Chiesa, D.M. Ceperley, R.M. Martin, M. Holzmann, Phys. Rev. Lett. 97, 076404 (2006). https://doi.org/10.1103/PhysRevLett.97.076404
7. "Spin and orbital ordering in double-layered manganites" R. Maezono, N. Nagaosa, Phys. Rev. B 61, 1825 (2000). https://doi.org/10.1103/PhysRevB.61.1825
8. "PyQMC: An all-Python real-space quantum Monte Carlo module in PySCF", W.A. Wheeler, S. Pathak, K.G. Kleiner, S. Yuan, J.N.B. Rodrigues, C. Lorsung, K. Krongchon, Y. Chang, Y. Zhou, B. Busemeyer, K.T. Williams, A. Munoz, C.Y. Chow, L.K. Wagner, J. Chem. Phys. 158, 114801 (2023) https://doi.org/10.1063/5.0139024
9. https://pyscf.org (URL confirmed on 2024.11)
10. "QMCPACK: an open source ab initio quantum Monte Carlo package for the electronic structure of atoms, molecules and solids", J. Kim, A.D. Baczewski, T.D. Beaudet, A. Benali, M.C. Bennett, M.A. Berrill, N.S. Blunt, E.J.L. Borda, M. Casula, D.M. Ceperley, J. Phys.: Condens. Matter 30, 195901 (2018), https://doi.org/10.1088/1361-648X/aab9c3
11. "Energy and variance optimization of many-body wave functions", C.J. Umrigar, C. Filippi, Phys. Rev. Lett. 94, 150201 (2005), https://doi.org/10.1103/PhysRevLett.94.150201
12. "TurboRVB: A many-body toolkit for ab initio electronic simulations by quantum Monte Carlo", K. Nakano, C. Attaccalite, M. Barborini, L. Capriotti, M. Casula, E. Coccia, M. Dagrada, C. Genovese, Y. Luo, G. Mazzola, A. Zen, S. Sorella, J. Chem. Phys. 152, 204121 (2020), https://doi.org/10.1063/5.0005037
13. https://trex-coe.eu/trex-quantum-chemistry-codes/qmc-chem (URL confirmed on 2024.11)
14. "Nexus: A modular workflow management system for quantum simulation codes" J.T. Krogel, Comput. Phys. Commun. 198, 154 (2016), https://doi.org/10.1016/j.cpc.2015.08.012
15. "TurboGenius: Python suite for high-throughput calculations of ab initio quantum Monte Carlo methods", K. Nakano, O. Kohulák, A. Raghav, M. Casula, S. Sorella, J. Chem. Phys. 159, 224801 (2023) https://doi.org/10.1063/5.0179003
16. "How to Build Your Own PC Cluster" (in Japanese), Ryo Maezono, Morikita Publishing (2017/12/14) ISBN:978-4627818217

17. "QMCkl: A Unified Approach to Accelerating Quantum Monte Carlo Codes", A. Scemama, https://github.com/Trex-CoE/qmckl (URL confirmed on 2024.11)
18. "TREXIO: A file format and library for quantum chemistry Special Collection: High Performance Computing in Chemical Physics", E. Posenitskiy, V.G. Chilkuri, A. Ammar, M. Hapka, K. Pernal, R. Shinde, E.J. Landinez Borda, C. Filippi, K. Nakano, O. Kohulak, S. Sorella, P. de Oliveira Castro, W. Jalby, P. Lopez Rios, A. Alavi, A. Scemama, J. Chem. Phys. 158, 174801 (2023), https://doi.org/10.1063/5.0148161

Chapter 10
Essence of Many-Body Electronic Correlation Theory

Ab initio QMC methods, particularly diffusion Monte Carlo (DMC), are regarded as the most reliable practical approach for describing quantum many-body interactions.[1] In this chapter, we explain why *ab initio* QMC methods are capable of achieving a higher level of descriptive accuracy, compared to other practical methods such as DFT or molecular orbital methods. We aim to deepen understanding of the relationship with other methods, as well as the differences in approach, thereby elucidating the **features of the method itself**.

This chapter may give the impression of being heavy with theoretical preliminaries, but the main argument begins in Sect. 10.5. The sections from Sect. 10.1 to Sect. 10.4 are preparatory, providing the necessary background for understanding the discussion in Sect. 10.5. It is also possible to begin reading from Sect. 10.5 and refer back to the earlier sections when one feels insufficiently prepared.

10.1 Decomposition of Quantum Many-Body Effects

In this section, we introduce several key concepts such as "exchange and correlation," "mean field and fluctuation," the pair distribution function, and the breakdown of self-interaction cancellation. These concepts will be examined in more detail in the following sections and will ultimately lead to the understanding of Fig. 10.10, which will be discussed later.

[1] If one limits the scope to "only small molecular systems" or "simple models," then highly reliable methods tailored by theorists can indeed be constructed in abundance. However, such methods are not easily applicable to realistic materials in practical materials science.

R. Maezono, *Ab initio Quantum Monte Carlo Tutorial*,
https://doi.org/10.1007/978-981-95-4134-8_10

10.1.1 What Are Quantum Many-Body Effects in Many-Electron Systems?

The *ab initio* QMC method is positioned as the most reliable method for describing the "effects arising from a large number of quantum mechanical particles interacting via Coulomb forces" (quantum many-body effects). In contrast to quantum many-body systems, the reference is a "system of classical particles with Coulomb interactions," and the difference from such systems defines what we refer to as *quantum* many-body effects. The features that distinguish quantum systems from classical ones include:

- Effects due to quantum statistics that do not arise in classical systems (most typically the Pauli exclusion principle),
- Unlike classical particles, particle positions follow probability distributions described by wavefunctions,

These aspects give rise to deviations from classical behavior. The former relates to exchange effects, while the latter relates to correlation effects. In what follows, we explain what constitutes **exchange effects** and what constitutes **correlation effects** (see Fig. 10.1).

10.1.2 Hartree–Fock Equation

To understand quantum many-body effects, especially electronic correlation, it is important to first revisit Hartree–Fock theory. This is because it serves as the reference point that separates exchange effects from correlation effects. A careful understanding of this theory is necessary in order to recognize clearly what additional effects must be incorporated beyond it. This foundation is essential for developing the arguments that follow with precision.

Meanfield + ***Fluctuations*** *(electronic correlation)*

= *[Classical meanfield]* + *[(Quantum statistics **(exchange)**)* + *(Electronic **correlation**)]*

= *[Classical meanfield]* + *[**Exchange-correlation** potentials]*

*** Quantum statistics ... Effects due to Pauli exclusion rule.*

*** Electronic correlation ... Effects of deformation in wavefunction by interactions*

Fig. 10.1 Separation of effects related to electron–electron interactions. The term "correlation" is often used loosely in various contexts due to its suggestive nuance, but it is important to understand the above classification properly

Starting from Hartree–Fock Is a Legitimate Approach, But...
When I began a lecture from Hartree–Fock theory, one professor remarked, "Why would you start from Hartree-Fock when discussing electron correlation?," expressing a sense of discomfort. However, after the lecture, he kindly told me, "I was impressed by the *fresh style* of building your argument from Hartree-Fock." Of course, this is not some *novel style* that I invented, rather, it is a thoroughly orthodox approach. Due to the somewhat seductive and self-propagating nature of the keyword "electronic correlation," the field of strongly correlated systems has been promoted for many years without always a clear distinction between correlation and exchange. It seems to me that the practice of starting with this proper separation has gradually been forgotten. If one discusses electron correlation too casually or intuitively, it can lead to widespread misinterpretation and misunderstanding, as discussed in Sect. 16.3.1.

Solving the many-body Schrodinger equation in Eq. (1.1),

$$\left[-\frac{1}{2}\sum_{j=1}^{N}\nabla_j^2 + V\left(\mathbf{r}_1,\cdots,\mathbf{r}_N\right)\right]\cdot\Psi\left(\mathbf{r}_1,\cdots,\mathbf{r}_N\right) = E\cdot\Psi\left(\mathbf{r}_1,\cdots,\mathbf{r}_N\right)$$

is realized as the minimization of the variational functional given in Eq. (1.3),

$$E = \frac{\int d\mathbf{R}\cdot\Psi^*\left(\mathbf{R}\right)\left[-\frac{1}{2}\sum_j\nabla_j^2 + V\left(\mathbf{R}\right)\right]\Psi\left(\mathbf{R}\right)}{\int d\mathbf{R}\cdot\Psi^*\left(\mathbf{R}\right)\cdot\Psi\left(\mathbf{R}\right)}$$

as per the variational principle. For a many-body potential with Coulomb interactions between particles,[2]

$$V\left(\mathbf{r}_1,\mathbf{r}_2,\cdots,\mathbf{r}_N\right) = \sum_j^N v_{\mathrm{ext}}\left(\mathbf{r}_j\right) + \sum_{1\le i<j\le N}\frac{1}{\left|\mathbf{r}_i-\mathbf{r}_j\right|},$$

we consider performing a variational minimization under the assumption that the many-body wave function is a Slater determinant:[3]

[2] Here, $v_{\mathrm{ext}}\left(\mathbf{r}\right)$ denotes the external potential commonly acting on all particles, such as the nuclear potential.

[3] Among the N particles, the first 1 through M are treated as spin-up, and $(M+1)$ through N as spin-down. Antisymmetry is enforced only within same-spin components via the determinant.

$$\Psi_{\mathrm{SD}}(\mathbf{r}_1,\mathbf{r}_2,\cdots,\mathbf{r}_N) = \begin{vmatrix} \psi_1(\mathbf{r}_1) & \cdots & \psi_1(\mathbf{r}_M) \\ \vdots & \ddots & \\ \psi_M(\mathbf{r}_1) & \cdots & \psi_M(\mathbf{r}_M) \end{vmatrix} \times \begin{vmatrix} \psi_{M+1}(\mathbf{r}_{M+1}) & \cdots & \psi_{M+1}(\mathbf{r}_N) \\ \vdots & \ddots & \\ \psi_N(\mathbf{r}_{M+1}) & \cdots & \psi_N(\mathbf{r}_N) \end{vmatrix} = \Psi_{\mathrm{SD}}\left[\{\psi_j(\mathbf{r})\}\right] .$$

In the present case, the variational degrees of freedom $\boldsymbol{\alpha}$ correspond to the variational relaxation of each orbital function $\psi(\mathbf{r})$. The Euler–Lagrange equation that determines this relaxation variationally yields the **Hartree-Fock equation**[1]:

$$\begin{aligned} &\left[-\frac{1}{2}\nabla^2 + v_{\mathrm{ext}}(\mathbf{r}) + \sum_{j\neq i}\int d\mathbf{r}'\frac{|\psi_j(\mathbf{r}')|^2}{|\mathbf{r}-\mathbf{r}'|}\right]\cdot\psi_i(\mathbf{r}) \\ &-\sum_{j\neq i}\left[\int d\mathbf{r}'\psi_j^*(\mathbf{r}')\frac{\delta_{\sigma_i\sigma_j}}{|\mathbf{r}-\mathbf{r}'|}\psi_i(\mathbf{r}')\right]\cdot\psi_j(\mathbf{r}) \\ &= \varepsilon_i\cdot\psi_i(\mathbf{r}) . \end{aligned} \tag{10.1}$$

The integral term in the first line is called the **Hartree term**, while the integral term in the second line is called the **Fock term**. These represent the classical electrostatic field and the exchange effect, respectively, as discussed below.

The term in the first line of Eq. (10.1), expressed as an integral, is the **Hartree term**. It takes the form: "the ith electron is affected by the sum of individual Coulomb interactions arising from the charge density $\rho_j(\mathbf{r}) = \left|\psi_j(\mathbf{r}')\right|^2$ produced by the jth electron." Thus, the Hartree term represents the contribution of "classical interactions due to $\rho_j(\mathbf{r})$." On the other hand, the **Fock term** in the second line of Eq. (10.1) has the orbital label in $\psi_j(\mathbf{r}')$ appearing as i instead of j. In the Hartree term, the index was j, so it could be combined with $\psi_j^*(\mathbf{r}')$ to yield $\left|\psi_j(\mathbf{r}')\right|^2$, but here, this exchange of indices renders the term no longer interpretable as a classical electrostatic field. Because of the factor $\delta_{\sigma_i\sigma_j}$, this term only contributes when the ith and jth electrons have the same spin. With the negative sign prefixed, this term is sometimes described pedagogically as "aligning spins leads to additional energy stabilization."[4] As discussed in the author's previous works [4], in the early theory [Hartree theory], where the trial wave function is taken without a determinant and

[4] There was a time when this was believed to be the origin of ferromagnetism and Hund's rules, but upon closer examination, it is now understood not to be the case [2, 3]. Nevertheless, since the explanation is intuitively appealing, this misunderstanding has proven difficult to dispel, and it still circulates erroneously in some contexts.

only with simple separation of variables,

$$\Psi_{\mathrm{H}}(\mathbf{r}_1, \mathbf{r}_2, \cdots, \mathbf{r}_N) = \psi_1(\mathbf{r}_1) \cdot \psi_2(\mathbf{r}_2) \cdots \psi_N(\mathbf{r}_N),$$

the exchange term does not appear. Once the Hartree theory is upgraded to the Hartree–Fock theory by taking the Pauli exclusion principle into account, an exchange hole emerges. Thus, the exchange hole reflects the exclusion principle. This can be understood directly in terms of the concept of the exchange hole, as discussed in the next section.

10.1.3 Exchange Hole and Correlation Hole

The Fock term is a purely quantum mechanical effect, reflecting the fact that the trial wave function incorporates antisymmetry in advance—that is, the property that same-spin electrons cannot come close to each other. This mutual avoidance behavior can be clearly observed by theoretically computing the **pair distribution function**. The pair distribution function describes how another particle is distributed around a given particle fixed at position $\mathbf{r}'$, normalized to unity. For a given many-body wave function $\Psi(\mathbf{r}_1, \mathbf{r}_2, \cdots, \mathbf{r}_N)$, it is defined as (see Sect. 17.3),

$$g(\mathbf{r}, \mathbf{r}') = \frac{N(N-1)}{n(\mathbf{r})\, n(\mathbf{r}')} \times \frac{\int |\Psi(\mathbf{r}_1, \cdots, \mathbf{r}_N)|^2 \delta(\mathbf{r} - \mathbf{r}_1)\, \delta(\mathbf{r}' - \mathbf{r}_2) \cdot d\mathbf{r}_1 \cdots d\mathbf{r}_N}{\int |\Psi(\mathbf{r}_1, \cdots, \mathbf{r}_N)|^2 d\mathbf{r}_1 \cdots d\mathbf{r}_N}.$$

The delta functions in the integrand serve to fix $\mathbf{r}_1 = \mathbf{r}$ and $\mathbf{r}_2 = \mathbf{r}'$, while summing over all possible values of the remaining coordinates $\mathbf{r}_j$. Among all $\mathbf{r}_j$, there are $N(N-1)$ possible ways to choose a pair to assign to $\mathbf{r}$ and $\mathbf{r}'$. However, due to the antisymmetry of $|\Psi|^2$ as in Eq. (4.3), the value remains the same regardless of the choice. Hence, we can take the case $\mathbf{r}_1 = \mathbf{r}$ and $\mathbf{r}_2 = \mathbf{r}'$ as a representative configuration and multiply it by $N(N-1)$. Such an integral represents the probability of finding two electrons at $\mathbf{r}$ and $\mathbf{r}'$. If $\mathbf{r}$ and $\mathbf{r}'$ are sufficiently far apart, they cease to influence each other, and the joint probability should converge to the product of independent single-electron probabilities: $n(\mathbf{r}) \cdot n(\mathbf{r}')$. Therefore, dividing by $n(\mathbf{r})\, n(\mathbf{r}')$ provides a normalized function that approaches 1 in the limit of large $r_{ij} = |\mathbf{r} - \mathbf{r}'|$ (see Fig. 10.2). For the Hartree–Fock many-body wave function of a three-dimensional electron gas, the pair distribution function can be evaluated analytically [5, 6], and its typical shape is shown in Fig. 10.2 (see also Sect. 10.4.1).

By examining the pair distribution function, one can obtain an image of how other electrons distribute themselves so as to avoid a spin-up electron fixed at the origin. If the avoiding electron also has spin-up, the distribution is described

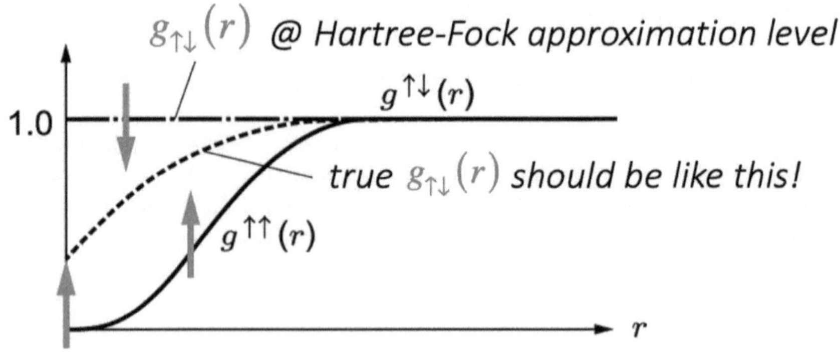

Fig. 10.2 Pair distribution function within the Hartree–Fock approximation. The initial rise in the functional form is depicted schematically in the figure; from the cusp condition (see Sect. 8.1), it follows that $g_{\uparrow\uparrow}$ [$g_{\uparrow\downarrow}$] rises quadratically [linearly]. This behavior is derived from the so-called Kimball condition

by $g^{\uparrow\uparrow}(r)$; if it has opposite spin, then $g^{\downarrow\uparrow}(r)$ captures the correlation. Since $g^{\sigma\sigma'}(r \to 0)$ corresponds to particle coalescence, the functional form of $g^{\sigma\sigma'}(r)$ is constrained by the cusp condition discussed in Sect. 8.1. As a result, it follows that $g_{\uparrow\uparrow}$ [$g_{\uparrow\downarrow}$] rises quadratically [linearly], a result known as the Kimball condition (Sect. 10.4.1).

When the avoidance occurs between electrons of the same spin, as r becomes small and another spin approaches the origin where an $\uparrow$ spin is fixed, the Pauli exclusion principle comes into effect. As a result of this avoidance, the distribution dips below 1.0 and must vanish completely at the origin. In the case of the Hartree–Fock approximation shown in Fig. 10.2, the assumed wavefunction explicitly imposes the **Pauli exclusion principle**, so the behavior of $g^{\uparrow\uparrow}(r)$ correctly satisfies this requirement. On the other hand, in the Hartree–Fock approximation, $g^{\downarrow\uparrow}(r)$ remains constantly 1.0, both near and far from the origin. This means that a $\downarrow$ spin does not perceive the presence of a fixed $\uparrow$ spin at the origin and makes no attempt to avoid it. This behavior does not reflect physical truth, but rather results from the assumption of using a Slater determinant for the many-body wavefunction, *i.e.*, the Hartree–Fock approximation, which leads to this incorrect consequence. From this, we see that the dip (or "hole") observed between same-spin electrons arises **not from Coulomb repulsion** but as a consequence of quantum statistics. If the hole were caused by Coulomb repulsion, it should also appear between opposite-spin electrons. In this sense, the hole in $g^{\uparrow\uparrow}(r)$ represented in the Hartree–Fock approximation is called the **exchange hole**. It is important to note again that the exchange hole appears even in the absence of interaction.

As emphasized above, if Coulomb repulsion is present, electrons with opposite spins should also avoid each other, resulting in a depletion (hole). This type of hole is called a **correlation hole**. Looking at $g^{\downarrow\uparrow}(r)$ in Fig. 10.2, we see that the Hartree–Fock theory does not exhibit a correlation hole. This theory implies that "only parallel-spin pairs avoid each other due to statistical effects, while antiparallel-spin pairs remain unaware of each other." The defining feature of the Hartree–Fock theory is that it does **not allow holes arising from interactions**.

It is important to note here that the Hartree–Fock theory does **not ignore interactions**.[5] The accurate description is that it uses a "many-body wave function constructed without interactions" to evaluate a "situation where interactions are included." When interactions are introduced into an originally non-interacting independent-particle system, two effects arise: "**(i)** the orbital functions forming the Slater determinant deform and relax under the influence of interactions," and "**(ii)** the functional form itself, being confined to that expressible by a Slater determinant, is no longer preserved and undergoes relaxation beyond that form." The Hartree–Fock theory gives an optimal solution, variationally minimized within a subspace of the total Hilbert space in which the many-body wave function resides, but only within the subspace accessible via effect (i) alone. When effect (ii) is included, there exists a more optimal exact solution beyond this subspace. The relaxation beyond the subspace expressible by Slater determinants corresponds to what is known as **electron correlation**.

In the Hartree–Fock theory, effect (i), that is, the individual distortion of orbital functions, is incorporated. However, if we examine how the interaction is included in the Hartree–Fock equation, Eq. (10.1), which determines such deformation, it becomes clear that the interaction is included only in a form based on the classical electrostatic field (Hartree term) arising from all electrons except the one under consideration.[6] When the variational minimization is restricted to the relaxation of individual orbitals only, the resulting interaction reflects only this **one-versus-all** mean-field-like description. While the Hartree–Fock theory can describe to some extent "how much the total energy changes due to the interaction," it fails to describe the **one-to-one avoidance behavior**, such as "how an electron placed at a certain location is consciously avoided by another," as reflected in the pair distribution function. This inadequacy is clearly illustrated in Fig. 10.2. **Electronic correlation** is defined as an effect that goes beyond the Hartree–Fock approximation. From this viewpoint, it is often interpreted symbolically as the **effect of higher-order fluctuations beyond the mean-field picture**. On the other hand, from the perspective of the mathematical structure in many-body wavefunction theory, it can be regarded as the **effect arising from the distortion of the many-body wavefunction beyond the Slater determinant form**.

To summarize the discussion so far on Hartree–Fock theory:

- **Classical effects** of interaction: represented by the Hartree term. These are effects that arise from Coulomb repulsion, not specific to quantum mechanics, and present also in classical physics.

[5] The statement that "the Hartree-Fock theory lacks electron *correlation*" is correct by definition, but occasionally one encounters the misinterpretation that "the Hartree-Fock theory does not include electron-electron *interactions*."

[6] The Fock term provides a correction.

- **Exchange effects** among quantum effects of interaction: represented by the Fock term. These are statistical avoidance effects, occurring regardless of the presence or absence of interaction.
- **Correlation effects** among quantum effects of interaction (electronic correlation): effects that cannot be incorporated within the Hartree–Fock equation. These arise when the influence of interaction leads to further relaxation beyond the space expressible by a Slater determinant.

10.1.4 Exchange-Correlation Potential

In the Hartree–Fock equation (10.1), let us write the Fock term as

$$-\sum_{j>i}\left[\int d\mathbf{r}'\psi_j^*\left(\mathbf{r}'\right)\frac{\delta_{\sigma_i\sigma_j}}{|\mathbf{r}-\mathbf{r}'|}\psi_i\left(\mathbf{r}'\right)\right]\cdot\underline{\psi_j\left(\mathbf{r}\right)}=:\hat{V}_X\cdot\underline{\psi_i\left(\mathbf{r}\right)}\ . \tag{10.2}$$

The subscript "X" denotes the "X" in e<u>x</u>change. Although the subscripts j and i in the underlined terms on the left- and right-hand sides of Eq. (10.2) differ, this is not a typographical error. Rather, it defines the operator $\hat{V}_X\cdot\psi_i\left(\mathbf{r}\right)$ formally as expressed in Eq. (10.2). For this reason, we attach the caret " ^ " to indicate that $\hat{V}_X$ is an operator. By relabeling the indices in this way, the Hartree–Fock equation can be reformulated in a clearer form as

$$\left[-\frac{1}{2}\nabla^2+v_{\mathrm{ext}}\left(\mathbf{r}\right)+\sum_{j\neq i}\int d\mathbf{r}'\frac{\left|\psi_j\left(\mathbf{r}'\right)\right|^2}{|\mathbf{r}-\mathbf{r}'|}+\hat{V}_X\right]\cdot\psi_i\left(\mathbf{r}\right)=\varepsilon_i\cdot\psi_i\left(\mathbf{r}\right)\ , \tag{10.3}$$

with improved readability. In this form, $\hat{V}_X$ can be interpreted as an "effective one-body potential representing exchange effects" (*i.e.*, exchange potential). However, this is merely a formal reinterpretation, defined as an operator as shown above. It should be noted that $\hat{V}_X$ is not a simple one-body potential in the usual sense.

The limitation of the Hartree–Fock theory lies in the fact that even when the many-body wave function is optimized within the subspace representable by a Slater determinant, correlation holes cannot be captured. In Eq. (10.3), only $\hat{V}_X$, which represents exchange effects, appears. To *symbolically* express a more accurate formulation that goes beyond the Hartree–Fock approximation and is capable of describing correlation holes, we write

$$\left[-\frac{1}{2}\nabla^2+v_{\mathrm{ext}}\left(\mathbf{r}\right)+\sum_{j\neq i}\int d\mathbf{r}'\frac{\left|\psi_j\left(\mathbf{r}'\right)\right|^2}{|\mathbf{r}-\mathbf{r}'|}+\hat{V}_X+\hat{V}_C\right]\cdot\psi_i\left(\mathbf{r}\right)=\varepsilon_i\cdot\psi_i\left(\mathbf{r}\right) \tag{10.4}$$

Here, $\hat{V}_C$ denotes the contribution from **correlation effects**, with C standing for "correlation." Alternatively, by combining $\hat{V}_X$ and $\hat{V}_C$, we can write

$$\left[-\frac{1}{2}\nabla^2 + v_{\text{ext}}(\mathbf{r}) + \sum_{j\neq i}\int d\mathbf{r}'\frac{\left|\psi_j(\mathbf{r}')\right|^2}{|\mathbf{r}-\mathbf{r}'|} + \hat{V}_{\text{XC}}\right]\cdot\psi_i(\mathbf{r}) = \varepsilon_i\cdot\psi_i(\mathbf{r})$$

where $\hat{V}_{\text{XC}}$ is referred to as the **exchange-correlation potential**.

The specific form of the exchange-correlation potential $\hat{V}_{\text{XC}}$ is unknown, and the above logical development is merely a *hopeful abstraction* that assumes the treatment beyond the Hartree–Fock approximation could be written in the form of a one-body problem for orbital functions, just like the Hartree–Fock equation. If so, then a term like $\hat{V}_{\text{XC}}$, which describes exchange and correlation effects, would be added here. The effects that cannot be captured by Hartree–Fock theory are, as discussed in the previous section, the consequences of distortions beyond the functional form of a Slater determinant. Whether such effects can truly be described by simply adding a correction potential to the framework of a one-body equation, which only allows for deformation within the constraint of the Slater determinant functional form, remains a highly *questionable assumption*. However, as will be outlined in Sect. 16.1.3, density functional theory provides a foundation for describing many-body correlation effects in the ground state exactly, using a one-body formalism of the type in Eq. (16.7) that incorporates $\hat{V}_{\text{XC}}$ (the Kohn–Sham equation). Of course, as suggested by the above concern about the validity of this approach, this result cannot be derived from a variational principle based on Slater determinant many-body wave functions alone, but rather emerges through a completely different and ingenious line of reasoning.[7]

The one-body form used to represent quantum many-body correlation in Eq. (10.4) is formally identical to an extension of the Hartree–Fock equation. However, as will be seen in the derivation in Appendix Sect. 16.1.3, the origin and interpretation of the orbital functions in that context are entirely different from those in the Hartree–Fock formalism. It should be reiterated that the introduction of the exchange-correlation potential $\hat{V}_{XC}$ in this section is a *coarse heuristic*, employed to skillfully circumvent the limitations of page count, time, or the reader's patience.

10.1.5 Non-local Nature of the Fock Term

The Hartree–Fock method may be regarded as the first approximation in dealing with electron–electron interactions. This may lead some to view it as an "overly simplistic approximation" (and thus computationally efficient?). On the other hand,

[7] In density functional theory as well, the explicit form of $\hat{V}_{\text{XC}}$ is unknown, and only its existence is proven.

among domain experts, one may also encounter expressions such as "Hartree-Fock treats exchange *seriously*" or "Hartree-Fock properly accounts for self-interaction cancellation" or even "if Hartree-Fock is done properly, one must deal with nonlocal terms, so it becomes computationally *expensive*." Such descriptions may give an opposite impression, that of a "rigorous treatment."

The term "nonlocal evaluation" refers to the situation in the Fock term of Eq. (10.2), where determining the value at a point $\mathbf{r}$ requires integration over the entire space with respect to $\mathbf{r}'$. That is, the value at a given point depends on the global spatial configuration. This is what is meant by the statement: "evaluating exchange effects requires **nonlocal evaluation**." The Hartree–Fock method entirely neglects correlation ($V_C = 0$), but provides a "rigorous" evaluation of exchange through proper nonlocal integration. In this sense, it can be said to handle exchange "seriously."

In contrast, practical implementations of DFT, which are widely used for real-world many-electron systems, include approximations for the correlation term ($V_C \neq 0$), but sometimes take shortcuts by localizing or simplifying the nonlocal exchange integral. In contrast to molecular quantum chemistry, Hartree–Fock calculations are rarely mentioned in *ab initio* calculations for solid-state (periodic) systems. This is because, in order to evaluate the exchange term properly, one must perform nonlocal integration over the entire space. While this may be tractable for isolated molecular systems with bounded spatial extent, it becomes computationally prohibitive in periodic systems where space extends indefinitely.

10.1.6 Balance of Exchange, Correlation, and Classical Effects

The interaction term in the Hartree–Fock equation (10.1), namely the sum of the Hartree and Fock terms,

$$
\begin{aligned}
V_{\text{int}}^{\text{HF}} &= \sum_{j\neq i} \int d\mathbf{r}' \frac{\left|\psi_j\left(\mathbf{r}'\right)\right|^2}{\left|\mathbf{r}-\mathbf{r}'\right|} \cdot \psi_i\left(\mathbf{r}\right) \\
&\quad -\sum_{j\neq i} \int d\mathbf{r}' \frac{\delta_{\sigma_i\sigma_j}\psi_j^*\left(\mathbf{r}'\right)\psi_i\left(\mathbf{r}'\right)}{\left|\mathbf{r}-\mathbf{r}'\right|} \cdot \psi_j\left(\mathbf{r}\right)
\end{aligned}
$$

contains the condition $j \neq i$, which expresses the natural physical requirement that electron i does not interact with itself. However, even if the $j = i$ term is included in both the Hartree and Fock terms for algorithmic convenience, these extra terms *cancel out* due to the opposite signs of the Hartree and Fock contributions. Thus, it is valid to write:

$$
V_{\text{int}}^{\text{HF}} = \sum_{j} \int d\mathbf{r}' \frac{\left|\psi_j\left(\mathbf{r}'\right)\right|^2}{\left|\mathbf{r}-\mathbf{r}'\right|} \cdot \psi_i\left(\mathbf{r}\right)
$$

$$
\begin{aligned}
& -\sum_{j} \int d\mathbf{r}' \frac{\delta_{\sigma_i \sigma_j} \psi_j^* (\mathbf{r}') \psi_i (\mathbf{r}')}{|\mathbf{r} - \mathbf{r}'|} \cdot \psi_j (\mathbf{r}) \\
& = \sum_{j} \int d\mathbf{r}' \frac{\rho (\mathbf{r}')}{|\mathbf{r} - \mathbf{r}'|} \cdot \psi_i (\mathbf{r}) \\
& -\sum_{j} \int d\mathbf{r}' \frac{\delta_{\sigma_i \sigma_j} \psi_j^* (\mathbf{r}') \psi_i (\mathbf{r}')}{|\mathbf{r} - \mathbf{r}'|} \cdot \psi_j (\mathbf{r})
\end{aligned}
$$

as a matter of computational convenience. By doing so, the Hartree term is expressed as a functional of the charge density $\rho(\mathbf{r})$, making it compatible with the framework of density functional theory. However, this formulation is only valid under the assumption that the Fock term is evaluated exactly as a non-local potential. In practical implementations of DFT, such as the local density approximation (LDA), the contribution of the Fock term is localized approximately as a function of the charge density $\rho(\mathbf{r})$. In this case, the above-mentioned cancellation of the $j = i$ term is incomplete, and as a result, electron i ends up contributing an **artifact of interacting with itself**. This is referred to as **the breakdown of self-interaction cancellation**. A well-known issue arising from the breakdown of self-interaction cancellation is the underestimation of the energy gap in solids when using DFT. It is important to note that this is a problem of exchange, not of correlation. These matters are discussed in more detail separately in Appendix Sect. 16.2.

The issue of the breakdown of self-interaction cancellation originates from the imbalance between the Hartree term (classical contribution) and the Fock term (correlation contribution). It is also known that various inconsistencies and artifacts can arise if the balance between exchange and correlation is not evaluated in a consistent manner.[8] In this sense, one must also be cautious about assuming that exchange effects evaluated faithfully in Hartree–Fock theory, which completely neglects correlation, can be taken as the true exchange. In an exact solution, once correlation is taken into account, the exchange must also be modified in a corresponding and consistent manner [6].

10.2 Overview of Representative Methods for Incorporating Correlation Effects

In this section, we provide a conceptual overview of the minimum necessary framework to understand the comparative positioning of the post-Hartree–Fock methods in the molecular orbital approach, density functional theory, and many-body perturbation theory, with respect to the *ab initio* QMC. The part on many-body

[8] This point is discussed in Sect. 10.5.1.

perturbation theory is somewhat lengthy, but this is due to the necessity of establishing a clear correspondence with it, given the historical development of many-body electronic theory largely under the banner of many-body perturbation theory.

10.2.1 *Molecular Orbital Method Incorporating Excited Orbitals*

In the Hartree–Fock approximation to the many-body eigenvalue problem

$$\hat{H} \cdot \Psi\left(\mathbf{r}_1, \cdots, \mathbf{r}_N\right) = E \cdot \Psi\left(\mathbf{r}_1, \cdots, \mathbf{r}_N\right)$$

a separable form assuming no interaction was postulated for the many-body wavefunction. By assuming a wavefunction $\Psi\left(\mathbf{r}_1, \cdots, \mathbf{r}_N\right)$ in the form of an N-body product constructed from orbital functions $\{\psi_k\left(\mathbf{r}\right)\}$ and substituting this into the equation, we obtain a one-body equation of the form

$$\hat{h} \cdot \psi_k\left(\mathbf{r}\right) = \varepsilon_k \cdot \psi_k\left(\mathbf{r}\right) \tag{10.5}$$

which corresponds to the Hartree–Fock equation. The lowest many-body eigenstate corresponds to the state in which the one-body eigenvalues $\{\varepsilon_k\}_{k=0}^{\infty}$ are taken for the N lowest levels in ascending order [7].

Let us write down M eigenfunctions of Eq. (10.5) as

$$\{\psi_1, \cdots, \psi_N, \psi_{N+1}, \cdots, \psi_M\}$$

with $M > N$. The Hartree–Fock method corresponds to the variational principle using as the trial wavefunction the Slater determinant constructed from the N lowest occupied orbitals,

$$\Psi\left(\mathbf{r}_1, \cdots, \mathbf{r}_N\right) = \det\left\{\psi_1, \cdots, \psi_N\right\}.$$

The expressive limitation of this approach lies, therefore, in the fact that it is constrained to the space expressible by a single Slater determinant. A natural extension beyond this limitation is to employ the variational principle with trial wavefunctions that go beyond a single Slater determinant.

In elementary courses on quantum mechanics, perturbation theory [8] teaches that the introduction of interactions opens channels between the ground and excited states, and through virtual transitions permitted by the uncertainty principle, further lowers the energy (*i.e.*, deforms the non-interacting ground state). In the Hartree–Fock method, the trial wavefunction for the many-body system is assumed to take a form valid for the non-interacting case. When interactions are present, the variational freedom associated with excited orbitals ($\psi_{N+1} \sim \psi_M$), which become involved through interactions, should also contribute. Following this line of reasoning, the molecular orbital method employs a more general many-body wavefunction constructed from molecular orbitals as

$$\Psi^{\mathrm{MO}}\left(\mathbf{r}_1, \cdots, \mathbf{r}_N\right) = \hat{\mathcal{A}}\left[\left\{\psi_1, \cdots, \psi_N, \psi_{N+1}, \cdots, \psi_M\right\}\right], \tag{10.6}$$

where $\hat{\mathcal{A}}$ denotes anti-symmetrization. This corresponds to a variational framework in which the energy functional

$$E_{\mathrm{est}} = \frac{\int d\mathbf{R} \cdot \Psi_T^*(\mathbf{R}) \cdot \hat{H}\Psi_T(\mathbf{R})}{\int d\mathbf{R} \cdot \left|\Psi_T(\mathbf{R})\right|^2} \tag{10.7}$$

is minimized using a trial function expressed in terms of molecular orbitals. $\hat{\mathcal{A}}$ denotes the operation of anti-symmetrization, which is introduced to ensure the anti-symmetric constraint given in Eq. (4.3):

$$\Psi_\alpha\left(\cdots, \mathbf{r}_j, \cdots, \mathbf{r}_i, \cdots\right) = (-) \cdot \Psi_\alpha\left(\cdots, \mathbf{r}_i, \cdots, \mathbf{r}_j, \cdots\right),$$

which must be satisfied for fermionic wavefunctions.

As the most elementary and well-known implementation of Eq. (10.6), we have the Configuration Interaction method (CI method):

$$\begin{aligned}\Psi^{\mathrm{CI}}\left(\mathbf{r}_1, \cdots, \mathbf{r}_N\right) = &\ \text{det(Prepare } N \text{ orbitals using only occupied orbitals)} \\ &+ c_1 \cdot \text{det(Replace one occupied orbital with an excited orbital)} \\ &+ c_2 \cdot \text{det(Replace two occupied orbitals with excited orbitals)} \\ &+ c_3 \cdot \text{det(Replace three occupied orbitals with excited orbitals)} \\ &+ \cdots .\end{aligned}$$

The second and subsequent terms in this expression represent "many-body excited states" expressed using excited orbitals,[9] and transitions between them and the first term occur via the interaction V. Such a sum of determinants is called a **multi-determinant**, and it is an extension of a Slater determinant (a single determinant) that preserves antisymmetry with respect to particle exchange. In the second and subsequent terms, such as "replacing only one occupied orbital with an excited orbital," there are various combinations possible depending on "which occupied orbital is replaced with which excited orbital." In general, this becomes a sum over all such possible combinations.[10] In this way, the "canvas (Hilbert space) on which

[9] These are referred to as virtual orbitals.

[10] In the case of small molecules, it is sometimes possible to take all possible combinations, and this is referred to as Full CI.

the wavefunction is drawn" is extended, and by using it as a trial function for the many-body wavefunction, the variational functional in Eq. (10.7) is minimized. This framework is called the configuration interaction (CI) method.

A serious drawback of the configuration interaction (CI) method is known: it violates size consistency.[11] To address this issue, the coupled cluster (CC) method has been developed and widely used. However, it is known that in the CC method, the variational principle no longer holds.

In the CI method, what are optimized according to the variational principle are, first and foremost, only the expansion coefficients of the multi-determinant. One may further choose whether to proceed to optimize the orbital functions as well. There also exist implementations that derive and solve equations to determine the orbital functions themselves, analogous to the Hartree–Fock theory, applied to the multi-determinant case. These include the MCSCF method (Multi Configuration SCF) and various related approaches [1].

10.2.2 Density Functional Theory

In the density functional theory (DFT), the foundational idea is the viewpoint that the fundamental variable describing the ground state of an interacting many-electron system is the charge density distribution $n(\mathbf{r})$.[12] This $n(\mathbf{r})$ is obtained by self-consistently solving the orbital equation in the form of a one-body problem, shown in Eq. (10.4) in Sect. 10.1.4,

$$\left[-\frac{1}{2}\nabla^2 + v_{\text{ext}}(\mathbf{r}) + \sum_{j\neq i}\int d\mathbf{r}'\frac{\left|\psi_j(\mathbf{r}')\right|^2}{|\mathbf{r}-\mathbf{r}'|} + \hat{V}_{\text{XC}}\right]\cdot\psi_i(\mathbf{r}) = \varepsilon_i\cdot\psi_i(\mathbf{r}) \tag{10.8}$$

and then defining $n(\mathbf{r}) = \sum_j \left|\psi_j(\mathbf{r})\right|^2$, from which the ground-state energy is obtained using the one-body eigenvalues $\left\{\varepsilon_j\right\}$ (see Appendix Chap. 16). This form

11 One of the important missions of electronic structure calculations is to evaluate various binding energy values, $\Delta E = E_{\text{A+B}} - (E_{\text{A}} + E_{\text{B}})$. In doing so, it is crucial that the combined system $E_{\text{A+B}}$, which is roughly twice the size of the individual systems, is evaluated with "comparable accuracy" to that of the separate systems E_{A} and E_{B}. This requirement is called size consistency. In many cases, the descriptive power for a doubled-size molecular system is inferior to that for the individual systems, which results in underestimation of binding energies (*i.e.*, size-inconsistent evaluation).

12 From the perspective of *ab initio* calculations, the ground state of a many-body system is determined by the geometry of the ions. According to the cusp condition equation described in Sect. 8.1.2, the charge density also exhibits cusps at the ionic positions, and the sharpness of these cusps reflects the atomic number Z of the ionic cores at those positions. If so, it is possible to capture the ionic geometry from the information of the charge density, and therefore, it is discerned that the ground state can be described solely from the charge density information.

closely resembles a *one-body approximation based on a mean-field*, but *the logical path to this conclusion is entirely different*, and the orbital functions and one-body eigenvalues are introduced only as mathematical conveniences. Details are given in Appendix Chap. 16; here, we extract only the conceptual outline leading to this conclusion.

Consider a system with an electron density $n(\mathbf{r})$, subjected to an external electrostatic potential $v_{\text{ext}}(\mathbf{r})$. The ground-state energy of the system, E_{GS}, can be written as

$$\begin{aligned} E_{\text{GS}} &= (\text{energy intrinsic to the interacting electron system}) \\ &\quad + \int d\mathbf{r} \cdot v_{\text{ext}}(\mathbf{r}) \cdot n(\mathbf{r}). \end{aligned}$$

If we denote the term "energy intrinsic to the interacting electron system" as F, then since this is determined by the many-body wave function $\Psi(\mathbf{r}_1, \cdots, \mathbf{r}_N)$, which is the fundamental quantity of the many-electron system, we can write:

$$\begin{aligned} E_{\text{GS}} & \left[\Psi_{\text{interacting}}(\mathbf{r}_1, \cdots, \mathbf{r}_N)\right] \\ &= F\left[\Psi_{\text{interacting}}(\mathbf{r}_1, \cdots, \mathbf{r}_N)\right] + \int d\mathbf{r} \cdot v_{\text{ext}}(\mathbf{r}) \cdot n(\mathbf{r}). \end{aligned} \tag{10.9}$$

For later convenience, we have attached the subscript "interacting" to the many-body wave function to indicate that it corresponds to an interacting system.

For this setting, the following foundational statements are established:

(1) F in Eq. (10.9) can be expressed as a functional $F[n(\mathbf{r})]$ of the charge density $n(\mathbf{r})$ (such an expression exists).[13]
(2) The dependence of $F[n(\mathbf{r})]$ on $n(\mathbf{r})$ is universal, independent of the specific system, as long as the type of interaction is the same.
(3) The exact ground-state energy of the system is given by minimizing

$$E_{\text{GS}}[n(\mathbf{r})] = F[n(\mathbf{r})] + \int d\mathbf{r} \cdot v_{\text{ext}}(\mathbf{r}) \cdot n(\mathbf{r})$$

with respect to $n(\mathbf{r})$.

Based on the above, Eq. (10.9) becomes

$$\begin{aligned} E_{\text{GS}}\left[\Psi_{\text{interacting}}(\mathbf{r}_1, \cdots, \mathbf{r}_N)\right] &= F[n(\mathbf{r})] + \int d\mathbf{r} \cdot v_{\text{ext}}(\mathbf{r}) \cdot n(\mathbf{r}) \\ &= E_{\text{GS}}[n(\mathbf{r})] \end{aligned}$$

[13] While it is standard to write functionals as $F[n]$, we intentionally write $F[n(\mathbf{r})]$ here. See the footnote in Appendix Chap. 16 for details.

and hence, the energy can ultimately be described solely in terms of the density $n(\mathbf{r})$. In other words, the theory asserts that the ground-state energy can be understood using a real-valued function $n(\mathbf{r})$ in 3D space, instead of the $3N$-dimensional complex-valued function $\Psi(\mathbf{r}_1, \cdots, \mathbf{r}_N)$. While descriptions in quantum mechanics typically emphasize the fundamental nature of complex functions and phase degrees of freedom, this result asserts that, for the ground-state energy of interacting electron systems, it is in fact sufficient to express it using a real function. Such a claim, which might have seemed unexpected before the 1960s, has been rigorously established through mathematical arguments using reductio ad absurdum and other techniques. It is now known as **density functional theory** [7].

The exchange-correlation potential $\hat{V}_{\mathrm{XC}}$ appearing in Eq. (10.8) can be grounded in the universal functional $F[n(\mathbf{r})]$ introduced above as follows: We adopt the convention of expressing $F[n(\mathbf{r})]$ in the form

$$F[n(\mathbf{r})] = T_{\mathrm{S}} + E_{\mathrm{H}} + E_{\mathrm{XC}}[n(\mathbf{r})]$$

which is "partitioned" in a manner similar to the Hartree–Fock formalism.[14] E_{H} denotes the Hartree term, and T_{S} is the "formal" kinetic energy defined with respect to the orbital functions in Eq. (10.8)—that is, while it takes the form of a kinetic energy term, it does not carry physical meaning as such. The exchange-correlation functional E_{XC} is introduced as a residual term, defined as the remainder of F after subtracting those parts whose character is well understood. Under this formulation, the minimization with respect to $n(\mathbf{r})$ in item (3) above leads to the requirement that Eq. (10.8) be solved (Appendix Chap. 16). The exchange-correlation potential $\hat{V}_{\mathrm{XC}}$ appearing therein is given by

$$\hat{V}_{\mathrm{XC}}[n(\mathbf{r})] = \frac{\delta E_{\mathrm{XC}}[n(\mathbf{r})]}{\delta n(\mathbf{r})} .$$

Equation (10.8), known as the Kohn–Sham equation, is not a relation obtained through approximation as in the Hartree–Fock theory, but rather is derived via a mathematically exact formal procedure. However, it does not provide an explicit form of the exchange-correlation functional $E_{\mathrm{XC}}[n]$; instead, it serves as a kind of existence theorem for the exact $V_{\mathrm{XC}}[n]$ that describes the true solution. This provides a guarantee that efforts to refine the exchange-correlation potential $V_{\mathrm{XC}}[n]$, in the pursuit of going beyond approximate solutions and approaching the exact solution, are not in vain. Thus, DFT constitutes an approach that seeks to describe correlation effects through the refinement of the exchange-correlation potential in a single-particle formalism.

[14] This approach is known as the Kohn–Sham scheme.

10.2.3 Overview of Methods Based on Many-Body Perturbation Theory

In DFT, a clever approach is taken in which the problem is mapped onto an "equivalent single-particle form" that exactly reproduces the ground state. Specifically, this involves the "Kohn-Sham orbitals and the associated exchange-correlation functional." In contrast, many-body perturbation theory adopts an approach based on the picture of "quasiparticles incorporating many-body interactions."

In the quasiparticle approach, the quantum many-body problem is treated within the framework of **quantum field theory**. When dealing with many-electron systems, the wavefunctions used in perturbation theory must satisfy the antisymmetric constraint given in Eq. (4.3). In Hartree–Fock theory, this condition was enforced by expressing the wavefunction as a linear combination of antisymmetrized products of orbital functions. However, if one explicitly represents the many-body wavefunction, perturbative expansions must be carried out using such extended function forms, making the derivation cumbersome. For many-electron systems, therefore, perturbative expansion is performed in the second-quantized formalism, in which the algebraic rules are designed to embed the antisymmetric constraint equivalently. Key concepts of second quantization are summarized in Appendix Sect. 17.1.

In quantum field theory, many-body interactions between particles are incorporated into a quantum field defined over three-dimensional space, and the creation of a particle is treated through an operator acting on this field. The properties imparted by the creation operator in the presence of many-body interactions are interpreted as the properties of quasiparticles—this approach is referred to as the quasiparticle method. When asking what specific quantities are handled to elaborate on the idea that "(properties imparted by the creation operator) = (properties of quasiparticles reflecting interactions)," the orthodox framework of many-body perturbation theory treats the following quantity, known as the **Green's function** or **propagator**:[15]

$$G\left(\mathbf{r}, t; \mathbf{r}', t'\right) := -i \left\langle T_t \left[\hat{\psi}\left(\mathbf{r}, t\right) \hat{\psi}^{\dagger}\left(\mathbf{r}', t'\right) \right] \right\rangle . \tag{10.10}$$

When this quantity is treated via Fourier transformation, the configuration of poles in the complex plane becomes the object of consideration. In the non-interacting case, these poles are shown to yield the **single-particle energy levels** (Appendix Chap. 18). When the interacting Green function (10.10) is perturbatively evaluated as

$$G_{\text{int}}\left(z\right) \sim \sum_k \frac{\rho_k}{z - \tilde{\varepsilon}_k},$$

as in Eq. (18.1), the poles

[15] Why this quantity is referred to as the Green's function, and in what context it was introduced, is explained starting from Sect. 17.2, with Sect. 17.2.6 as the goal.

$$\tilde{\varepsilon}_k = \varepsilon_k + \Sigma_k \tag{10.11}$$

are regarded as quantities that reduce to the single-particle levels in the limit of vanishing interaction. These poles can be interpreted as **energy levels of quasiparticles**, which are generalizations of one-electron levels in the presence of many-body interactions. Σ_k is called the **self-energy**; its real part gives the energy-level shift, while the imaginary part determines the lifetime of the quasiparticle. ρ_k, which is unity in the non-interacting case, acquires the meaning of quasiparticle weight under interaction.

In the context of electronic correlation analysis, discussions are more commonly carried out using **pair distribution functions** rather than quasiparticle spectra. The pair distribution function is proportional to the product of charge densities,

$$g\left(\mathbf{r},\mathbf{r}'\right) \propto \left\langle \hat{\rho}\left(\mathbf{r}\right)\hat{\rho}\left(\mathbf{r}'\right)\right\rangle - \delta\left(\mathbf{r}'-\mathbf{r}_j\right)\left\langle \hat{\rho}\left(\mathbf{r}\right)\right\rangle , \tag{10.12}$$

as discussed in (Sect. 17.3). Since interparticle interactions are governed by the product of particle densities, they can be expressed in terms of the pair distribution function or its Fourier transform, the structure factor,

$$S^{\sigma\sigma'}\left(\mathbf{q}\right) := \frac{1}{N}\left\langle \rho_{-q\sigma}\cdot\rho_{q\sigma'}\right\rangle .$$

In fact, a relationship in terms of a functional derivative with respect to the energy change ΔE due to the introduction of interparticle interaction is given by

$$\frac{\delta\Delta E}{\delta v\left(q\right)} = \frac{1}{2}\left[n\cdot S^{\sigma\sigma'}\left(\mathbf{q}\right) - \delta_{\sigma\sigma'}\cdot n^{\sigma}\right], \tag{10.13}$$

as derived in Appendix Chap. 18.

In the analysis of electronic correlation, the main focus is on **how the shape of the pair distribution function changes** upon the introduction of interparticle interactions. This is handled by computing the right-hand side from the left-hand side in the above equation. The left-hand side, ΔE, is obtained by evaluating the thermodynamic potential Ω from the partition function $\Xi = \exp\left(-\beta\Omega\right)$ using the perturbative-diagrammatic method. If one can determine how the partition function Ξ changes under the introduction of interparticle interactions via diagrams, then the change in the thermodynamic potential, $\Delta\Omega = \Delta E$, can be derived. After analytically expressing this variation ΔE in terms of diagrams, one performs the functional derivative in the above equation to evaluate the structure factor $S^{\sigma\sigma'}\left(\mathbf{q}\right)$. Taking the Fourier transform of this quantity reveals how the introduction of interparticle interactions modifies the shape of the pair distribution function (Sect. 10.5.1).

10.3 Treating Interaction Effects Using Diagrams

In Sect. 10.2.3, we outlined the general framework of many-body perturbation theory, wherein interaction effects are estimated by examining the change in Green functions and thermodynamic potentials due to the inclusion of interactions as perturbation processes. This idea is implemented by representing perturbative contributions from interactions using diagrams called **Feynman diagrams**, and summing over diagrams corresponding to each underlying process. In this section, we provide an overview of this procedure.

Why Explaining Many-Body Perturbation Theory?
As noted in the column in Sect. 8.2.2, there is a historical background in which "many-body electronic theory" has been developed primarily through many-body perturbation theory. If one does not properly understand the terminology and concepts of perturbation theory, it becomes difficult to conduct discussions in the field. It is not uncommon to be asked slightly off-the-mark questions such as: "What kind of diagrams does your *ab initio* QMC theory use?" or "What sort of truncation approximation does your method assume?" Even when one explains that "it is a variational method, not a perturbative one," some people cannot separate themselves from the mindset of "quantum many-body theory equals many-body perturbation theory," and fail to grasp the parallel positioning of perturbative and variational methods. This has often made it difficult to engage in an open-minded conversation.

10.3.1 Diagrammatic Representation of Many-Body Perturbation

Whether for the Green function or for the thermodynamic potential,[16] these quantities are expressed in terms of the field operator

$$\hat{\psi}(\mathbf{r}, t) = e^{i\hat{H}^{(2)}t/\hbar} \cdot \hat{\psi}(\mathbf{r}) \cdot e^{-i\hat{H}^{(2)}t/\hbar} \tag{10.14}$$

[16] Since the Green function can be computed from the thermodynamic potential (Sect. 18.1.2), it is possible to unify the discussion in a textbook-style manner by simply stating, "Let us evaluate the thermodynamic potential" [5]. However, for beginners, it is often difficult to see "where the argument is heading." In order to clarify the goal, namely, what is to be evaluated in the quasiparticle picture, we take a pedagogical approach that starts from the Green function.

where the time dependence in the Heisenberg representation reflects the Hamiltonian.[17] Moreover, the average value expressed by the notation $\left\langle \hat{A}(t) \right\rangle$, which appears in the formulas defining the Green function and the thermodynamic potential, is a statistical mechanical thermal average defined as

$$\left\langle \hat{A} \right\rangle = \frac{1}{\Xi}\mathrm{Tr}\left[\exp\left[-\beta\hat{H}^{(2)}\right]\cdot\hat{A}\right] \quad , \quad \Xi = \mathrm{Tr}\left[\exp\left[-\beta\hat{H}^{(2)}\right]\right] . \tag{10.15}$$

Here too, the Hamiltonian is incorporated in the formulation. These Hamiltonian terms are treated as a perturbation with respect to the interaction $\hat{H}_I$,[18] as in

$$\hat{H} = \hat{H}_0 + \hat{H}_I$$

and the explicit implementation of many-body perturbation theory consists of evaluating the effect of the variation induced by $\hat{H}_I$. Let us now take a closer look at this process.

For the time dependence in Eq. (10.14), we will, for the sake of simplicity (for the time being),[19] treat it using imaginary time $\tau = it/\hbar$, so that

$$\hat{\psi}(\mathbf{r},\tau) = e^{\tau\hat{H}}\cdot\hat{\psi}(\mathbf{r})\cdot e^{-\tau\hat{H}} \tag{10.16}$$

and proceed with manipulations expressed in terms of τ. By treating the Hamiltonian perturbatively, the time (imaginary time) evolution of the field operator becomes

$$\begin{aligned}\hat{\psi}(\mathbf{r},\tau) &= e^{\tau\hat{H}}\hat{\psi}(\mathbf{r})\,e^{-\tau\hat{H}} \\ &= e^{\tau\hat{H}}\cdot e^{-\tau\hat{H}_0}e^{\tau\hat{H}_0}\cdot\hat{\psi}(\mathbf{r})\cdot e^{-\tau\hat{H}_0}e^{\tau\hat{H}_0}\cdot e^{-\tau\hat{H}} \\ &= e^{\tau\hat{H}}\cdot e^{-\tau\hat{H}_0}\left[e^{\tau\hat{H}_0}\cdot\hat{\psi}(\mathbf{r})\cdot e^{-\tau\hat{H}_0}\right]e^{\tau\hat{H}_0}\cdot e^{-\tau\hat{H}} \\ &= \hat{U}^{\dagger}(\tau)\cdot\hat{\psi}_0(\mathbf{r},\tau)\cdot\hat{U}(\tau)\ .\end{aligned} \tag{10.17}$$

In this way, using the unperturbed field operator $\hat{\psi}_0(\mathbf{r},\tau)$ as the reference, one can develop the perturbative expansion that accounts for changes due to the interaction $\hat{H}_I$.[20] Likewise, the statistical mechanical average defined in (10.15) can be re-

[17] For field operators, see Sect. 17.1. For the notation with superscripts such as "(2)," see the explanation in Sect. 17.1.2.

[18] From this point onward, since only the second-quantized form of the Hamiltonian is relevant, the superscript (2) on $\hat{H}^{(2)}$ will be omitted.

[19] The relation to the thermal Green function is explained in Sect. 17.2.7.

[20] The time dependence denoted with the subscript "0," such as $\hat{\psi}_0(\mathbf{r},\tau)$, corresponds to what might be more aptly called the "non-interacting representation in the interacting system." However, it is traditionally referred to by the somewhat misleading name "interaction representation." In this terminology, the time dependence in the Heisenberg representation is governed by the full Hamiltonian $\hat{H}$, while that in the interaction representation is governed by the unperturbed

expressed as

$$\langle A \rangle = \frac{\mathrm{Tr}\left[e^{-\beta \hat{H}} \cdot \hat{A}\right]}{\mathrm{Tr}\left[e^{-\beta \hat{H}}\right]} = \frac{\mathrm{Tr}\left[e^{-\beta \hat{H}_0} \cdot e^{\beta \hat{H}_0} \cdot e^{-\beta \hat{H}} \cdot \hat{A}\right]}{\mathrm{Tr}\left[e^{-\beta \hat{H}_0} \cdot e^{\beta \hat{H}_0} \cdot e^{-\beta \hat{H}}\right]} = \frac{\mathrm{Tr}\left[e^{-\beta \hat{H}_0} \cdot U(\beta) \cdot \hat{A}\right]}{\mathrm{Tr}\left[U(\beta)\right]}$$
$$= \frac{\Xi_0^{-1} \cdot \mathrm{Tr}\left[e^{-\beta \hat{H}_0} \cdot U(\beta) \cdot \hat{A}\right]}{\Xi_0^{-1} \cdot \mathrm{Tr}\left[U(\beta)\right]} = \frac{\left\langle U(\beta) \cdot \hat{A}\right\rangle_0}{\langle U(\beta)\rangle_0} \tag{10.18}$$

and thus reduced to an evaluation using the non-interacting average $\langle \cdots \rangle_0$.

In the rewriting to non-interacting evaluation, the operator

$$\hat{U}(\tau) = e^{\tau \hat{H}_0} \cdot e^{-\tau \hat{H}} \tag{10.19}$$

emerges. Its time evolution can be written as

$$\frac{\partial \hat{U}(\tau)}{\partial \tau} = e^{\tau \hat{H}_0}\left(\hat{H}_0 - \hat{H}\right)e^{-\tau \hat{H}} = e^{\tau \hat{H}_0}\hat{H}_I e^{-\tau \hat{H}}$$
$$= e^{\tau \hat{H}_0} e^{-\tau \hat{H}} \cdot e^{\tau \hat{H}}\hat{H}_I e^{-\tau \hat{H}} = \hat{U}(\tau)\hat{H}_I(\tau)$$

From this differential form, using iterative substitution, $\hat{U}(\tau)$ can be solved as

$$\hat{U}(\tau) = \sum_{n=0}^{\infty} \frac{(-)^n}{n!} \int_0^{\tau} \cdots \int_0^{\tau} d\tau_1 \cdots d\tau_n \cdot T_\tau \left[\hat{H}_I(\tau_1) \cdots \hat{H}_I(\tau_n)\right] \tag{10.20}$$

as shown in [5, 9–11]. By substituting this into Eq. (10.17) and also using Eq. (10.18), the Green function, (10.10), can be evaluated as

$$G\left(\mathbf{r}, \tau; \mathbf{r}', \tau'\right) \sim \left\langle T_\tau \left[\hat{\psi}^\dagger(\mathbf{r}, \tau)\, \hat{\psi}\left(\mathbf{r}', \tau'\right)\right]\right\rangle$$
$$= \left\langle T_\tau \left[U(\tau)\, \hat{\psi}_0^\dagger(\mathbf{r}, \tau)\, U^\dagger(\tau) \cdot U^\dagger\left(\tau'\right) \hat{\psi}_0\left(\mathbf{r}', \tau'\right) U\left(\tau'\right)\right]\right\rangle$$
$$= \left\langle \sum_{n=0}^{\infty} \frac{(-)^n}{n!} \int \cdots \int d\tau_1 \cdots d\tau_n \right.$$
$$\left. \times\, T_\tau \left[\hat{\psi}_0(\mathbf{r}, \tau)\, \hat{\psi}_0^\dagger\left(\mathbf{r}', \tau'\right) \hat{H}_I(\tau_1) \cdots \hat{H}_I(\tau_n)\right]\right\rangle_0 \tag{10.21}$$

as derived in [5, 9–11].

The interaction $\hat{H}_I$ in second quantization takes the form (Eq. 17.16),

---

Hamiltonian $\hat{H}_0$. In this text, we adopt the subscript "0" notation, though it is more common in literature to use "I" (for "interaction"). This usage can be confusing for beginners, as it suggests dependence on the perturbation part H_I, despite being based on H_0.

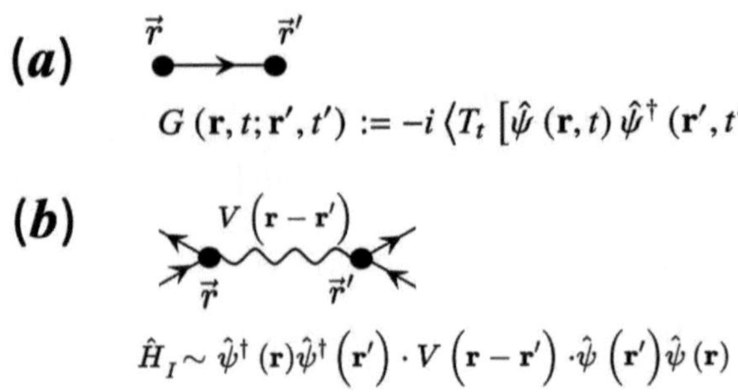

Fig. 10.3 The Green function (propagator) $G(\mathbf{r}, t; \mathbf{r}', t')$ is interpreted as "annihilating a particle at $\mathbf{r}'$ by $\hat{\psi}(\mathbf{r}')$ and creating one at $\mathbf{r}$ by $\hat{\psi}^{\dagger}(\mathbf{r})$." Accordingly, it is represented diagrammatically as an arrow pointing from $\mathbf{r}'$ to $\mathbf{r}$, as illustrated in panel (**a**). The interaction term in Eq. (10.22), involving $V(\mathbf{r}-\mathbf{r}')$, is drawn using a wavy line connecting $\mathbf{r}$ and $\mathbf{r}'$. The operator $\hat{\psi}(\mathbf{r})$, which annihilates a particle at point $\mathbf{r}$, is represented as an incoming arrow, while $\hat{\psi}^{\dagger}(\mathbf{r})$, which creates a particle at $\mathbf{r}$, is drawn as an outgoing arrow. Together, these graphical elements are shown in panel (**b**)

$$\hat{H}_I = \frac{1}{2}\int d^3x d^3y \cdot \hat{\psi}^{\dagger}(\mathbf{r})\,\hat{\psi}^{\dagger}(\mathbf{r}')\,V(\mathbf{r}-\mathbf{r}')\,\hat{\psi}(\mathbf{r}')\,\hat{\psi}(\mathbf{r})\ . \qquad (10.22)$$

This interaction is diagrammatically represented as shown in Fig. 10.3b. Meanwhile, the propagator (Green function)

$$G(\mathbf{r}, t; \mathbf{r}', t') := -i\left\langle T_t\left[\hat{\psi}(\mathbf{r}, t)\,\hat{\psi}^{\dagger}(\mathbf{r}', t')\right]\right\rangle$$

can be interpreted as annihilating a particle at $\mathbf{r}$ by $\hat{\psi}(\mathbf{r})$ and creating one at $\mathbf{r}'$ by $\hat{\psi}^{\dagger}(\mathbf{r}')$. It is represented by a directed arrow from $\mathbf{r}'$ to $\mathbf{r}$, as depicted in Fig. 10.3a.

By substituting Eq. (17.29) into the general term of the Green function expanded perturbatively with respect to the interaction $\hat{H}_I$ in Eq. (10.21),

$$G(\mathbf{r}, \tau; \mathbf{r}', \tau') \sim \left\langle T_\tau\left[\hat{\psi}_0(\mathbf{r}, \tau)\,\hat{\psi}_0^{\dagger}(\mathbf{r}', \tau')\,\hat{H}_I(\tau_1)\cdots\hat{H}_I(\tau_n)\right]\right\rangle_0 ,$$

we obtain an expression where the thermal expectation value $\langle\cdots\rangle_0$ involves a sequence of creation and annihilation operators:[21]

$$\left\langle\cdots\left(\psi_3^{\dagger}\psi_4^{\dagger}V_{34}\psi_4\psi_3\right)\left(\psi_1^{\dagger}\psi_2^{\dagger}V_{12}\psi_2\psi_1\right)\cdots\psi_r^{\dagger}\psi_{r'}\right\rangle_0 .$$

---

[21] The subscripts such as ψ_k correspond to dummy variables $\mathbf{r}_k$ that appear in the integrals as arguments. The subscript "0," previously used to indicate the non-interacting representation of field operators, is omitted from here on.

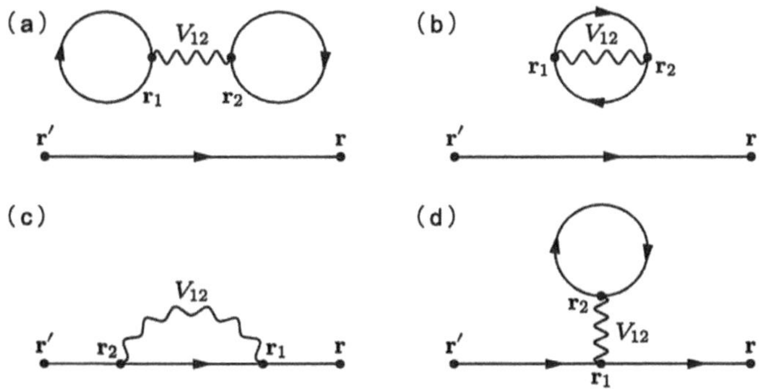

Fig. 10.4 When expressing the first-order perturbation term involving a single interaction line (wavy line) as a product of two-body terms using the Bloch–De Dominicis theorem, four types of variations, labeled **(a)**–**(d)**, arise depending on how the index pairs are selected within each propagator

Within the expectation value $\langle \cdots \rangle$, there are an equal number of creation operators $\hat{\psi}^\dagger$ and annihilation operators $\hat{\psi}$, and the total number of operators is even ($2N$). It can be shown that this expression can be decomposed into a product of N propagators, each consisting of a pair of operators, such as

$$V_{12}\left\langle \psi_r^\dagger \psi_1 \right\rangle \left\langle \psi_1^\dagger \psi_2 \right\rangle \left\langle \psi_2^\dagger \psi_{r'} \right\rangle =: D^{(c)},$$

which is a computationally convenient result established by a formal theorem [5, 9–11]. At absolute zero, this decomposition is known as the **Wick's theorem**, whereas at finite temperature, it is called the **Bloch–De Dominicis theorem**. The above expression corresponds to a first-order perturbation term involving a single instance of $\hat{H}_I$, but there are also other possible forms, such as

$$V_{12}\left\langle \psi_2^\dagger \psi_2 \right\rangle \left\langle \psi_1^\dagger \psi_1 \right\rangle \left\langle \psi_r^\dagger \psi_{r'} \right\rangle =: D^{(a)}$$

$$V_{12}\left\langle \psi_2^\dagger \psi_1 \right\rangle \left\langle \psi_1^\dagger \psi_2 \right\rangle \left\langle \psi_r^\dagger \psi_{r'} \right\rangle =: D^{(b)}$$

$$V_{12}\left\langle \psi_2^\dagger \psi_2 \right\rangle \left\langle \psi_r^\dagger \psi_1 \right\rangle \left\langle \psi_1^\dagger \psi_{r'} \right\rangle =: D^{(d)}$$

These are alternative decompositions within the same first-order perturbative framework. These differ from one another in terms of the variation in how the index pairs are assigned within each propagator. When attention is paid to the connectivity of the index pairs, they can be illustrated in diagrammatic form as shown in Fig. 10.4.

The graphical representation of perturbation terms according to such rules is referred to as a **Feynman diagram**. The propagators connected to the vertices $\mathbf{r}$ and $\mathbf{r}'$ are those that originally appeared in Eq. (10.21) and not those emerging from the interaction $\hat{H}_I$. These are called external lines. Depending on the pattern of index

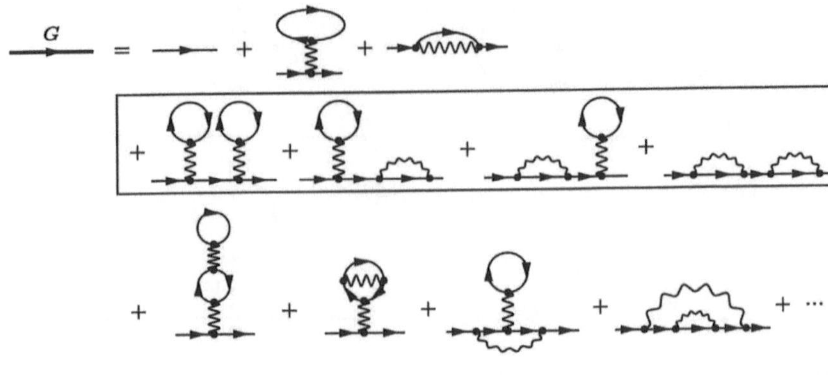

Fig. 10.5 When we enumerate the diagrams that appear in the perturbative expansion, they can be classified into two types: those that can be separated into multiple parts by cutting through a propagator line (the *improper diagrams*, shown as boxed groups), and those that cannot be divided into smaller components by cutting any single propagator line (the *proper diagrams*)

connections, general terms in the expansion may include diagrammatic structures such as those in Fig. 10.4a,b, which are *disconnected from the external lines*, or those in Fig. 10.4c,d, which are *connected to the external lines*. It can be shown that the diagrammatic structures *disconnected from the external lines* cancel out with contributions arising from the denominator in Eq. (10.18) and thus *do not contribute to the perturbative counting*. This result is known as the **linked cluster theorem** [5, 9–11].

Let us briefly summarize the discussion so far. By introducing the effects of the inter-particle interaction $\hat{H}_I$ perturbatively, the Green function (propagator)

$$
\begin{aligned}
G\left(\mathbf{r},\tau;\mathbf{r}',\tau'\right) &\sim \left\langle T_\tau\left[\hat{\psi}^\dagger\left(\mathbf{r},\tau\right)\hat{\psi}\left(\mathbf{r}',\tau'\right)\right]\right\rangle \\
&= \left\langle T_\tau\left[U\left(\tau\right)\hat{\psi}_0^\dagger\left(\mathbf{r},\tau\right)U^\dagger\left(\tau\right)\cdot U^\dagger\left(\tau'\right)\hat{\psi}_0\left(\mathbf{r}',\tau'\right)U\left(\tau'\right)\right]\right\rangle \\
&= \left\langle\sum_{n=0}^{\infty}\frac{(-)^n}{n!}\int\cdots\int d\tau_1\cdots d\tau_n \right. \\
&\quad \left. \times\, T_\tau\left[\hat{\psi}_0\left(\mathbf{r},\tau\right)\hat{\psi}_0^\dagger\left(\mathbf{r}',\tau'\right)\hat{H}_I\left(\tau_1\right)\cdots\hat{H}_I\left(\tau_n\right)\right]\right\rangle_0
\end{aligned}
\tag{10.23}
$$

can be evaluated. Each term of this expansion can be diagrammatically classified and represented as shown in Fig. 10.5. By collecting only the connected diagrams and summing their contributions, the propagator can be systematically evaluated.

10.3.2 Diagrammatic Expansion of the Partition Function

The partition function Ξ appearing in Eq. (10.15) can also be expanded perturbatively as follows:

$$\begin{aligned}\Xi &= \mathrm{Tr}\left[e^{-\beta H}\right] = \mathrm{Tr}\left[e^{-\beta H_0}e^{\beta H_0}e^{-\beta H}\right]\\ &= \Xi_0 \cdot \mathrm{Tr}\left[\frac{e^{-\beta H_0}}{\Xi_0}e^{\beta H_0}e^{-\beta H}\right] =: \Xi_0 \cdot \left\langle e^{\beta H_0}e^{-\beta H}\right\rangle_0\\ &=: \Xi_0 \cdot \langle U(\beta)\rangle_0 \end{aligned} \tag{10.24}$$

Then, inserting the operator $U(\tau)$ of Eq. (10.19),

$$\hat{U}(\tau) = \sum_{n=0}^{\infty}\frac{(-1)^n}{n!}\int_0^{\tau}\cdots\int_0^{\tau} d\tau_1\cdots d\tau_n \cdot T_\tau\left[\hat{H}_I(\tau_1)\cdots\hat{H}_I(\tau_n)\right]$$

one can evaluate the partition function accordingly. As a result, comparing with the expression for evaluating the Green function in Eq. (10.23),

$$G\left(\mathbf{r},\tau;\mathbf{r}',\tau'\right) \sim \left\langle\hat{\psi}(\mathbf{r},\tau)\,\hat{\psi}^\dagger\left(\mathbf{r}',\tau'\right)\hat{H}_I(\tau_1)\cdots\hat{H}_I(\tau_n)\right\rangle$$

$$\Xi \sim \left\langle\hat{H}_I(\tau_1)\cdots\hat{H}_I(\tau_n)\right\rangle$$

we find that they take essentially the same form. In the case of the partition function Ξ, there are no external lines like $\hat{\psi}(\mathbf{r},\tau)\,\hat{\psi}^\dagger\left(\mathbf{r}',\tau'\right)$, and the expression is closed within the $\langle\cdots\rangle$ average. Thus, the diagrammatic representation consists solely of propagators originating from the interaction $\hat{H}_I$, such as in Fig. 10.3b. In the case of first-order perturbation, diagrams such as Fig. 10.6c and d appear. If one makes a **cut** through a single propagator in these diagrams, one notices that they correspond to the Feynman diagrams of the Green function shown in Fig. 10.4c and d.

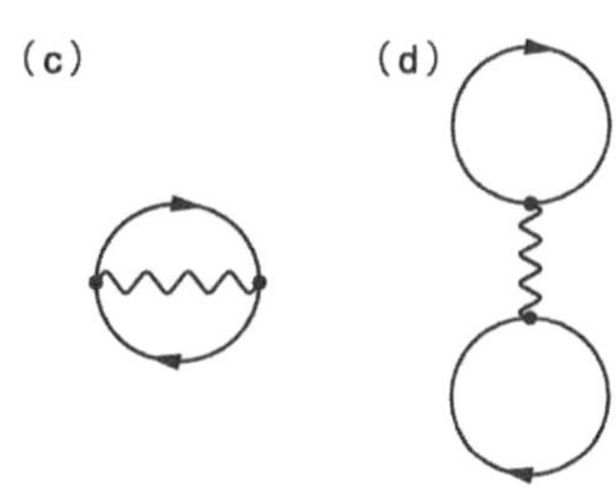

Fig. 10.6 The perturbative expansion terms of the partition function consist of closed and connected diagrams with no external lines. By making a single **cut** through one of the propagators in the diagrams (**c**) and (**d**) correspond to the labels in Fig. 10.4, one finds that they correspond to the Feynman diagrams for the Green function shown in Fig. 10.4c and **d**

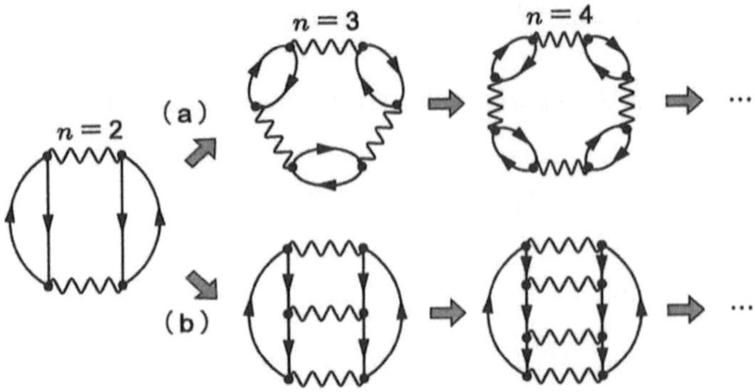

Fig. 10.7 To incorporate higher-order contributions, one may construct diagrams by connecting many interaction lines (wavy lines). There are two possible strategies: one is to consume them in series by arranging them in a ring, as shown in (**a**); the other is to consume them in parallel by arranging them in a ladder, as shown in (**b**)

In this way, the diagrammatic series for the perturbative expansion of the Green function can be derived by **making cuts** in the diagrammatic series of the partition function. Accordingly, it is common practice to discuss **how many diagrams to include** in the expansion of the partition function and **which diagrams correspond to which physical processes**.

In the nth-order perturbative expansion, diagrams that contain n interaction lines (wavy lines) appear. As the order n increases beyond $n = 2$, diagrams such as those shown in Fig. 10.7 emerge. As illustrated in Fig. 10.7, when the interaction order n increases, two representative classes of diagrams appear: **ring-type** diagrams, which extend in the direction of arrow (a), and **ladder-type** diagrams, which extend in the direction of arrow (b). In reality, diagrams that contain mixtures of ring- and ladder-type structures also appear. Techniques have been developed to treat the operation of "summing up infinitely many diagrams" as a recursive procedure and to express it in terms of geometric series.[22] These techniques are formalized in frameworks such as the **Dyson equation** and the **Bethe-Salpeter equation** (Sect. 18.1.3).

The diagrams contributing to the first-order ($n = 1$) perturbative expansion in the interaction are Fig. 10.6(1)c and d. It is known that truncating the expansion at this level corresponds to the Hartree–Fock theory.[23] Rather than truncating the

[22] The logic goes: "Suppose we could sum an infinite series, and call the result Γ; then we can express the series itself using Γ. Writing this relation as an equation and solving it, we obtain $\Gamma = \cdots$." See Appendix Chap. 18.

[23] To express the Hartree–Fock method as a variational approach within the framework of second quantization, one rewrites the Hamiltonian in terms of field operators and then reduces the four-body terms to effective two-body terms by taking expectation values [12]. The two diagrams correspond to this procedure. Furthermore, the process by which the Hartree–Fock equations are properly derived from the diagrammatic form of the thermodynamic potential, via the self-energy, is presented in several references [13, 14].

perturbative expansion at a finite order, one of the first successful attempts to incorporate all orders through a recursive technique was the method of summing only the ring diagrams in Fig. 10.7 as a geometric series. This approach is known as the **random phase approximation** (RPA). It was revealed that the RPA describes physical processes such as the **screening effect**, where a single electron is surrounded by many others that screen the interaction, as well as plasma oscillations (Sect. 18.2.2).

The discussion regarding diagrams has become a bit long, but let us now return to Eq. (10.13):

$$\frac{\delta \Delta E}{\delta v\left(q\right)} = \frac{1}{2}\left[n \cdot S^{\sigma\sigma'}\left(\mathbf{q}\right) - \delta_{\sigma\sigma'} \cdot n^{\sigma}\right] .$$

To summarize what has been discussed thus far, the left-hand side ΔE is obtained analytically from the partition function evaluated using many-body perturbation theory as $\Xi = \exp\left(-\beta\Omega\right) \sim \exp\left(-\beta\Delta E\right)$. Taking the functional derivative of this expression yields the structure factor $S^{\sigma\sigma'}\left(\mathbf{q}\right)$. Its Fourier transform then allows us to determine how the pair distribution function is modified depending on which diagrams have been included.

10.4 Analyzing Electron Correlation via Pair Correlation Functions

10.4.1 Hartree–Fock Theory and RPA Theory

The definition of the **electron correlation effect** is "the effect that cannot be described by Hartree-Fock theory." If we first understand how the **pair correlation function in Hartree–Fock theory** behaves, we can then identify how higher-order diagrams affect the shape of the pair correlation function. In the Hartree–Fock approximation, the pair correlation function of the uniform electron gas is given, with $r = |\mathbf{r}_1 - \mathbf{r}_2|$, as

$$\begin{aligned} g_{\mathrm{HF}}^{\uparrow\uparrow}\left(r\right) &= \left(\frac{2}{n}\right)^2 \cdot \int \frac{d^3p}{(2\pi)^3} \int \frac{d^3p'}{(2\pi)^3} f\left(\mathbf{p}\right) \cdot f\left(\mathbf{p}'\right) \cdot \frac{1}{2} \cdot \left|1 - e^{i(\mathbf{p}-\mathbf{p}')\cdot\mathbf{r}}\right|^2 \\ &= 1 - \left\{\frac{3 \cdot j_1\left(p_{\mathrm{f}}r\right)}{p_{\mathrm{f}}r}\right\}^2 \\ g_{\mathrm{HF}}^{\uparrow\downarrow}\left(r\right) &= 1 \end{aligned}$$

as can be derived [5, 6]. By expanding the spherical Bessel function $j_1\left(x\right)$ in the above expression, we obtain the behavior

$$g_{\mathrm{HF}}^{\uparrow\uparrow}\left(r\right) \sim r^2$$

which rises as the square of the inter-electron distance r. This corresponds to the result shown in Fig. 10.2 in Sect. 10.1.3. That is, under this approximation, the exchange hole $[g^{\uparrow\uparrow}(r)]$ opens with a quadratic rise in r, whereas the correlation hole $[g^{\uparrow\downarrow}(r)]$ remains flat.

On the other hand, the **exact behavior of the pair distribution function near the origin** is governed by the electron–electron cusp condition, which yields

$$g^{\uparrow\downarrow}(r) = g^{\uparrow\downarrow}(0) \cdot \left(1 + \frac{r}{a_0} + \cdots\right)$$

$$g^{\uparrow\uparrow}(r) = \frac{1}{2} \cdot \left.\frac{d^2 g^{\uparrow\uparrow}(r)}{dr^2}\right|_{r=0} \cdot r^2 \cdot \left(1 + \frac{r}{2a_0} + \cdots\right) \tag{10.25}$$

as the resulting expression.[24] That is to say, in the exact case, $g^{\uparrow\downarrow}(r)$ should rise as a "constant plus a term linear in r," and $g^{\uparrow\uparrow}(r)$ should start from zero and rise quadratically in r. Therefore, within the Hartree–Fock approximation, it is evident that the contributions from the r-linear term and beyond in the expansion factor $(1 + \cdots)$ cannot be captured.[25]

In the Hartree–Fock approximation, we had $g_{\text{HF}}^{\uparrow\downarrow}(r) = 1$, which means that the correlation hole could not be described. Going beyond the Hartree–Fock approximation that considers only diagrams (c) and (d) of Fig. 10.6(1), the approximation known as the random phase approximation (RPA) incorporates only the ring-type diagrams found in the upper row along arrow (a) of Fig. 10.7 to all orders. As explained in Sect. 18.2.2, this inclusion of part of the electron correlation allows the description of screening effects. Within the range of the RPA approximation, if one evaluates the pair distribution function from the structure factor using Eq. (10.13), the resulting shape of the correlation hole is shown in Fig. 10.8a. It can be seen that the correlation hole, which could not be described in the Hartree–Fock approximation, is indeed captured with a shape that rises as "a constant plus linear in r."

Here, let us recall that the pair distribution function represents a probability of existence (Sect. 10.1.3). Therefore, the pair distribution function must be non-negative. However, in Fig. 10.8a, it can be seen that the RPA-based pair distribution function yields negative values when the parameter r_s is large. This suggests that the inadequacies of the theory become pronounced as r_s increases. The parameter r_s corresponds to the average distance between electrons, and a larger r_s indicates a more dilute electronic system. RPA theory describes the process in which a given electron is surrounded by many other electrons that collectively screen the interaction (Sect. 18.2.2). That the RPA description deteriorates in the dilute regime can be attributed to the lack of a sufficient electron density to ensure the presence of enough screening electrons, which is a fundamental premise of the RPA framework.

[24] This result is known as the **Kimball** condition. Here, a_0 denotes the Bohr radius.

[25] In this expansion, the coefficient of a_0 differs by a factor of two between spin-parallel and spin-antiparallel pairs. This originates from the same reason as the corresponding difference in the cusp condition, as seen in Eq. (8.8).

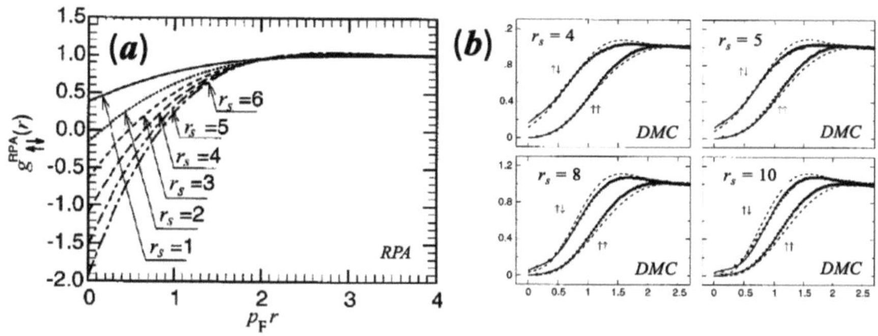

Fig. 10.8 The pair distribution function between antiparallel spins evaluated within the RPA approximation is shown in panel (**a**) (reproduced from [5]). Panel (**b**) shows the pair distribution function evaluated using the first-principles quantum diffusion Monte Carlo (DMC) method (reproduced from [15]). The bold curves represent the DMC results, while solid and dashed lines indicate results from other theoretical approaches for comparison. It is evident that the correct power-law behavior near the origin, as described by Eq. (10.25), is successfully reproduced

To correctly restore the correlation hole, overly deep and negative as a result of moving from the Hartree–Fock approximation to RPA, to a physically meaningful, positive-valued form, it becomes necessary to incorporate not only the ring-type diagrams but also the ladder-type diagrams. This will be discussed in the following subsection.

10.4.2 Short-Range and Long-Range Correlations

In the vicinity of the origin of the correlation hole, the repulsion due to interactions leads to a **locally diluted** region. Recalling the Fourier transform, describing such localized spatial variations requires the inclusion of many wavevectors **q** simultaneously.[26] On the other hand, recalling the diagrammatic structure of the RPA, the exchanged wavevector **q** in such diagrams is limited to a "single **q**" as illustrated in Fig. 10.9a. As a result, it fails to capture such localized spatial variations. This is the reason why RPA provides a poor description in the vicinity of the correlation hole.

In response, instead of limiting the number of interaction lines bridging the ring to just one, we are led to the idea of allowing multiple lines to connect in a ladder-like fashion, as illustrated in Fig. 10.9b, so that a variety of **q** values are involved to capture spatially localized variations. In the dilute regime near the correlation hole, the interacting particle pair is well-defined, and it is precisely the multiple scattering

[26] The Fourier spectrum of a Dirac delta function, which represents the extreme case of spatial localization, is constant in momentum space. This implies that accurately describing strong localization necessitates all wavevector components being equally involved.

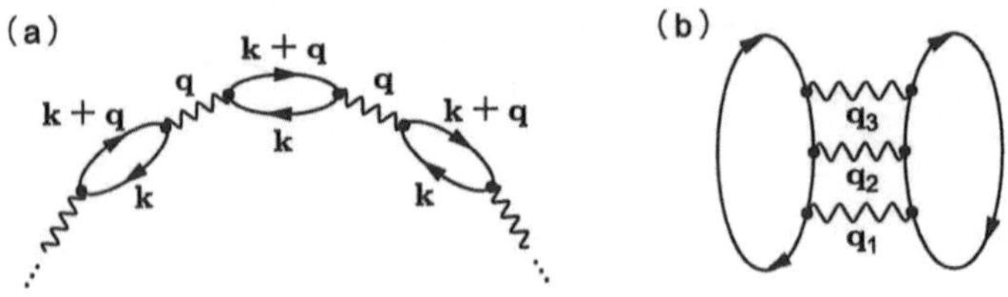

Fig. 10.9 Higher-order terms in the perturbation expansion are represented as closed diagrams constructed with multiple interaction lines (wavy lines). Among these, in the contribution from ring-type diagrams (panel (**a**)), the diagrammatic constraints restrict the involved wavevector $\mathbf{q}$ to a single value. In contrast, the contribution from ladder-type diagrams (panel (**b**)) allows various distinct wavevectors $\mathbf{q}_1, \mathbf{q}_2, \cdots$ to be involved

processes between this pair of particles that become important. The ladder diagrams provide a physical picture of such processes.

In the scheme for analyzing how "diagrams representing the partition function" affect the pair distribution function (Eq. 10.13), the way in which ladder-type diagrams contribute can be understood through the following line of reasoning: It has been shown that ladder-type diagrams contribute to the structure factor S in the large-q asymptotic regime as [16, 17]

$$\bar{S}^{\uparrow\uparrow}_{\mathrm{HF+ladd}}(\mathbf{q}) - 1 \sim q^{-6} \quad , \quad \bar{S}^{\uparrow\downarrow}_{\mathrm{ladd}}(\mathbf{q}) \sim q^{-4} \, . \tag{10.26}$$

Now, suppose $f(r)$ is a function depending only on the radial coordinate in d-dimensional space. If its Fourier transform $F(q)$ has the following asymptotic form:

$$F(q) \sim q^{-\beta} \quad (\text{as} \quad q \to \infty)$$

then it follows that $f(r)$ behaves asymptotically as $r \to 0$ according to[27]

$$f(r) \sim C_d(\beta) \cdot r^{\beta - d} + \cdots \quad (\text{as} \quad q \to \infty/0)$$

$$C_d(\beta) = \frac{(4\pi)^{d/2} \cdot \Gamma\left(\dfrac{d}{2} - \dfrac{\beta}{2}\right)}{2^{\beta} \cdot \Gamma\left(\dfrac{\beta}{2}\right)} \, .$$

Since the pair distribution function $g(|\mathbf{r}|)$ is the Fourier transform of the structure factor $S(\mathbf{q})$, the correction contributions from ladder diagrams can be evaluated, based on Eq. (10.26), as

[27] By expanding the exponential function contained in the Fourier transform into Bessel functions and rewriting the integral into the form known as a Hankel transform, one can evaluate the asymptotic behavior by using the asymptotic form of the Bessel functions [18].

$$\Delta g_{\mathrm{HF+ladd}}^{\uparrow\uparrow}(\mathbf{q}) \sim r^{6-3} \propto r^3 \quad , \quad \Delta g_{\mathrm{ladd}}^{\uparrow\downarrow}(r) \sim r^{4-3} \propto r \,.$$

These terms serve as the correction terms with the coefficient $1/a_0$ in the Kimball condition given in Eq. (10.25). In this way, the cusp condition,[28] that is, **electron correlation associated with particle coalescence**, can be understood to be properly described only when the ladder-type diagrams are included beyond the RPA (ring diagrams), which account for **electron correlation describing screening**. Accordingly, ladder-type diagrams are characterized as those describing **short-range correlations**, while RPA, in a complementary manner, is regarded as describing **long-range correlations**.

Another important intuition to develop is the sense of "balance between exchange and correlation." In order to achieve a correct description of short-range correlation (cusp condition), a key point was that $\bar{S}_{\mathrm{HF+ladd}}^{\uparrow\uparrow}(\mathbf{q})$ has an expansion whose leading term behaves as q^{-6}. The leading term from the ladder diagrams alone actually appears as q^{-4}. However, this q^{-4} contribution is exactly canceled out by the q^{-4} leading term that arises from the exchange part within the Hartree–Fock approximation. As a result, the q^{-6} term becomes the leading behavior, thus recovering the correct cusp condition. This indicates the **importance of incorporating exchange and correlation diagrams on an equal footing**. If this balance is lacking, a linear r term appears in the parallel-spin pair distribution function, which leads to an incorrect description of the exchange hole, which should rise quadratically in r.

10.5 Electron Correlation Description by DMC

10.5.1 Short-Range and Long-Range Correlation

The pair distribution function of the uniform electron gas, evaluated using DMC, is shown in Fig. 10.8b. In contrast to the unphysical result seen in Fig. 10.8a, where the pair distribution function becomes negative in the low-density regime with large r_s under RPA, the function remains positive even in the dilute regime. Moreover, the behavior near the origin, *i.e.*, the cusp condition, is correctly reproduced. In Sect. 10.4.2, we stated that, in order to reproduce the correct behavior of the cusp condition as short-range correlation, it is necessary to incorporate the diagrams representing exchange and correlation with accurate balance. The DMC method reflects the fact that this balance is automatically considered as a numerical variational method.

The r_s-dependence in the Hamiltonian of an electronic system can be easily understood: the kinetic energy term scales as $1/r_s^2$, while the electron–electron interaction term scales as $1/r_s$ [10].[29] In the *high-density regime*, where the average

[28] In terms of the pair distribution function, this is known as the Kimball condition.

[29] This is referred to as the r_s-expansion.

electron distance r_s is small, the kinetic energy dominates. Thus, approaches that treat electron–electron interactions as a perturbative expansion are well justified. In the high-density regime, where electrons are always surrounded by other electrons, the electron–electron interactions are well screened. Therefore, the "perturbed single-particle picture" is expected to hold effectively. Even within the RPA theory shown in Fig. 10.8a, no negative values appear in the high-density regime. This suggests that the perturbation expansion based on ring diagrams remains valid in such conditions.

On the other hand, in the **low-density regime** with large r_s, the validity of treating the electron–electron interaction term as a perturbation breaks down. In the approach based on many-body perturbation theory, as discussed in Sect. 10.4.2, it becomes necessary to go beyond RPA and incorporate ladder diagrams. However, as the system becomes increasingly dilute, even such considerations lose their validity, ultimately leading to unphysical conclusions. On the other hand, the DMC method can readily describe the pair distribution function even in the dilute regime. This remarkable capability stems from the fact that DMC is a variational method. The **strength of the variational approach** lies in its guarantee that proceeding in a way that lowers the variational energy leads to a solution closer to the exact one, which proves effective in non-perturbative regimes. Moreover, the balance between exchange and correlation is also automatically incorporated, as long as the calculation proceeds according to the principle of variational minimization.

Magnetism, which involves the description of localized magnetic moments, remains a particularly challenging subject even for *ab initio* calculations and is considered a highly nontrivial problem in many-body electronic theory. The reason can be understood in terms of the delicate balance between exchange and correlation required to describe local electronic correlations, as discussed above. When advancing many-body electronic theory via a many-body perturbation approach, it is important to handle the diagrams carefully in terms of the exchange-correlation balance; otherwise, one may fail to capture the correct physics.[30] In this respect, the promise of *ab initio* QMC in the field of magnetism and spintronics lies in the strengths of numerical variational approaches, which automatically fulfill such subtle electronic correlation balances.[31]

[30] I once heard from a distinguished scholar that although theories explaining anomalous phenomena can be crafted by selectively considering specific diagrams, they often result in "false theories" that violate conservation laws, gauge invariance *etc.* those are managed to be guaranteed by diagrammatic balance.

[31] However, as mentioned in Sect. 8.3.3, there exists a serious obstacle: the difficulty of constructing accurate pseudopotentials for transition metal elements has hindered the application of these methods in the field of magnetism.

- ***Nodes*** *or* ***Amplitude*** *of many-body wavefunction*
- ***Exchange*** *or* ***Correlation*** *... 'Quantum statistics' or 'Deformation of wavefunction'*
- ***Long-range*** *or* ***Short-range***
  (Screening) *(Cusp)*
  Ring*-type diagrams* $\sim q^{-4}$ ***Ladder****-type diagrams* $\sim q^{-6}$
  ... in terms of ***pair-correlation function****, ...*
  digging exchange-hole... *setting exchange-hole back to positive value..*
- ***Static*** *or* ***Dynamic*** *correlation*
  (multi-reference) *(Coalescence)*

Fig. 10.10 Electron correlation refers to the contribution that goes beyond the description provided by Hartree–Fock theory. There are, however, various ways to classify and divide the underlying causes of this contribution. It is important not to confuse these different perspectives with one another

10.5.2 Static and Dynamical Correlation

In the previous section, we discussed the classification of electronic correlation into long-range correlation (ring-type) and short-range correlation (ladder-type). Let us again review the effects of electron–electron interactions. They are first separated into classical effects (Hartree term) and quantum effects. The quantum effects are further divided into exchange effects (Fock term) and correlation effects. The correlation effects can, in turn, be classified into long-range correlation and short-range correlation. As another perspective for classifying correlation effects, a division into **dynamical correlation** and **static correlation** is often made (see Fig. 10.10). Just like the term "electron correlation," "dynamical correlation" and "static correlation" are potentially misleading in nuance, and **special care must be taken not to develop a mistaken image**. We explain this point below.

Returning once again to the language of many-body wave functions, Hartree–Fock theory is a many-body wave function theory that employs a single Slater determinant as the trial function. This serves as the reference point at which the description of electron correlation is considered zero. Accordingly, electron correlation can be understood as an effect that emerges from multi-determinant representations.[32] A representative example of such a situation is a degenerate **multi-reference** ground state. For instance, consider an isolated boron atom (with five electrons). As electrons fill the lower energy levels, we encounter an equivalent choice for placing the fifth electron—either in p_x, p_y, or p_z. A complete description requires a linear combination of three Slater determinants, corresponding to configurations where the fifth electron occupies p_x, p_y, or p_z. Only then can a variationally satisfactory representation be obtained. Such electron correlation effects, which are difficult to describe with a single configuration and arise from degeneracy or near-degeneracy, are referred to as geometric correlation or **static correlation**.

[32] This discussion assumes that no corrections such as Jastrow factors are employed.

On the other hand, **dynamic correlation** refers to the correlation effects associated with electron coalescence or scattering processes. In the context of this book, this corresponds to the description of the cusp condition related to coalescence. The cusp refers to the singularity that the many-body wave function exhibits as the interparticle distance $r_{ij} \to 0$. Since a Slater determinant is composed of linear combinations of products of one-body functions $\psi\left(\mathbf{r}_j\right)$, it is difficult for a single determinant to represent such non-analytic features like a cusp. It should be kept in mind that the role of the "second and subsequent terms" in the multi-determinant expansion beyond the Hartree–Fock level includes not only the description of multi-reference effects but also amplitude corrections necessary for representing electron cusps. In VMC discussed in this book, instead of using multi-determinants, the electron–electron cusp was efficiently represented by appending a Jastrow factor to a single determinant. Moreover, in the case of DMC, amplitude corrections including the cusp description were automatically restored via imaginary-time evolution (→ Fig. 10.11).

Should We Overlook Mild Misuses?
If one casually uses terminology like "dynamic correlation" based on the superficial impression that it refers to "correlation arising from electron dynamics" (incorrect!), it can lead to a telephone game-like spread of misinterpretation. In serious discussions within many-body electron theory, such terminology can cause confusion, one might respond, "Wait, isn't that actually static correlation?." In large research funding efforts, it is often considered important to coin appealing and impactful keywords. Due to such circumstances, research projects labeled with phrases like "dynamic electronic correlation," for example, those exploring new possibilities of electronic phases via real-time observation of electron dynamics made possible by cutting-edge experimental techniques, have yielded significant results. While the use of the term "dynamic correlation" in these contexts may not align with its traditional technical definition, it would be rather tactless to harshly nitpick such successful endeavors for terminological inaccuracy. In fact, one might argue instead that the terminology within specialist communities, whether for electronic correlation or for dynamic and static correlation, is confusing and should be revised. That said, when misuse of technical terms becomes widespread, precise communication suffers. For that reason, occasional course correction remains necessary.

Dynamic correlation is a matter of cusp representation and thus concerns the description of the wavefunction amplitude. The amplitude correction toward the exact solution can be captured perfectly using the fixed-node DMC. The fixed-node DMC generally succeeds remarkably well, and it is actually more difficult to find problems where the fixed-node approximation clearly breaks down. Such cases, in

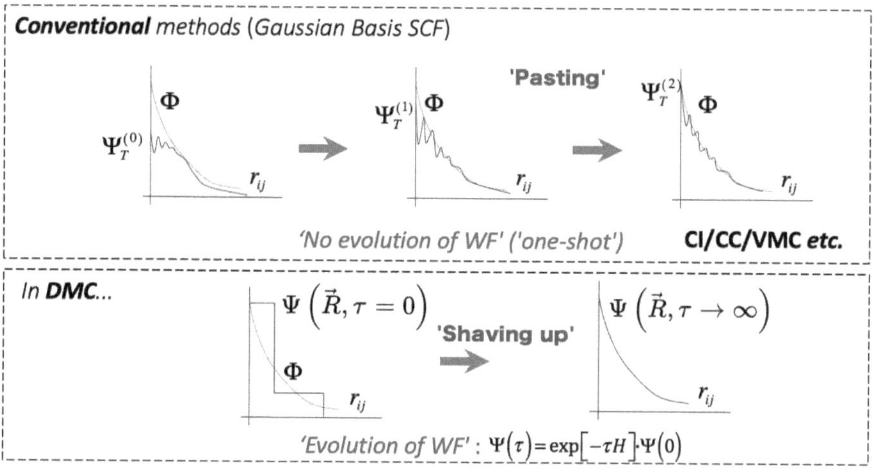

Fig. 10.11 In *ab initio* QMC, one begins with an initial trial wavefunction, reasonably guessed, and refines it through (imaginary) time evolution, gradually sculpting it toward the exact solution. In contrast, conventional molecular orbital methods do not employ such refinement via time evolution, requiring instead a more accurate wavefunction shape to be prepared as a one-shot construct from the outset

most situations, are problems involving static correlation with quasi-degeneracy. In communities that have traditionally studied electronic correlation, such as electron gas theory and quantum chemistry, there is a strong awareness that "open shell systems are particularly difficult problems in electronic correlation." It is therefore important to develop the intuition that **static correlation tends to be more problematic**. Due to the superficial impression given by the terms, one might naively think that "dynamic" sounds more difficult, and this is a source of misunderstanding surrounding the terminology of dynamic and static correlation. In the case of simple metals, it is often assumed that, compared to strongly correlated systems like Mott insulators, "electronic correlation must be easier to deal with." However, these simple-metal systems with large open shells are actually regarded as systems where the treatment of electronic correlation in the traditional sense is highly nontrivial. In standard textbooks, one learns that the Hartree–Fock theory fails in describing properties such as the specific heat of metals [5, 6]. For insulators, on the other hand, the existence of a finite gap mitigates this breakdown and allows the Hartree–Fock theory to serve as a good starting point [5, 6].

10.5.3 The Distinction Between Nodes and Amplitudes

From the viewpoint of handling DMC, it is useful to classify electronic correlations into those that can be captured solely through modulation of the many-body wave function amplitude and those that require nodal modulation. In the former case, the fixed-node approximation implemented with a Slater–Jastrow-type trial wave function can handle the problem relatively easily. In contrast, for the latter, it becomes necessary to further refine the trial wave function in order to modulate the nodes appropriately.

An intriguing perspective on nodal structures in many-body systems is offered by one-dimensional systems: More generally, in d dimensions, the nodal surface of a many-body wave function, defined by the implicit condition $\Psi(\mathbf{r}_1, \cdots, \mathbf{r}_N) = 0$, forms a space of dimension $(d \cdot N - 1)$. Among such nodal surfaces, those where particles coincide are of particular interest. When the position of some $\mathbf{r}_i$ is made to coincide with that of $\mathbf{r}_j$, the Pauli exclusion principle dictates that the wave function must vanish: $\Psi\left(\mathbf{r}_1, \cdots, \mathbf{r}_i = \mathbf{r}_j, \cdots, \mathbf{r}_N\right) = 0$. Such **trivial nodal surfaces** are called coalescence nodes. Since one particle coordinate is removed, these coalescence nodes have dimension $d \cdot (N - 1) = (d \cdot N - d)$. This is lower than the dimension $(d \cdot N - 1)$ of a general nodal surface, implying that coalescence nodes form a subspace within the more general nodal structure. The fact that many more unknown general nodal surfaces exist beyond these trivial coalescence nodes is what makes the many-body fermionic problem so difficult.

Here, one may notice that in the one-dimensional case ($d = 1$), the dimension of the trivial coalescence nodal surface $[(d \cdot N - d)]$ coincides with that of the general nodal surface $[(d \cdot N - 1)]$. This implies that all nodal surfaces can be identified by particle coalescence alone. There are no unknown nodal surfaces, and thus, from the perspective of the "sign problem associated with nodal surfaces" in DMC, this case is considered a "tractable problem." In this sense, the description of electronic correlation in one-dimensional systems is relatively simple. However, many-body electronic systems in lower dimensions are generally known to exhibit enhanced interaction effects due to restricted motion and are considered prototypical systems for realizing strongly correlated electrons [14]. In one dimension, exotic states such as the Tomonaga–Luttinger liquid are realized, and these systems are regarded as "challenging problems" in quantum many-body physics.[33] This stands in contrast to the relatively "tractable" impression suggested from the perspective of many-body nodal structures.

10.6 Descriptive Advantage of DMC

With the foundational knowledge developed up to this section in this chapter, including the "classification of electron correlations" and the "diagrammatic interpretation based on many-body perturbation theory," we can gain a deeper understanding of the descriptive advantage of *ab initio* DMC.

[33] It should also be recalled that in one-dimensional systems, exact solutions can be obtained via the Bethe ansatz [6].

10.6.1 Explanation Using the Result of the Hydrogen Molecule

The result of the DMC for the ground-state energy of the hydrogen molecule is shown in Fig. 10.12. The hydrogen molecule is a two-electron system, and with appropriate techniques, it can be solved exactly in an analytical manner. Therefore, the exact ground-state energy value is known as "Exact Value; − 1.1745" [hartree]. The Hartree–Fock energy is shown as "HF; − 1.1330", and the difference from − 1.1745 corresponds to the "electron correlation energy." The figure compares how much percentage of the correlation energy each method (VMC/DMC/CCSD/CISD) is able to reproduce.

The Hartree–Fock value is listed as − 1.1330, but this value was obtained using a moderately accurate Gaussian basis set, namely 6-311G.... Since improving the basis set expands the search space for optimal values, the energy can potentially be lowered further. With sufficiently large basis sets, one can estimate the lowest energy achievable by Hartree–Fock calculations [1]. For the hydrogen molecule, a literature value of the HF-limit has been estimated as "Hartree-Fock limit: − 1.1336"[19]. This value should be adopted as the more accurate Hartree–Fock reference.

The ground state of the hydrogen molecule is a two-electron spin singlet, and therefore, the many-body wave function has no node. Since the fixed-node approximation, normally the sole source of error in DMC, is not present in this case, the amplitude of the many-body wave function converges to the exact solution through the simulation. As a result, the computed ground-state energy [DMC: − 1.1746(2)] matches the exact value. At first glance, one might exclaim, "Isn't − 1.1746 lower than the exact value − 1.1745?! What happened to the variational principle?" However, " − 1.1746" is a statistical estimate and agrees within the error bar given by "(2)".

The competing methods indicated as "CCSD" and "CISD" are quantum chemical approaches based on configuration interaction, as discussed in Sect. 10.2.1. The value " − 1.1723" reproduces 94.5% of the electron correlation energy. Since the hydrogen molecule is a small system, it is possible to exhaustively consider all determinant combinations in the configuration interaction method. This allows one to perform a "Full CI" calculation that uses a trial wave function capable of describing the function space completely. Although Full CI should in principle provide the exact solution, its performance here is limited to 94.5%, due to limitations of the employed basis

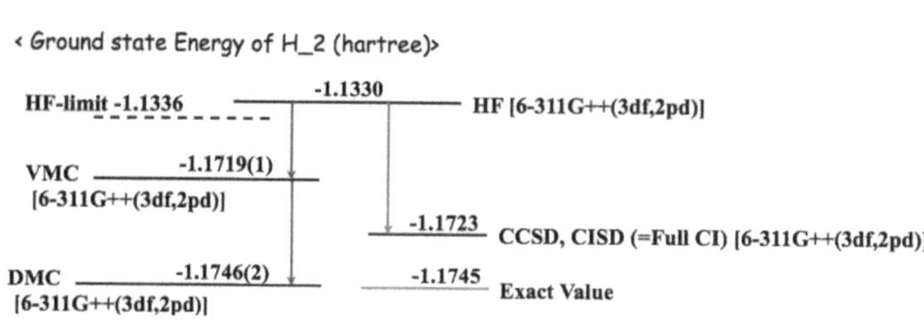

Fig. 10.12 Ground-state energy of the hydrogen molecule. Since all methods (HF, CCSD, DMC, ...) obey the variational principle, one can compare how closely each approaches the exact solution (Exact) and thereby discuss the relative merits of each method (The data are from computations by Prof. Kenta Hongo.)

set [6-311G++(3df,2pd)]. Gaussian basis set calculations approximate the cusp of the exact wave function using a superposition of Gaussian functions with no cusp but sufficiently narrow width. If one refines the Gaussian basis set to a higher-level one (Fig. 10.11), then in the case corresponding to Full CI, the computed value would approach the exact solution. However, in order to achieve high accuracy solely with quantum chemical methods, not only must one improve the basis set but also carefully select the active space and design the convergence strategy. Basis set correction schemes often involve complex procedures, and bringing such calculations to a reliable publication level requires considerable effort.

From this viewpoint, a particularly powerful feature of DMC is that even with a poor basis set, the amplitude is automatically repaired through imaginary time evolution (Fig. 10.11). The value "VMC: $-1.1719(1)$" in Fig. 10.12 is the result from variational Monte Carlo (VMC), which serves as a precursor to DMC. The correlation energy that can be described using a Jastrow factor reaches only about 93.5% of the total. This limitation also depends on the basis set used [6-311G++(3df,2pd)]. Using this basis set, the correlation description achieved by "a single determinant plus Jastrow function (VMC)" is 93.5%, while that of "a multi-determinant expansion (CISD)" reaches 94.5%, showing the latter performs better.[34] However, if the seemingly poor "single determinant plus Jastrow function" description is followed by the time evolution in DMC, the amplitude is self-corrected, and the full correlation energy is recovered.

Not Just About Cusp Description, of Course

When I explained the feature that even with a poor basis set, the cusp can be automatically carved out, a certain professor persistently objected, saying, "Your method only describes the cusp, but electronic correlation is not just about the cusp." I tried to clarify, saying, "That is just one of its features. Later I also discussed adjusting nodal degrees of freedom and using trial wave functions with multi-determinants...," but he kept insisting, "Electronic correlation is not just about the cusp," over and over, and it was quite frustrating. There is a certain kind of researcher who, once a topic close to their specialty arises, gets so eager to voice their own opinion that the rest of the talk's logical structure never reaches their ears.

Topics such as the effort to describe cusps using Gaussian basis functions, Jastrow factors, and multi-determinants have appeared. Let us recall that the cusps to be described include two types: cusps due to electron–electron coalescence, and cusps due to electron–nucleus coalescence. Figure 10.13 shows how the charge density cusp,

[34] See the column in Sect. 8.1.4. In the past, I once struggled to convince a senior expert and their followers, who dismissed any method using a Jastrow factor as "an inferior technique they disliked," despite numerous attempts to explain the role of the Jastrow in DMC, which differs from that in VMC.

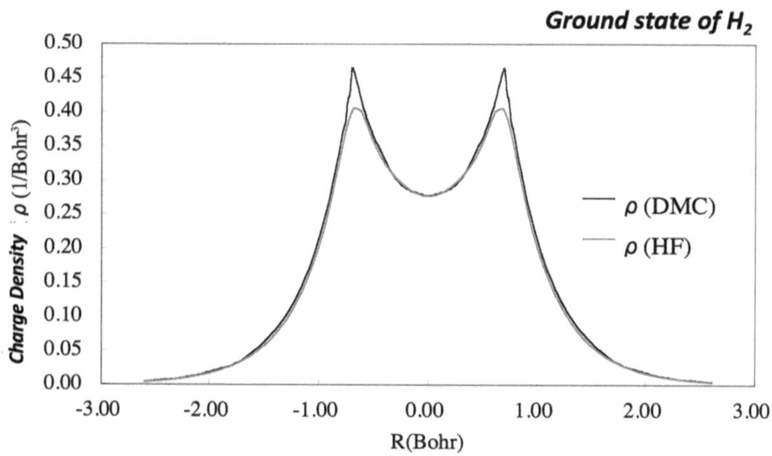

Fig. 10.13 Charge density of the hydrogen molecule. According to the cusp condition (see Sect. 8.1), the charge density should exhibit a cusp (a sharp point) at the position of the ionic core. However, at the level of Hartree–Fock (HF) theory used for the trial function, the cusp is not adequately described. Even starting from such a poor initial estimate, the cusp is well reproduced in *ab initio* DMC calculations through shape relaxation under imaginary-time evolution (calculated by Prof. Kenta Hongo)

corresponding to electron–nucleus coalescence, is gradually restored in DMC. In the figure, the Hartree–Fock (HF) calculation indicated as "HF" shows an insufficient description of the electron–nucleus cusp, due to the limitations of the Gaussian basis set. Even when starting from such a poorly described trial wave function, the numerical projection operation of the DMC restores the amplitude and reproduces the cusp, as illustrated in the conceptual diagram in Fig. 10.11.

It was previously noted that pursuing an accurate description of electron correlation using quantum chemical methods requires considerable expertise. In contrast, the element of expertise corresponding to this in the context of *ab initio* QMC lies in variational optimization. If one does not proceed to DMC and attempts to conclude with only VMC, significant skill is required in both setting the Jastrow factor according to the characteristics of periodic or isolated systems and numerically optimizing it. Many graduate students have struggled with the painstaking task of tuning dozens of parameters to achieve "93% correlation energy captured," or "94.2% achieved," and so forth. However, when VMC is performed for the modern purpose of providing a trial wave function for DMC, the role of the Jastrow factor is not to describe electron correlation itself, but rather to stabilize the computation and accelerate convergence. In such cases, it suffices to use a somewhat universal functional form that does not strongly depend on the target system and to optimize it only moderately. This greatly reduces the required level of expertise.

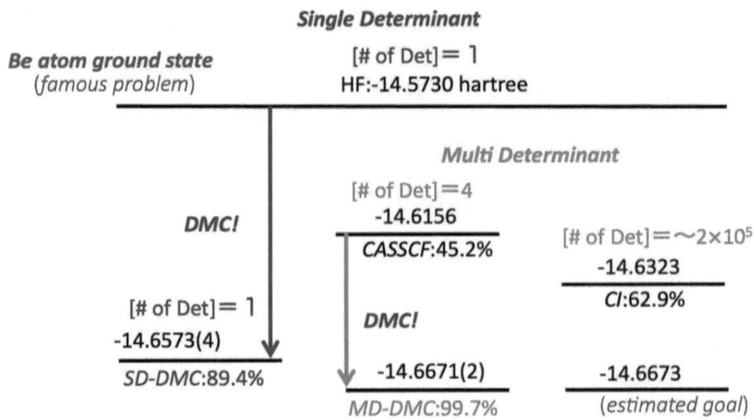

Fig. 10.14 Ground-state energy calculation for the beryllium atom. Just as with the hydrogen molecule, all methods (HF, CASSCF, CI, DMC) adhere to the variational principle. Therefore, their relative performance can be discussed by comparing how close each method approaches the lowest energy (Figure based on calculations by Prof. Kenta Hongo). The arrows for SD-DMC and MD-DMC indicate the origin of the respective initial trial wave functions used in each DMC calculation. Each arrow points to the method used to generate the corresponding trial function

10.6.2 Explanation Using the Result of the Beryllium Atom

Figure 10.14 shows the results of ground-state energy calculations for the beryllium atom. The interpretation of the figure is the same as in the case of the hydrogen molecule presented in the previous section (Fig. 10.12). In this system, unlike the hydrogen molecule, an analytic exact solution cannot be obtained. However, based on the accumulation of previous theoretical studies, the estimated ground-state energy has been established as -14.6673 hartree [20]. The label on the left of the figure, "SD-DMC: 89.4%," represents the result of a standard DMC calculation using a trial function composed of a single Slater determinant plus a Jastrow factor. In typical DMC for isolated atomic systems, nearly 99% of the correlation energy is usually recovered. However, in the case of the beryllium atom, only about 89.4% is reproduced, which highlights the difficulty of correlation description in this system. This difficulty arises because the low-energy excitation levels are relatively close to the ground state, and these excitations originate from the degeneracy among p_x, p_y, and p_z orbitals. This leads to static correlation involving multiple reference configurations. Since the exact ground state is formed by a hybridization of nearly degenerate multi-electron excited configurations, describing such a ground state with a single determinant poses a significant challenge.

Beyond the single-determinant level, the results of quantum chemistry calculations using multi-determinant expansions are shown as "MD: -14.6156" and "MD: -14.6323". Even with a four-term expansion achieving 45.2%, and a 200,000-term expansion, only about 63% of the correlation energy is recovered.[35] This illustrates why the ground-state description of the beryllium atom is known as a textbook-common difficult problem. A four-term multi-determinant expansion, which by itself reproduces only 45.2% of the correlation energy and is clearly insufficient, can nonetheless be used as a trial wave function in a DMC that astonishingly recovers as much as 99.7% of the correlation energy, easily surpassing even quantum chemistry calculations involving 200,000 multi-determinant terms. In quantum chemistry calculations, both the nodal structure and the amplitude of the wave function must be described by the original four determinants, limiting the achievable accuracy to 45.2%. In contrast, within the fixed-node DMC method, the nodal structure of the many-body wave function is constructed from just four determinants, while the amplitude degrees of freedom are automatically refined toward the exact form, regardless of how poor the initial guess or basis functions may be.

The results in Fig. 10.14 suggest that, while a single determinant is clearly insufficient to describe the many-body nodal structure, using only about four determinants may already be sufficient. This stands in stark contrast to quantum chemistry calculations, where even with as many as 200,000 determinants, only about 63% of the correlation energy can be captured– because both the nodal structure and the amplitude must be represented solely by the determinant expansion, without any self-repair. In such quantum chemical approaches, the vast majority of the 200,000 determinants are used to strengthen the amplitude component, whereas fixed-node DMC does not require all of them. This implies that the nodal structure alone may be described with far fewer determinants than conventionally thought. This insight is reminiscent of ideas in diagrammatic approaches, such as proper diagrams or skeleton diagrams.[36]

References

1. "Introduction to Computational Chemistry, Third Edition" (Wiley, 2017), F. Jensen, ISBN: 978-1118825990
2. "Lecture Notes on Electron Correlation and Magnetism" (World Scientific ,1999) Patrik Fazekas, ISBN: 978-9810224745
3. "Interpretation of Hund's multiplicity rule for the carbon atom", K. Hongo, R. Maezono, Y. Kawazoe, H. Yasuhara, M.D. Towler, and R.J. Needs, J. Chem. Phys. 121, 7144 (2004), https://doi.org/10.1063/1.1795151
4. "Ab initio Calculation Tutorial: For Materials Analysis, Informatics and Design", (Springer Nature, 2023), R. Maezono, ISBN: 978-9819909216
5. "Tatai Mondai" (in Japanese, Asakura Publishing, 1999) Y. Takada, ISBN: 978-4254136791
6. "Many-Particle Physics, Third Edition", (Springer, 2000/10/31), G.D. Mahan, ISBN: 978-0306463389

[35] It is useful to note that modern quantum chemistry calculations can easily handle multi-determinant expansions with tens of thousands of terms for small molecular systems.

[36] See Sect. 18.1.2.

7. “Density-functional Theory of Atoms And Molecules” (Oxford University Press, 1989), R.G. Parr, W. Yang ISBN: 978-0195092769
8. “Lectures on Quantum Mechanics, Second Edition” (Cambridge University Press, 2015), S. Weinberg, ISBN: 978-1107111660
9. “Bussei-butsuri no tame no Bano Riron, Green Kansu (Second Edition)” (in Japanese, Saiensu-Sha Co.,Ltd, 2024) M. Ogata, ISBN: 978-4781916132
10. “Quantum Theory of Many-Particle Systems” (Dover Publications, 2003), A.L. Fetter, J.D. Walecka, ISBN: 978-0486428277
11. “Quantum Many-particle Systems” (Perseus, 1998), J.W. Negele, H. Orland, ISBN: 978-0738200521
12. “Quantum Monte Carlo Approaches for Correlated Systems” (Cambridge University Press, 2017), F. Becca, S. Sorella, ISBN: 978-1107129931
13. “Tatai Mondai” (in Japanese, Baifukan Co., Ltd, 1975) F. Takano, ISBN: 978-4563024185
14. “Quantum Theory of the Electron Liquid” (Cambridge University Press, 2005), G. Giuliani, G. Vignale, ISBN: 978-0521821124
15. “Electronic Structure: Basic Theory and Practical Methods” (Cambridge University Press, 2004), R.M. Martin, ISBN: 978-0521782852
16. “Electron Correlations and Dielectric Function of Electron Gas at Metallic Densities” H. Yasuhara, J. Phys. Soc. Jpn. 36, 361 (1974), https://doi.org/10.1143/JPSJ.36.361
17. “On the short-range correlation in electron gas”, H. Yasuhara, Physica 78, 420 (1974), http://dx.doi.org/10.1016/0031-8914(74)90372-3
18. “Asymptotic Behavior of Fourier Transform”, Y. Fukui and K. Hiroike, Technology Reports, Tohoku Univ., Vol. 32, No. 1 (1967).
19. “The partial-wave self-consistent-field method for diatomic molecules: Computational formalism and results for small molecules”, E. A. McCullough, Jr., J. Chem. Phys. 62, 3991-3999 (1975) https://doi.org/10.1063/1.430322
20. “Beryllium atom reinvestigated: A comparison between theory and experiment” A-M Martensson-Pendrill, S.A. Alexander, L. Adamowicz, N. Oliphant, J. Olsen, P. Oster, H.M. Quiney, S. Salomonson, D. Sundholm, Phys. Rev. A 43, 3355 (1991). https://doi.org/10.1103/PhysRevA.43.3355

Chapter 11
Appendix A: Terminal Setting (For Macintosh)

We will set up a "Linux OS environment" on a Macintosh machine. Since Mac computers actually run on a Unix-based OS, the necessary environment for this book can be configured using the native[1] terminal environment. Within this setup, we will install the ab initio quantum Monte Carlo package "CASINO" along with various utilities required for its use.

- Using the terminal on Mac
- Online installation via the package manager homebrew

are assumed as prior knowledge. For details on these topics, refer to the author's previous book, "Ab initio Calculation Tutorial" [1], specifically: §A.2 "Setting Up the Terminal Environment" and §A.3.1 "Installing Homebrew."

11.1 Installation of CASINO

11.1.1 Obtaining the Tutorial Materials

First, prepare a "working directory" on your PC for the exercises. Execute the following commands in the terminal:

```
% cd
% mkdir work
% cd work
% pwd
```

[1] This refers to functionality that is built-in without requiring additional preparation.

R. Maezono, *Ab initio Quantum Monte Carlo Tutorial*,
https://doi.org/10.1007/978-981-95-4134-8_11

After executing these commands,[2] verify that the displayed directory path matches "/Users/maezono/work". In all examples in this book where "maezono" is written, replace it with "your own PC's username".[3] Following Linux conventions, we will use the notation:

```
'/Users/maezono/work' = '~/work'
```

Here, the part replaced by " ~/" corresponds to "/Users/maezono/", which serves as the starting point of the user's directory structure and is referred to as the **home directory**.

Next, we will download the complete set of tutorial materials from the web. These files are hosted on the authors' GitHub server and can be obtained using the git command. First, navigate to the "working directory" prepared in the previous section by executing the following commands:

```
% cd
% cd work
```

Then, verify that the directory path matches "/Users/maezono/work" by executing:

```
% pwd
```

Now, execute the following command to clone the repository:[4]

```
% git clone https://github.com/rmaezono/qmcTutorials.git
```

If the following output appears on the screen, the download is complete:

```
Cloning into 'qmcTutorials'...
remote: Enumerating objects: 19, done.
remote: Counting objects: 100% (19/19), done.
remote: Compressing objects: 100% (15/15), done.
remote: Total 19 (delta 1), reused 18 (delta 0), pack-reused 0
Unpacking objects: 100% (19/19), done.
```

Finally, confirm that the directory "qmcTutorials/" exists by running:

```
% ls
```

This directory contains the complete set of tutorial materials.

Next, navigate to the " ~/work/qmcTutorials" directory:

```
% cd
% cd work/qmcTutorials
% pwd
```

---

[2] Instead of typing each command from scratch, you can press the up arrow key in the terminal to recall the previously executed command. By modifying only the last argument and re-executing, you can save time typing.

[3] Many beginners encounter errors by forgetting to make this substitution, so reminders will be included throughout the text.

[4] Replace "maezono" with "your own PC's username".

Then, execute "`% ls`" to list the contents of the folder. You should see a directory named "`casino`." If this folder is present, the extraction has completed successfully.

The command sequences appearing in the subsequent installation steps are also documented in:

```
~/work/qmcTutorials/docs/ForMac.txt
```

Beginners who are concerned about typing errors can open this file and copy-paste commands directly from it to ensure accuracy.

11.1.2 Installing the Required Environment

We will use Homebrew to install the necessary packages required for compilation and practical exercises.[5]

```
% brew install open-mpi
% brew install emacs
% brew install gnuplot
% brew install grace
% brew install --cask xquartz
```

The `open-mpi` library is required for parallel computing, and CASINO cannot be installed without it. `emacs` is a text editor used for modifying input files, while `gnuplot` and `grace` are tools for visualizing results. For these tools to function properly, `xquartz` must be installed.

After installing `xquartz`, a system restart is required to apply the changes. Be sure to **restart** your Mac before proceeding to the next step.

11.1.3 Appending to .bash_profile

This book assumes that the Mac command line is used with **bash**. By default, macOS uses **zsh**, so switch to bash using:

```
chsh -s /bin/bash
```

before proceeding. Next, use redirection to append the following lines to your `.bash_profile`:[6]

---

[5] For instructions on installing Homebrew on Mac, refer to the author's previous book [1] or check the latest information available online. If Homebrew is not recognized in your terminal, ensure that the following line: `eval "$(/opt/homebrew/bin/brew shellenv)"` is added to your `~/.bash_profile` or `~/.zprofile`, and activate it by running the `source` command.

[6] Since these commands involve lengthy text input, beginners are prone to typing errors. It is recommended to copy and paste from `~/work/qmcTutorials/docs/ForMac.txt`. **Do not**

```
% cd
% echo 'export CASINO_ARCH=macos-gnu-parallel' >> .bash_profile
% echo 'export PATH=‘‘$PATH’’:$HOME/work/qmcTutorials/
    [Do not insert a line break; continue on the next line]
    casino/bin_qmc/$CASINO_ARCH/opt' >> .bash_profile
% echo 'export PATH="$PATH":$HOME/work/qmcTutorials/
    [Do not insert a line break; continue on the next line]
    casino/bin_qmc' >> .bash_profile
% echo 'export PATH="$PATH":$HOME/work/q-e/bin' >> .bash_profile
% echo 'export PLOTTER=/opt/homebrew/bin/xmgrace' >> .bash_profile
```

The variables written in uppercase, `CASINO_ARCH`, `PATH`, and `PLOTTER`, are all environment variables. Each time a new terminal session is started, the contents of `.bash_profile` are executed, assigning these variables the specified values. The system then follows these environment variable settings to execute the desired operations.

- `CASINO_ARCH`: This variable is referenced when installing the `CASINO` package to identify the architecture of the target machine.
- `PATH`: This ensures that the commands related to the newly installed `CASINO` package are added to the command path, allowing them to be executed from anywhere in the terminal.
- `PLOTTER`: This variable specifies the path to `xmgrace`, the plotting tool used by `CASINO`, ensuring that it can be accessed properly.

After executing the `echo` command sequence, check that the appended lines have been correctly added to `.bash_profile` by running:

```
% cd
% cat .bash_profile
```

If the lines appear as expected, exit the terminal session with:

```
% exit
```

Then, restart the terminal to apply the changes.

11.1.4 Installing CASINO

Next, we proceed with the installation of the **CASINO** package, which is used for ab initio quantum Monte Carlo calculations. This installation is done by compiling from source.

Navigate to the following directory and confirm that the listed files are present:

---

copy from this PDF file, as copying from a PDF can cause quotation marks to change from actual symbols to font-based representations, leading to errors.

```
% cd
% cd work/qmcTutorials/casino
% ls
    Makefile   arch/ ...
```

Then, execute the compilation process as follows:[7]

```
% make -j 4 opt
```

If the compilation completes successfully, the following file should be present:

```
% pwd
    /Users/maezono/work/qmcTutorials/casino/bin_qmc/macos-gnu-parallel/opt/
% ls
    casino
```

If the compilation fails,[8] reset the compilation process by running:

```
% make clean
```

This will remove any partially compiled files, allowing you to restart the process.

This concludes the CASINO compilation. Proceed to Sect. 2.1.

11.2 Installing Other Simulation Packages

11.2.1 Installing QuantumEspresso

In the latter part of this book, we will use the density functional theory (DFT) package **QuantumEspresso**. This package must be installed by compiling from source.

First, install the necessary packages for compilation: **gcc**, **fftw**, and **openblas** using `brew`.

- **gcc**: Compiler for C/C++ and Fortran.
- **fftw**: Library for performing Fast Fourier Transforms (FFT).
- **openblas**: Numerical computation library, including matrix diagonalization algorithms.

Install these packages with the following command:[9]

---

[7] The option "-j 4" of the make command means "compile in parallel with 4 cores." Adjust the specification "4 cores" appropriately according to how many cores are available on your own CPU.

[8] Users who have already installed and are using Anaconda's `open-mpi` may encounter issues due to conflicting command paths. In such cases, locate the section in `.bash_profile` where conda is initialized, comment it out, and retry the compilation.

[9] This installation can be executed from any directory.

```
% brew install gcc
% brew install fftw
% brew install openblas
```

Next, obtain the source code for **QuantumEspresso**.[10] It is distributed on the **GitHub** server, which specializes in source code distribution, at the following URL:

https://github.com/QEF/q-e

Although it is possible to download the source code via a web browser, we will use a dedicated command (`git`) to retrieve it. Navigate to the `work` directory (`% cd work`) and execute the following commands:[11]

```
% pwd
    /Users/maezono/work
% git clone https://github.com/QEF/q-e
    Cloning into 'q-e'...
    remote: Enumerating objects: 2763, done.
    ...
```

After the download completes, confirm the existence of the `q-e` directory inside `work` to verify a successful download.

Next, navigate into the `q-e` directory. Executing `% ls` should reveal the file **configure**. This script automatically collects information about the compiler and required libraries for the installation. It must be executed before compilation. However, we need to explicitly specify the **C language preprocessor** (`cpp`) installed via Homebrew. To check the available preprocessors, run `cpp` in the terminal and press the **Tab** key several times. This will display the installed preprocessors:

```
% cpp
    cpp      cpp-14
```

Here, `cpp` is the preprocessor that comes with macOS by default, while `cpp-14` is the one installed via Homebrew. For the upcoming QuantumEspresso installation, we will use `cpp-14`.[12]

Next, execute the `configure` script, paying close attention to the details outlined in the footnote,[13] execute the following command:

---

[10] As mentioned in the footnote of Sect. 5.1.3, "-pw2casino" seems to be no longer supported in versions later than v7.4. See "Version Information" at the beginning of this book.

[11] Pay attention to the meaning of `% pwd` (as explained in the Acronyms section). Replace `maezono` with your own username.

[12] The number 14 indicates the version of `cpp`, which is the latest as of 2024. With future updates, this number may increase (e.g., `cpp-15`). Check the support page mentioned at the beginning of this book for the latest information.

[13] Ensure that there is **no space** after the dot and slash (`./`). However, **a space** must be included before `-fallow-...`. `cpp-14` refers to the latest version at the time of writing. As noted in the previous footnote, adjust accordingly if a newer version is available. Since the command is quite long and prone to typing errors, copying and pasting from the `.../qmcTutorials/docs/ForMac.txt` file is recommended. Avoid copying directly from this PDF, as quotation marks and other symbols may become corrupted.

```
% pwd
    /Users/maezono/work/q-e
% ./configure FC=mpif90 CC=mpicc CPP=cpp-14 FFLAGS= @\
    "-I/usr/local/include/ -fallow-argument-mismatch"
```

Here, the option `FFLAGS="-I/usr/local/include/"` informs the system that the FFTW function libraries installed via `brew` are located in `/usr/local/include`. While `configure` is designed to automatically gather necessary information, it is not flawless, and users may need to manually specify missing paths.[14]

Once the `configure` process is complete, you can proceed with compilation using the **make** command based on the collected information:

```
% pwd
    /Users/maezono/work/q-e
% make pwall
```

After compilation, run `% ls bin` to check if the executable `pw.x` has been created. If it exists, the compilation was successful. To verify proper execution, run the following command:

```
% pwd
    /Users/maezono/work/q-e/bin
% ./pw.x
     Program PWSCF v.7.1 starts on  7Dec2022 at 16:59: 1
     ...
     Waiting for input...
```

If you see the message "Waiting for input...," the program is running correctly.[15] At this point, you can terminate the program using "Ctrl+C."

Reference

1. "Ab initio electronic structure calculation" (in Japanese, Second edition, Morikita publishing Co. Ltd, 2023), R. Maezono, T. Ichibha, ISBN: 978-4627170322

[14] Such detailed configurations typically require trial and error, along with extensive online research. One of the primary purposes of tutorials like this book is to consolidate and provide such information in a structured manner.

[15] On the first execution, it may take about a minute for the response to appear. However, subsequent runs should return output immediately.

Chapter 12
Appendix B: Terminal Environment Setup (Windows Version)

We will set up an environment where “Linux OS” runs on a Windows machine. A program that enables the execution of “Ubuntu,” a distribution of Linux OS, is available on the Microsoft Store, which we will use in this book. Alongside setting up this basic operational environment, we will also proceed with the installation of the ab initio quantum Monte Carlo package “CASINO” and other necessary utilities. The following sections provide step-by-step instructions for this process.[1]

Before Starting the Setup

Ensure that your OS is upgraded to the latest version. Otherwise, the setup process during the setup may trigger an automatic upgrade, which could take up to 30 minutes in some cases.

Additionally, some users may already have a “WSL1” environment installed on their PC. It has been confirmed that this may interfere with the setup described in this book. To proceed with the environment setup without issues, it is necessary to **uninstall WSL1** beforehand.

12.1 Preparing the Linux OS Environment

For machines running Windows OS (Windows 10 or later), an “Ubuntu OS emulator” is available, which allows the creation of a “Linux OS environment.” Beginners may find this terminology somewhat intimidating. An emulator, as the

[1] Since these procedures depend on the latest OS updates, please check the support page referenced at the beginning of this book for the most up-to-date information.

R. Maezono, *Ab initio Quantum Monte Carlo Tutorial*,
https://doi.org/10.1007/978-981-95-4134-8_12

term "emulate" suggests, means "to imitate or replicate." Thus, an Ubuntu emulator does not run an actual Linux system but provides an environment that behaves as if it were Linux. From a practical user perspective within the scope of this book, this can effectively be considered a "Linux environment."

12.1.1 Setting Up the WSL Environment

Before installing the Ubuntu emulator, it is necessary to set up WSL (Windows Subsystem for Linux), which provides an execution environment for running Linux on Windows. Follow the steps below to install and configure WSL.

First, update Windows OS to the latest version by running "Windows Update." Click on the search bar at the bottom of the screen and type "Windows Update." Next, click on "Windows Update Settings" and then "Check for updates." Once the update is complete, your PC may prompt you to restart. Follow the instructions to restart the system.

Now, we are about to "install WSL on the PC." This installation is performed using PowerShell, the "command-line interface of Windows OS." Follow the steps below to open Windows PowerShell in administrator mode. Click on the search bar at the bottom of the screen and type "Windows PowerShell." When the Windows PowerShell option appears, click on "Run as administrator" to open PowerShell **in administrator mode**.

Enter the following command in PowerShell:

```
> wsl --install --no-distribution
```

Then, press Enter to execute it.[2] Once the WSL installation is complete, you will be prompted to restart your PC. Follow the instructions and restart your system.

Hacking Your PC Environment
The steps described in this book essentially boil down to "1: Installing Ubuntu," "2: Downloading the tutorial files," and "3: Installing various applications." However, along the way, additional steps such as "Installing WSL" and "Linking directories with 'ln -s'" are included. As of the time of writing,[3] the default settings of the OS do not allow Ubuntu to modify the administrator

(continued)

[2] If an error occurs at this step, WSL1 may already be installed. You will need to uninstall WSL1 and retry the installation.

[3] This is based on the OS environment in 2024.

permissions of the Windows file system, which can cause issues during the tutorials in this book. The **hacking** techniques detailed in this appendix serve as workarounds for these limitations. Depending on the OS environment at the time of working, some of these steps may become unnecessary, or adjustments may be required. When the OS is updated, incompatibilities with linked libraries can arise, necessitating new hacking solutions (For the latest information, please refer to the support page mentioned at the beginning of this book.). For beginners, these small yet essential hacks often become the biggest practical obstacles. Gaining experience through repeated successes in hacking is what ultimately transitions a beginner into a proficient user.

Nowadays, simply copying and pasting error messages into AI tools like ChatGPT can sometimes lead to quick solutions. Rather than merely lamenting that "an error occurred and it won't work," it is crucial to calmly analyze the cause of the error, understand the underlying issue, and search for workaround solutions online. However, without knowing the appropriate terminology or concepts that describe the root cause, efficiently searching for solutions becomes significantly more difficult. Gaining experience in troubleshooting errors is the fastest way to master these concepts.

12.1.2 Setting Up the Emulator

Linux has numerous "distributions" or variants. For Windows PCs, an "emulator" of the Ubuntu OS (pronounced "Oo-boon-too") is available online. This Linux environment can be downloaded for free from the Microsoft Store. At the **taskbar** located at the bottom of the screen (which will appear frequently in later sections), click on the "shopping bag icon." The "Microsoft Store" window will open. Click on the magnifying glass icon in the upper right corner, which will allow you to type into the search box. Type "ubuntu" in the search box, and several candidate apps will appear. Select the one labeled simply as "ubuntu" and click on it. Then, an "Install" button will be displayed. Click this button to begin the installation.

Once the installation is complete, the button that previously displayed "Install" will change to a button labeled "Open." Clicking this will launch the Ubuntu terminal.[4] Next, you will be prompted to set a username and password for Ubuntu. Here,

[4] In some cases, an error code "0x80070050" may be displayed, preventing the terminal from launching. If this happens, reinstalling Ubuntu may resolve the issue. Click the search bar at the bottom of the screen, type "Add or remove programs," and uninstall Ubuntu from there. Then, reinstall Ubuntu again. For the latest information, please refer to the support page mentioned at the beginning of this book.

carefully enter "`maezono`" for "`Enter new UNIX username`," and "`mammamia`" for both "`Enter new UNIX password`" and "`Retype new UNIX password`".[5]

Note that the username "`maezono`" and password set here are different from the "PC's actual username." The username "`maezono`" set here is for a "**virtual** Linux machine" installed on your PC. In the next section, settings related to the "PC's actual username" will be introduced. Be careful **not to confuse** the "virtual Linux machine username" set here with the "PC's actual username." Once the setup is complete, close the Ubuntu application for now.

Close Applications Regularly
To ensure optimal PC performance, make it a habit to close applications that are not in use. There was once a student who left dozens of applications and nearly a hundred PDF files open. When the author attempted to work on that student's PC, the response was significantly slow, revealing the issue for the first time.

On a Windows machine, running applications are displayed as icons at the bottom of the screen. Right-clicking on the respective application's icon will reveal the option "Close all windows." Clicking this option will terminate the application.[6]

12.1.3 Logging into the Terminal Environment

Type "Ubuntu" into the search bar in the taskbar, and a result labeled "Ubuntu App" will appear. Click the "Open" button below it to launch the Ubuntu emulator environment. A black screen will appear, but unlike the previous instance, you will not be asked to set a username or password. The cursor prompt will be blinking.[7] At the prompt, type **pwd** and press Enter. It should return `/home/maezono`. Here again, remember that "maezono" is not the "PC's actual username" but the "username of the virtual Linux machine."

Once the above has been confirmed, the terminal environment has been set up correctly. At this point, close the terminal application for now.

---

[5] Replace "maezono" with your own user name, and replace "mammamia" with the password you set yourself. If you happen to set the wrong password, you can reset it by executing the command "passwd." For details, refer to information about using Linux commands, which can be found on the Internet and other sources.

[6] On a Windows machine, closing all windows terminates the application.

[7] A prompt means "to encourage." It indicates that input is expected at that location.

12.1.4 Preparing the Working Directory

Next, we prepare a "working directory" on the PC for the tutorials. This is essentially "reserving a space" for learning Linux command operations, which must be done using commands before actually learning them. For absolute beginners, this might seem like a kind of "jargon operation".[8] Be very careful to avoid mistakes while typing.[9]

First, execute the following command:

```
% ls /mnt/c/Users/
```

to check what is set as **your own PC username**. A word of caution: some people mistakenly try to type the leading "%" as well. However, this symbol only represents the "terminal input prompt," indicating "Type here!" as a prompt. Therefore, when instructed to enter the above command, you should type only the part starting from "`ls`" onward.

Now, the command executed above will return **your own PC username**. In the following explanation, we will use "rmaez" as an example. However, wherever "rmaez" appears in the next few lines of command descriptions, you should **replace it with your own PC username**. To emphasize again, the "username set for the Ubuntu emulator" (maezono) and the "username of your Windows PC itself" (rmaez) are **different**. Please proceed carefully, keeping this distinction in mind.

Next, execute the following commands:[10]

```
% cd
% mkdir /mnt/c/Users/rmaez/work
```

Then, carefully execute the following commands:[11,12,13]

---

[8] Like reciting an unfamiliar prayer or chant, where one simply types what is written without understanding its meaning.

[9] After mastering Linux commands, reviewing this section again should make it clear what was being done.

[10] Beginners should make use of **tab completion** to avoid typos.

[11] Be particularly careful with this operation. A mistyped command may require extra steps to recover. Pay close attention to the placement of spaces, as most beginner mistakes stem from incorrect spacing. For those who feel uncertain, it may be advisable to first type the command into a text editor, such as Notepad, and verify multiple times that there are no mistakes before copying and pasting it into the terminal.

[12] From this point onward, instead of typing everything from scratch, you can use the up arrow key in the terminal to recall the previously executed command. By utilizing this feature and modifying only the last argument, you can reduce the amount of typing required.

[13] In Windows, opening the "home directory on the Ubuntu emulator" is oddly cumbersome, and the following setup is designed to address this issue. The reason for this inconvenience is that "`~/work` on the emulator is created on a separate device in Windows OS. This is a key difference between "Mac, where UNIX-OS runs natively" and "Windows, where Linux is

```
% cd
% ln -s /mnt/c/Users/rmaez/work
% ln -s /mnt/c/Users/rmaez/Downloads
```

Next, try entering the commands below and confirm that the displayed directory name matches "`/home/maezono/work`":

```
% cd
% cd work
% pwd
```

For further details, refer to the Linux tutorial "appendix_linux.pdf," which you will download later. This document explains that the notation:

`'/home/maezono/work' = '~/work'`

is used. The portion " ~/" replaces "`/home/maezono/`," which serves as the starting point of the user's directory structure and is called the **home directory**.

12.2 Installing CASINO

12.2.1 Obtaining and Placing the Tutorial Set

Before downloading the tutorial files, you need to shut down and restart WSL (Windows Subsystem for Linux), which was launched in Sect. 12.1.1.[14] First, close the Ubuntu application. Then, instead of running PowerShell as an administrator, open it in normal user mode, as explained in Sect. 12.1.1. From the command line in the opened PowerShell, execute:

```
> wsl --shutdown
```

After that, restart Ubuntu.

With this, you are now ready to download the tutorial files from the internet. The tutorial files include: a basic Linux operations tutorial (used in the web appendix) and a set of input files for a series of tutorials used from Chap. 3 onward. These files are publicly available on the authors' GitHub server and can be downloaded using the "`git`" command as follows.

constructed virtually as an emulator." The command "`% ln -s ...`" is used to link the `work` directory, which is created on a separate device, to a location where it appears as if it is a `work` directory visible from Windows. This operation is intended to make "the directory structure in terminal operations on Windows machines" appear more Linux-like. The descriptions in this book assume that these steps have been performed, allowing the instructions to be written in a way that applies equally to both Mac and Windows environments.

[14] Otherwise, Ubuntu will not be able to modify file ownership permissions, which has been confirmed to cause failures in downloading the tutorial files via git, as described later.

First, execute the following commands to navigate to the "working directory" prepared in Sect. 12.1.4.

```
% cd
% cd work
```

Next, enter "`% pwd`" and confirm that the displayed directory name matches "`/home/maezono/work`". Then, execute the following command:

```
% git clone https://github.com/rmaezono/qmcTutorials.git
```

If messages such as:

```
    Cloning into 'qmcTutorials'...
    remote: Enumerating objects: ...
...
```

appear on the screen, the download is complete. Next, execute the "`% ls`" command. You should see the directory "`qmcTutorials`" listed. The tutorial set is stored within this directory.

The commands appearing in the subsequent installation steps are also listed in the downloaded file:

```
~/work/qmcTutorials/docs/ForWin.txt
```

Beginners who are unsure about typing accuracy can open this file and copy-paste the commands as needed to avoid errors.

12.2.2 Appending to the .bash_profile File

Next, use redirection to append the following content to the .bash_profile file:[15]

```
% cd
% echo 'export CASINO_ARCH=linuxpc-gcc-parallel' >> .bash_profile
% echo 'export PATH="$PATH":$HOME/work/qmcTutorials/
    [Do not insert a line break; continue on the next line]
    casino/bin_qmc/$CASINO_ARCH/opt' >> .bash_profile
% echo 'export PATH="$PATH":$HOME/work/qmcTutorials/
    [Do not insert a line break; continue on the next line]
    casino/bin_qmc' >> .bash_profile
```

The uppercase variables, such as CASINO_ARCH and PATH%/PLOTTER, are called environment variables. Each time a new terminal session is started, the

[15] Since the number of characters to be typed is long, beginners are more likely to make mistakes. It is safer to copy and paste from ~/work/qmcTutorials/docs/ForWin.txt, which was downloaded as part of the tutorial. Be careful **not to copy and paste from this PDF file**. Copying from a PDF may alter quotation marks from "actual symbols" to "font-specific representations," which can cause issues.

contents of .bash_profile are executed, assigning these environment variables the specified values. The system then uses these environment variables appropriately to perform the desired operations. CASINO_ARCH is referenced when installing the CASINO package to identify the target machine's architecture. The PATH variable is configured to ensure that the commands related to the newly installed CASINO are included in the command path.

After executing the above echo commands, check the contents of the .bash_profile file by running:

```
% cd
% cat .bash_profile
```

Verify that the expected lines have been appended correctly. If everything looks correct, exit the terminal with:

```
% exit
```

Then, restart the terminal.

12.2.3 Installing the Required Environment

Next, install the necessary packages for compilation and exercises using the package management tool apt.

```
% sudo apt update
% sudo apt install libopenmpi-dev openmpi-bin gfortran
% sudo apt install emacs
% sudo apt install make
% sudo apt install g++
% sudo apt install gnuplot-x11
% sudo apt install grace
```

The packages listed in the line starting with "libopenmpi-dev..." include the compiler and libraries for parallel computation, which are essential for installing CASINO. Emacs is a text editor for editing input files, while gnuplot-x11 and grace are tools for visualizing results.

12.2.4 Installing CASINO

Finally, install the ab initio quantum Monte Carlo package "CASINO" by compiling it from source. Navigate to the following directory and verify that the required files are present:

```
% cd
% cd work/qmcTutorials/casino
% ls
    Makefile   arch/ ...
```

Then, execute the compilation as follows:[16]

```
% make -j 4 opt
```

If the compilation completes successfully, you should find the following file:

```
% pwd
    /Users/maezono/work/qmcTutorials/casino/bin_qmc/linuxpc-gcc-parallel/opt/
% ls
    casino
```

With this, the compilation of CASINO is complete. Return to Sect. 2.1 in the main text and proceed with the next steps.

12.3 Installation from an Emulator

12.3.1 Installing QuantumEspresso

Next, install the electronic structure calculation package **QuantumEspresso**, which is used in this book, by downloading it via Linux's **apt** package management system.[17]

```
% sudo apt install -y quantum-espresso
```

After the installation is complete, executing "`pw.x`" should display the following message if the installation was successful:

```
% pw.x
     Program PWSCF v.6.0 (svn rev. 13079) starts on 30Oct2019 at 18:59:17
        ...
     Parallel version (MPI), running on     1 processors
     Waiting for input...
```

After confirming this message, exit by pressing "Ctrl+C" (refer to the web appendix).

[16] The option "-j 4" of the make command means "compile in parallel with 4 cores." Adjust the specification "4 cores" appropriately according to how many cores are available on your own CPU.

[17] As mentioned in the footnote of Sect. 5.1.3, "-pw2casino" seems to be no longer supported in versions later than v7.4. See "Version Information" at the beginning of this book.

Chapter 13
Appendix C: Derivation of the Diffusion Equation from Random Walk

This chapter provides detailed mathematical derivations related to the diffusion equation, separated into an appendix for clarity. Section 13.1 is guided by the main text Sect. 7.1. Sections 13.2.1 and 13.2.2 are guided by the main text. Section 13.2.3 is guided by the main text Sect. 7.3.1. Section 13.2.4 is guided by the main text Sect. 8.3.4. Section 13.2.5 is guided by the main text Sect. 9.1.1.

13.1 Random Walk and the Diffusion Equation

13.1.1 Derivation of the Moyal Expansion

To derive the differential form of the Chapman–Kolmogorov equation, Eq. (7.6),

$$P\left(x_3, t_3; x_1, t_1\right) = \int dx_2 \cdot P\left(x_3, t_3; x_2, t_2\right) \cdot P\left(x_2, t_2; x_1, t_1\right) \quad \cdots \left(t_1 < t_2 < t_3\right) \tag{13.1}$$

we consider evaluating $P\left(x_2, t_1 + \Delta t; x_1, t_1\right)$ and taking the limit $\Delta t \to 0$. To first order in Δt, we write

$$P\left(x_2, t_1 + \Delta t; x_1, t_1\right) = F \cdot \delta\left(x_2 - x_1\right) + \Delta t \cdot W\left(x_1 \to x_2\right) \tag{13.2}$$

The first term represents the contribution for $\Delta t = 0$, where no transition occurs and $x_2 = x_1$, hence the appearance of $\delta\left(x_2 - x_1\right)$. The coefficient F can be determined from the normalization condition over possible destinations of the transition:

$$\int dx_2 \cdot P(x_2, t_1 + \Delta t; x_1, t_1) = 1$$

R. Maezono, *Ab initio Quantum Monte Carlo Tutorial*,
https://doi.org/10.1007/978-981-95-4134-8_13

Substituting Eq. (13.2) into the above gives

$$F = 1 - \Delta t \int dx' \cdot W(x_1 \to x')$$

Substituting this result into Eq. (13.2) yields

$$P(x_2, t_1 + \Delta t; x_1, t_1) = \delta(x_2 - x_1) \left\{ 1 - \Delta t \int dx' \cdot W(x_1 \to x') \right\} + \Delta t \cdot W(x_1 \to x_2) \tag{13.3}$$

In the Chapman–Kolmogorov equation (Eq. 13.1), we set $t_3 = t_2 + \Delta t$ and write

$$P(x_3, t_2 + \Delta t; x_1, t_1) = \int dx_2 \cdot P(x_3, t_2 + \Delta t; x_2, t_2) \cdot P(x_2, t_2; x_1, t_1)$$

Then, by substituting Eq. (13.3) for $P(x_3, t_2 + \Delta t; x_2, t_2)$, we obtain

$$\begin{aligned} &P(x_3, t_2 + \Delta t; x_1, t_1) \\ &= \int dx_2 \cdot P(x_3, t_2 + \Delta t; x_2, t_2) \cdot P(x_2, t_2; x_1, t_1) \\ &= \int dx_2 \cdot \begin{pmatrix} \delta(x_3 - x_2)\left\{1 - \Delta t \int dx' \cdot W(x_2 \to x')\right\} \\ + \Delta t \cdot W(x_2 \to x_3) \end{pmatrix} \cdot P(x_2, t_2; x_1, t_1) \\ &= \left\{ 1 - \Delta t \int dx' \cdot W(x_3 \to x') \right\} P(x_3, t_2; x_1, t_1) \\ &\quad + \Delta t \int dx_2 \cdot W(x_2 \to x_3) \cdot P(x_2, t_2; x_1, t_1) \end{aligned}$$

By relabeling the integration variable x_2 in the second term as x' and organizing the terms, we obtain

$$\begin{aligned} &P(x_3, t_2 + \Delta t; x_1, t_1) - P(x_3, t_2; x_1, t_1) \\ &= -\Delta t \int dx' \cdot W(x \to x') \cdot P(x_3, t_2; x_1, t_1) \\ &\quad + \Delta t \int dx' \cdot W(x' \to x_3) \cdot P(x', t_2; x_1, t_1) \end{aligned}$$

Then, rewriting $(x_3, t_2) \to (x, t)$, we arrive at

$$\frac{P(x, t + \Delta t; x_1, t_1) - P(x, t; x_1, t_1)}{\Delta t}$$

$$= -\int dx' \cdot W(x \to x') \cdot P(x, t; x_1, t_1) + \int dx' \cdot W(x' \to x) \cdot P(x', t; x_1, t_1)$$

Taking the limit $\Delta t \to 0$, we obtain

$$\frac{\partial P(x, t; x_1, t_1)}{\partial t} = -\int dx' \cdot W(x \to x') \cdot P(x, t; x_1, t_1) + \int dx' \cdot W(x' \to x) \cdot P(x', t; x_1, t_1)$$

This yields the evolution equation for $P(x, t; x_1, t_1)$. This equation is called the **master equation**, and it describes the balance between the inflow (second term) and outflow (first term) of probability at position x.

Let us now revisit the origin of $W(x_1 \to x_2)$ as introduced in Eq. (13.2). For $x_1 < x_2$, the displacement $\xi = x_2 - x_1$ is considered as a distance advanced during Δt, and we assume $\xi \ll 1$. With this in mind, we rewrite the equation as

$$\frac{\partial P(x, t; x_1, t_1)}{\partial t} = -\int d\xi \cdot [W(x \to x + \xi) \cdot P(x, t; x_1, t_1) - W(x - \xi \to x) \cdot P(x - \xi, t; x_1, t_1)]$$

Defining

$$G(x) := W(x \to x + \xi) \cdot P(x, t; x_1, t_1),$$

we obtain the following form:

$$\frac{\partial P(x, t; x_1, t_1)}{\partial t} = -\int d\xi \cdot [G(x) - G(x - \xi)]$$

Applying the Taylor expansion, we write

$$G(x - \xi) = G(x) + \sum_{n=1}^{\infty} \frac{1}{n!} \left(-\xi \frac{d}{dx}\right)^n G(x)$$

Substituting this into the previous expression, we obtain

$$\begin{aligned} \frac{\partial P(x, t; x_1, t_1)}{\partial t} &= -\int d\xi \cdot \left[-\sum_{n=1}^{\infty} \frac{1}{n!} \left(-\xi \frac{d}{dx}\right)^n G(x)\right] \\ &= \int d\xi \cdot \sum_{n=1}^{\infty} \frac{1}{n!} \left(-\xi \frac{d}{dx}\right)^n \{W(x \to x + \xi) \cdot P(x, t; x_1, t_1)\} \end{aligned}$$

$$= \sum_{n=1}^{\infty} \frac{1}{n!} \left(-\frac{d}{dx}\right)^n \left\{ \left(\int d\xi \cdot \xi^n \cdot W(x \to x+\xi) \right) \cdot P(x, t; x_1, t_1) \right\}$$

Introducing the definition

$$\alpha_n(x) = \int d\xi \cdot \xi^n \cdot W(x \to x + \xi) \tag{13.4}$$

we finally obtain

$$\frac{\partial P(x, t; x_1, t_1)}{\partial t} = \sum_{n=1}^{\infty} \frac{1}{n!} \left(-\frac{d}{dx}\right)^n \{\alpha_n(x) \cdot P(x, t; x_1, t_1)\}$$

Equation (13.4) takes the form of the n-th moment, but it is expressed in terms of the intermediate quantity $W(x_1 \to x_2)$ that was introduced along the way. By using Eq. (13.3), the expression for $\alpha_n(x)$ can be rewritten in terms of the original quantity $P(x_2, t_2; x_1, t_1)$. By setting $(x_1, t_1) \to (x, t)$ and $x_2 \to x+\xi$ in Eq. (13.3), we have

$$P(x + \xi, t + \Delta t; x, t) = \delta(\xi) \left\{ 1 - \Delta t \int dx' \cdot W(x \to x') \right\} + \Delta t \cdot W(x \to x + \xi)$$

Multiplying both sides by ξ^n and integrating yields

$$\begin{aligned} \int d\xi \cdot \xi^n \cdot P(x+\xi, t+\Delta t; x, t) &= \int d\xi \cdot \xi^n \cdot \delta(\xi) \left\{ 1 - \Delta t \int dx' \cdot W(x \to x') \right\} \\ &\quad + \Delta t \int d\xi \cdot \xi^n \cdot W(x \to x + \xi) \\ &= \Delta t \int d\xi \cdot \xi^n \cdot W(x \to x + \xi) = \Delta t \cdot \alpha_n(x) \end{aligned}$$

Thus, we obtain

$$\alpha_n(x) = \frac{\int d\xi \cdot \xi^n \cdot P(x + \xi, t + \Delta t; x, t)}{\Delta t}$$

In the integral appearing in the numerator, since all arguments of $P(x + \xi, t + \Delta t; x, t)$ other than ξ are fixed, we can write

$$\int d\xi \cdot \xi^n \cdot P(x + \xi, t + \Delta t; x, t) = \int d\xi \cdot \xi^n \cdot P_\xi(\xi) = \langle \xi^n \rangle \tag{13.5}$$

This shows that the integral represents the "n-th moment with respect to the distribution function $P_\xi(\xi)$ of the transition displacement ξ". Since $\xi = \Delta x$, we can write

$$\alpha_n(x) = \lim_{\Delta t \to 0} \frac{\langle \Delta x^n \rangle}{\Delta t}$$

13.1.2 Derivation of the Fokker–Planck Equation

Starting from the Langevin equation, Eq. (7.10),

$$\frac{dx}{dt} = F(x,t) + \gamma(t) \tag{13.6}$$

we integrate it to obtain

$$\begin{aligned}\Delta x &= \int_t^{t+\Delta t} F\left(x,t'\right) + \gamma\left(t'\right) \cdot dt' \\ &= F(x,t) \cdot \Delta t + \int_t^{t+\Delta t} \gamma\left(t'\right) \cdot dt'\end{aligned}$$

Here, we write

$$\underline{\Delta x} = F(x,t) \cdot \Delta t + \int_t^{t+\Delta t} \underline{\gamma\left(t'\right)} \cdot dt' \tag{13.7}$$

with underlines indicating which terms are stochastic variables. Keeping in mind the normalization condition of the distribution function $p(\gamma)$, we find

$$\begin{aligned}\langle A + \underline{B} \rangle &= \int d\gamma \cdot p(\gamma) \cdot \left(A + \underline{B(\gamma)}\right) \\ &= A \int d\gamma \cdot p(\gamma) + \int d\gamma \cdot p(\gamma) \cdot \underline{B(\gamma)} \\ &= A + \langle \underline{B} \rangle\end{aligned}$$

Now, applying Eq. (14.2) from white Gaussian noise (Sect. 14.1),

$$\langle \gamma(t) \rangle = 0 \quad , \quad \langle \gamma(t)\, \gamma\left(t'\right) \rangle = \sigma^2 \cdot \delta\left(t - t'\right) \tag{13.8}$$

we find that since $\langle \gamma \rangle = 0$, it follows that

$$\langle \underline{\Delta x} \rangle = F(x,t) \cdot \Delta t + \int_t^{t+\Delta t} \langle \underline{\gamma\left(t'\right)} \rangle \cdot dt' = F(x,t) \cdot \Delta t\ .$$

Next, consider $\langle \underline{\Delta x^2} \rangle$. Since $\langle \gamma \rangle = 0$, we have

$$
\begin{aligned}
\langle \underline{\Delta x^2} \rangle &= \left\langle \left(F(x,t) \cdot \Delta t + \int_t^{t+\Delta t} \underline{\gamma(t')} \cdot dt' \right) \right\rangle^2 \\
&= \left\langle F^2(x,t) \cdot \Delta t^2 + 2F(x,t) \cdot \Delta t \int_t^{t+\Delta t} \underline{\gamma(t')} \cdot dt' \right. \\
&\quad \left. + \int_t^{t+\Delta t} \underline{\gamma(t')} \cdot dt' \int_t^{t+\Delta t} \underline{\gamma(t'')} \cdot dt'' \right\rangle \\
&= F^2(x,t) \cdot \Delta t^2 + \iint_t^{t+\Delta t} \underline{\langle \gamma(t')\gamma(t'') \rangle} \cdot dt' dt''
\end{aligned}
$$

Using Eq. (13.8), we obtain

$$
\begin{aligned}
\langle \underline{\Delta x^2} \rangle &= F^2(x,t) \cdot \Delta t^2 + \sigma^2 \iint_t^{t+\Delta t} \delta(t' - t'') \cdot dt' dt'' \\
&= F^2(x,t) \cdot \Delta t^2 + \sigma^2 \cdot \Delta t
\end{aligned}
$$

13.1.3 Derivations in the Multi-dimensional and Multi-particle Case

When the particle position is extended to an M-dimensional space, the arguments of the transition probability become

$$
P\left(\mathbf{x}', t'; \mathbf{x}, t\right) = P\left(x_1', x_2', \cdots x_M', t'; \mathbf{x}, t\right)
$$

Accordingly, the Taylor expansion takes the form

$$
P(\mathbf{x}+\Delta\mathbf{x}, t; \mathbf{x}_0, t_0) = P(\mathbf{x}, t; \mathbf{x}_0, t_0) + \sum_i \frac{\partial P}{\partial x_i} \Delta x_i + \frac{1}{2} \sum_{ij} \frac{\partial^2 P}{\partial x_i \partial x_j} \Delta x_i \Delta x_j + \cdots
$$

In correspondence with this, the Moyal expansion in Eq. (7.9) becomes

$$
\begin{aligned}
\frac{\partial \rho(\mathbf{x},t)}{\partial t} &= \sum_{n=1}^{\infty} \frac{1}{n!} \left(-\frac{\partial^n}{\partial x_1 \cdots \partial x_n} \right) \left[\rho(\mathbf{x},t) \langle \Delta x_1 \cdots \Delta x_n \rangle \right] \\
&= \sum_i \frac{-\partial \rho(\mathbf{x},t)}{\partial x_i} \Delta x_i + \frac{1}{2} \sum_{ij} \frac{\partial^2 \rho(\mathbf{x},t)}{\partial x_i \partial x_j} \langle \Delta x_i \Delta x_j \rangle + \cdots
\end{aligned}
$$

Next, we evaluate Δx_i by specifying a concrete model, and substitute the result into the expression above. Assume that the particle positions evolve in time according to the Langevin equation Eq. (7.10) for each coordinate component in the M-dimensional space:

$$\frac{dx_k}{dt} = F_k(\mathbf{x}, t) + \gamma_k(t) \quad , \quad (k = 1, \cdots, M)$$

Then, the multi-dimensional version of Eq. (7.13) becomes

$$\underline{\langle \Delta x_j \rangle} = F_j(x, t) \cdot \Delta t \quad , \quad \underline{\langle \Delta x_i \Delta x_j \rangle} \sim \sigma^2 \cdot \Delta t \cdot \delta_{ij} \tag{13.9}$$

Substituting this result and using the identification from Eq. (7.16),

$$D = \frac{\sigma^2}{2} \quad , \quad \tilde{F} = \frac{F}{D}$$

we obtain the multi-dimensional version of the Fokker–Planck equation:

$$\frac{\partial \rho\left(\{x_k\}_{k=1}^M, t\right)}{\partial t} = D \sum_{m=1}^{M} \frac{\partial}{\partial x_m}\left[\frac{\partial}{\partial x_m} - \tilde{F}_m\left(\{x_k\}_{k=1}^M, t\right)\right] \rho\left(\{x_k\}_{k=1}^M, t\right) . \tag{13.10}$$

The corresponding random walk process, by the same reasoning as that used to derive Eq. (7.18), is given by

$$\underline{\Delta x_k} = D \cdot \tilde{F}_k(\mathbf{x}, t) \cdot \Delta t + \chi_k^{\mathrm{Gauss}}[0, 2D\Delta t] \ , \quad (k = 1, \cdots, M) . \tag{13.11}$$

When there are N particles in three-dimensional space, the system becomes $M = 3N$ dimensional, and we define

$$\mathbf{R} := \begin{pmatrix} x_{k=1} \\ \vdots \\ x_{k=3N} \end{pmatrix} = \begin{pmatrix} \mathbf{x}_{j=1} \\ \vdots \\ \mathbf{x}_{j=N} \end{pmatrix} \quad , \quad \tilde{\mathbf{F}}^{(3N)} = \begin{pmatrix} \tilde{F}_{k=1} \\ \vdots \\ \tilde{F}_{k=3N} \end{pmatrix} = \begin{pmatrix} \tilde{\boldsymbol{F}}_{j=1} \\ \vdots \\ \tilde{\boldsymbol{F}}_{j=N} \end{pmatrix} \tag{13.12}$$

Then, for the summation index m in Eq. (13.10), we rewrite it by separating into particle index j and spatial component index α as

$$a_m = a_{(j,\alpha)} = a_\alpha^{(j)} .$$

Then the inner product can be written as

$$\sum_{m=1}^{3N} a_m b_m = \sum_{j=1}^{N}\left(\sum_{\alpha=1}^{3} a_\alpha^{(j)} b_\alpha^{(j)}\right) = \sum_{j=1}^{N}\left(\mathbf{a}^{(j)} \cdot \mathbf{b}^{(j)}\right) = \sum_{j=1}^{N}\left(\mathbf{a}_j \cdot \mathbf{b}_j\right) .$$

Using this, the Fokker–Planck equation becomes

$$\frac{\partial \rho\left(\{x_k\}_{k=1}^{3N}, t\right)}{\partial t} = D \sum_{j=1}^{N} \nabla_j \left[\nabla_j - \tilde{\boldsymbol{F}}_j\left(\{x_k\}_{k=1}^{3N}, t\right)\right] \rho\left(\{x_k\}_{k=1}^{3N}, t\right)$$

or equivalently,

$$\frac{\partial \rho\left(\{\mathbf{x}_j\}_{j=1}^{N}, t\right)}{\partial t} = D \sum_{j=1}^{N} \nabla_j \left[\nabla_j - \tilde{\boldsymbol{F}}_j\left(\{\mathbf{x}_j\}_{j=1}^{N}, t\right)\right] \rho\left(\{\mathbf{x}_j\}_{j=1}^{N}, t\right)$$

Alternatively, using the compact configuration snapshot notation $\mathbf{R} = (\mathbf{x}_1, \cdots, \mathbf{x}_N) \in \mathcal{R}^{3N}$, it can be written as

$$\frac{\partial \rho(\mathbf{R}, t)}{\partial t} = D \sum_{j=1}^{N} \nabla_j \left[\nabla_j - \tilde{\boldsymbol{F}}_j(\mathbf{R}, t)\right] \rho(\mathbf{R}, t) \quad , \quad \mathbf{R} \in \mathcal{R}^{3N} .$$

The corresponding random walk from Eq. (13.11) can be written as "random walk of the j-th particle in 3D space":

$$\underline{\boldsymbol{\Delta x}}_j = D \cdot \tilde{\mathbf{F}}_j(\mathbf{R}, t) \cdot \Delta t + \chi_{\text{Gauss}}[0, 2D\Delta t] \quad , \quad (j = 1, \cdots, N)$$

or, collectively as "random walk of the configuration snapshot $\mathbf{R}$ in $3N$-dimensional space":

$$\underline{\boldsymbol{\Delta R}} = D \cdot \tilde{\mathbf{F}}^{(3N)}(\mathbf{R}, t) \cdot \Delta t + \chi_{\text{Gauss}}^{(3N)}[0, 2D\Delta t]$$

Here, the notation $\chi_{\text{Gauss}}[0, 2D\Delta t]$ denotes a multi-dimensional vector whose components are independent white Gaussian noises with mean 0 and variance $2D\Delta t$.

13.2 Topics Related to Importance Sampling

13.2.1 Equilibrium Solution of the Fokker–Planck Equation

Let the equilibrium solution be denoted by $\bar{\rho}(\mathbf{R})$. By setting the time derivative in the left-hand side of Eq. (7.22) to zero, we obtain

$$0 = \sum_j D \nabla_j \left[\nabla_j - \mathbf{F}_j(\mathbf{R})\right] \bar{\rho}(\mathbf{R})$$

$$\therefore \qquad \nabla_j^2\,\bar{\rho}\,(\mathbf{R}) - \boldsymbol{\nabla}_j\left[\mathbf{F}_j\,(\mathbf{R})\,\bar{\rho}\,(\mathbf{R})\right] = 0$$

$$\therefore \qquad \nabla_j^2\,\bar{\rho}\,(\mathbf{R}) = \bar{\rho}\,(\mathbf{R})\,\boldsymbol{\nabla}_j\mathbf{F}_j\,(\mathbf{R}) + \mathbf{F}_j\,(\mathbf{R})\,\boldsymbol{\nabla}_j\bar{\rho}\,(\mathbf{R})$$

We now consider the problem of choosing $\mathbf{F}_j\,(\mathbf{R})$ such that the above identity holds. If we try setting the first term on the right-hand side $\mathbf{F}_j\,(\mathbf{R})$ as

$$\mathbf{F}_j\,(\mathbf{R}) = \frac{1}{\bar{\rho}}\boldsymbol{\nabla}_j\bar{\rho}\,(\mathbf{R})$$

we may expect that this will lead to a form resembling $\nabla_j^2\,\rho\,(\mathbf{R})$ on the left-hand side. Let us substitute this and evaluate the right-hand side:

$$\begin{aligned}
(\text{RHS}) &= \bar{\rho}\cdot\boldsymbol{\nabla}_j\left[\frac{1}{\bar{\rho}}\boldsymbol{\nabla}_j\bar{\rho}\right] + \frac{1}{\bar{\rho}}\boldsymbol{\nabla}_j\bar{\rho}\boldsymbol{\nabla}_j\bar{\rho} \\
&= \bar{\rho}\cdot\boldsymbol{\nabla}_j\left[\frac{1}{\bar{\rho}}\right]\cdot\boldsymbol{\nabla}_j\bar{\rho} + \bar{\rho}\cdot\frac{1}{\bar{\rho}}\boldsymbol{\nabla}_j\left[\boldsymbol{\nabla}_j\bar{\rho}\right] + \frac{1}{\bar{\rho}}\boldsymbol{\nabla}_j\bar{\rho}\boldsymbol{\nabla}_j\bar{\rho} \\
&= \bar{\rho}\cdot\frac{-1}{\bar{\rho}^2}\boldsymbol{\nabla}_j\bar{\rho}\cdot\boldsymbol{\nabla}_j\bar{\rho} + \nabla_j^2\cdot\bar{\rho} + \frac{1}{\bar{\rho}}\boldsymbol{\nabla}_j\bar{\rho}\boldsymbol{\nabla}_j\bar{\rho} = \nabla_j^2\cdot\bar{\rho}
\end{aligned}$$

Thus, both sides agree, confirming that the equality indeed holds.

13.2.2 *Time-Evolution Equation Mapped to f*

The time-evolution equation in DMC, Eq. (3.15), is

$$\frac{\partial\Phi\,(\mathbf{R},\tau)}{\partial\tau} = \left(\frac{1}{2}\sum_{j=1}^{N}\nabla_j^2 + E_T - V\,(\mathbf{R})\right)\Phi\,(\mathbf{R},\tau)$$

We will rewrite this more compactly as

$$-\,\partial_\tau\Phi = -\frac{1}{2}\nabla^2\Phi + v\cdot\Phi$$

$$v = V\,(\mathbf{R}) - E_T \quad , \quad \nabla^2 = \sum_{j=1}^{N}\nabla_j^2$$

Multiplying both sides by Y, we define

$$f\,(\mathbf{R},\tau) = Y\,(\mathbf{R})\cdot\Phi\,(\mathbf{R},\tau)$$

and proceed to rewrite the equation as

$$-\partial_\tau \Phi Y = -\frac{1}{2}\left(\nabla^2\Phi\right)Y + v\cdot\Phi Y$$

$$\therefore \quad -\partial_\tau f = -\frac{1}{2}\left(\nabla^2\Phi\right)Y + v\cdot f \tag{13.13}$$

Rewriting the first term on the right-hand side:

$$\begin{aligned}\left(\nabla^2\Phi\right)Y &= \nabla^2(\Phi Y) - \Phi\left(\nabla^2 Y\right) - 2(\nabla Y)(\nabla\Phi)\\ &= \nabla^2 f + \Phi\left(\nabla^2 Y\right) - 2\nabla\left(\Phi(\nabla Y)\right)\end{aligned}$$

Substituting this into Eq. (13.13) yields

$$\begin{aligned}-\partial_\tau f &= -\frac{1}{2}\left(\nabla^2 f - 2\nabla\left(\Phi(\nabla Y)\right) + \Phi\nabla^2 Y\right) + v\cdot f\\ &= -\frac{1}{2}\nabla\left(\nabla f - 2\Phi(\nabla Y)\right) + v\cdot f - \frac{1}{2}\Phi\nabla^2 Y\end{aligned}$$

Now, substituting $\Phi = Y^{-1}f$ gives

$$\begin{aligned}-\partial_\tau f &= -\frac{1}{2}\nabla\left(\nabla f - 2Y^{-1}f\cdot(\nabla Y)\right) + v\cdot f - \frac{1}{2}Y^{-1}f\nabla^2 Y\\ &= -\frac{1}{2}\nabla\left(\nabla - \frac{2}{Y}\nabla Y\right)f + \left(v - Y^{-1}\frac{\nabla^2}{2}Y\right)f\end{aligned}$$

Returning to the full notation and summarizing the result, we obtain the time-evolution equation satisfied by $f(\mathbf{R},\tau)$ as:

$$\begin{aligned}\frac{\partial f(\mathbf{R},\tau)}{\partial\tau} &= \frac{1}{2}\sum_{j=1}^{N}\nabla_j\left(\nabla_j - 2\frac{\nabla_j Y(\mathbf{R})}{Y(\mathbf{R})}\right)f(\mathbf{R},\tau)\\ &\quad - \left(V(\mathbf{R}) - E_T - Y^{-1}\cdot\frac{1}{2}\sum_{j=1}^{N}\nabla_j{}^2\cdot Y\right)f(\mathbf{R},\tau)\end{aligned}$$

13.2.3 Derivation of the Propagator of the Fokker–Planck Equation

From Eq. (7.33), we have

$$G(x,t;x_0) = \frac{1}{2\pi}\int dp\cdot\exp\left[ip(x_0-x)+tD\cdot ip\left[ip-\tilde{F}(x,t)\right]\right]$$
$$= \frac{1}{2\pi}\int dp\cdot\exp\left[-tD\cdot\left(p^2+p\cdot i\left(\tilde{F}(x,t)-\frac{(x_0-x)}{tD}\right)\right)\right].$$

By completing the square for the exponent,

$$p^2+p\cdot i\left(\tilde{F}(x,t)-\frac{(x_0-x)}{tD}\right)$$
$$=\left[p+\frac{i}{2}\left(\tilde{F}(x,t)-\frac{(x_0-x)}{tD}\right)\right]^2+\frac{1}{4}\left(\tilde{F}(x,t)-\frac{(x_0-x)}{tD}\right)^2$$

and using the Gaussian integral formula that holds even for complex variables [1],

$$\int dx\cdot\exp\left[-a(x+\xi)^2\right]=\sqrt{\frac{\pi}{a}}$$

we obtain

$$G(x,t;x_0)=\frac{1}{2\pi}\int dp\cdot\exp\left[-tD\left[p+\frac{i}{2}\left(\tilde{F}(x,t)-\frac{(x_0-x)}{tD}\right)\right]^2\right]$$
$$\times\exp\left[-\frac{tD}{4}\left(\tilde{F}(x,t)-\frac{(x_0-x)}{tD}\right)^2\right]$$
$$=\frac{1}{2\pi}\sqrt{\frac{\pi}{tD}}\cdot\exp\left[-\frac{1}{4tD}\left(tD\tilde{F}(x,t)-(x_0-x)\right)^2\right]$$
$$=\sqrt{\frac{1}{2\pi\cdot 2tD}}\cdot\exp\left[-\frac{\left[x-\left(x_0-tD\tilde{F}(x,t)\right)\right]^2}{4tD}\right].$$

13.2.4 Importance-Sampled DMC for Non-local Potentials

The time evolution equation (7.29) for $f(\mathbf{R},\tau)=\Psi(\mathbf{R},\tau)\cdot\Phi(\mathbf{R})$, obtained via importance sampling, is given by

$$\frac{\partial f(\mathbf{R},\tau)}{\partial \tau} = \frac{1}{2}\sum_{j=1}^{N} \nabla_j \left(\nabla_j - 2\frac{\nabla_j \Psi_\alpha(\mathbf{R})}{\Psi_\alpha(\mathbf{R})} \right) f(\mathbf{R},\tau)$$
$$- \left(E_L(\mathbf{R}) - E_T \right) f(\mathbf{R},\tau)$$

Here, the local energy E_L appears in the second term under the assumption that the potential V is a c-number. In this case, we can write $V = \Psi^{-1} V \Psi$, so that

$$\left(V + \Psi^{-1}\hat{T}\Psi \right) \cdot f = \left(\Psi^{-1} V \Psi + \Psi^{-1}\hat{T}\Psi \right) \cdot f$$
$$= \Psi^{-1}\left(V + \hat{T} \right) \Psi \cdot f = E_L \cdot f \tag{13.14}$$

and the kinetic energy $\hat{T}$ and potential V can be jointly expressed in terms of E_L. When the pseudopotential is a non-local operator such that $\hat{V} \neq \Psi^{-1}\hat{V}\Psi$, one must return to the expression

$$\left(\hat{V} + \Psi^{-1}\hat{T}\Psi \right) \cdot f$$
$$\to \left(\hat{V}\Phi \right)\Psi + \Psi^{-1}\hat{T}\Psi \cdot \Phi\Psi =: I$$

and perform the evaluation from that point.[1] This I can be evaluated as[2]

$$I = \left[\Psi^{-1}\hat{V}\Psi \cdot \Phi\Psi - \Psi^{-1}\hat{V}\Psi \cdot \Phi\Psi \right] + \left(\hat{V}\Phi \right)\Psi + \Psi^{-1}\hat{T}\Psi \cdot \Phi\Psi$$
$$= E_L \cdot f + \left(\hat{V}\Phi \right)\Psi - \Psi^{-1}\hat{V}\Psi \cdot \Phi\Psi = E_L \cdot f + \delta\ .$$

The remainder δ becomes

$$\delta = \left(\hat{V}\Phi \right)\Psi - \Psi^{-1}\hat{V}\Psi \cdot \Phi\Psi = \frac{\hat{V}\Phi}{\Phi}\Phi\Psi - \Psi^{-1}\hat{V}\Psi \cdot \Phi\Psi$$
$$= \left(\frac{\hat{V}\Phi}{\Phi} - \frac{\hat{V}\Psi}{\Psi} \right) \cdot \Phi\Psi$$

and thus Eq. (8.13) is obtained.

[1] For the term $\left(\hat{V}\Phi \right)\Psi$, recall Eq. (13.13). Originally, in the Schrodinger equation for Φ, we had $\hat{V}\Phi$, and we introduced $f = \Phi\Psi$ by multiplying both sides by $Y = \Psi$. Accordingly, $\hat{V}$ acts only on Φ, not on Ψ.

[2] In the first line on the right-hand side, the term $0 = [\cdots]$ is added. The first term in this bracket is combined with $\hat{T}$ into E_L, and the remaining term is expressed as the remainder δ.

13.2.5 Choice of Time Step

The time step $\delta\tau$ refers to the time interval associated with one step of advancement by a sampling particle (walker). Before entering the statistical accumulation phase, these sampled configurations must be sufficiently spread throughout the system. Otherwise, fair sampling-based evaluation cannot be achieved. When L_{sys} denotes the characteristic length scale of the system (e.g., Wigner–Seitz length for solids,[3] molecular size, *etc.*), the diffusion length L_{diff} of the sampling particles determined by the time step $\delta\tau$ must satisfy the relation $L_{\mathrm{diff}} > L_{\mathrm{sys}}$. Ensuring that this condition is met by sufficiently diffusing the walkers throughout the system is the role of the equilibration process (see Sect. 3.3.2).

When the number of steps required for equilibration is denoted by N_{eq}, the total diffusion time for the equilibration process is given by $\delta T = \delta\tau \cdot N_{\mathrm{eq}}$. The relationship between diffusion length and diffusion time in the diffusion process is[4]

$$L_{\mathrm{diff}} = \sqrt{\delta T}\ .$$

Therefore,

$$L_{\mathrm{diff}} = \sqrt{\delta T} = \sqrt{\delta\tau \cdot N_{\mathrm{eq}}} > L_{\mathrm{sys}}$$

is the required condition described above. If we denote the average inter-electron distance derived from the charge density of the target system as r_s, then in a D-dimensional system, the average volume occupied by one electron is $\sim r_s^D$. With N_{el} electrons, the characteristic system length L_{sys} is related by

$$L_{\mathrm{sys}}^D = N_{\mathrm{el}} \cdot r_s^D$$

Substituting into the previous condition gives

$$N_{\mathrm{eq}} > \frac{N_{\mathrm{el}}^{2/D} \cdot r_s^2}{D \cdot d\tau}$$

This shows that the choice of equilibration step count N_{eq} and the time step $d\tau$ in DMC are interrelated. It is essential to set these parameters while imagining: "Will the walkers be able to sufficiently diffuse throughout the system after N_{eq} steps with this $d\tau$?."

Although it is now hidden from the user, a time step parameter "dtvmc" also exists in VMC. Since the introductory explanation of VMC given in Sect. 2.2

[3] This corresponds to what is called the Wigner–Seitz length.

[4] See the footnote just before Eq. (7.18).

contains no mention of a time step, this may confuse beginners. To explain the concept of importance sampling in VMC, a simplified discussion based on the Metropolis algorithm was provided in Sect. 2.2.3. On the other hand, in Sect. 7.2.2, it was explained that importance sampling can also be achieved by using the equilibrium solution of the Fokker–Planck equation.

In practical package implementations, this Fokker–Planck-based importance sampling is also adopted in VMC, and the discretized time step used there is what is called "dtvmc." An appropriate initial value is set for dtvmc, and the simulation is run while adjusting it so that the acceptance ratio of Monte Carlo updates approaches around 50%. The resulting value, opt_dtvmc, is written in the out file. As for the $d\tau$ ("dtdmc") used in DMC, empirical guidelines exist within package development communities that suggest choosing it as a certain percentage of the opt_dtvmc value.

Reference

1. "Mathematical Methods for Physics and Engineering, Third Edition" (Cambridge University Press, 2006), K.F. Riley, M.P. Hobson, S.J. Bence, ISBN: 978-0521679718

Chapter 14
Appendix D: Supplementary Remarks on Mathematical Topics

This chapter provides supplementary explanations, as an appendix, on mathematical topics that appear in the main text. Section 14.1 is guided by the main text Sect. 7.1.3. Section 14.2 is guided by the main text Sect. 7.3.1. Section 14.3 is guided by the main text Sect. 8.2.4.

14.1 White Noise and White Gaussian Process

For a signal $\gamma(t)$, consider the following integral:

$$g(\tau) = \int^{T} dt \cdot \gamma(t)\,\gamma(t+\tau) \tag{14.1}$$

The integrand here gives a positive contribution when the time-shifted signal $\gamma(t+\tau)$ has the same sign as $\gamma(t)$, and a negative contribution when the signs are opposite. For a random noise signal, there is no correlation between signal values separated by time τ, and over a sufficiently long interval T, the positive and negative contributions largely cancel out, leading to an expected value near zero. In this sense, $g(\tau)$ serves as an indicator of whether the signal is temporally correlated and is called the **time correlation function** (auto-correlation).

In typical cases, the value of $\gamma(t)$ at a given time retains some "memory" for a while, and a realistic form for $g(\tau)$ is $g(\tau) \sim \exp(-\kappa\tau)$. An idealized noise model in which the signal loses its memory instantaneously as soon as any time has passed would then be represented as

$$g(\tau) = \int^{T} dt \cdot \gamma(t)\,\gamma(t+\tau) \propto \delta(\tau)$$

R. Maezono, *Ab initio Quantum Monte Carlo Tutorial*,
https://doi.org/10.1007/978-981-95-4134-8_14

Recalling that "the Fourier transform of a delta function is a constant," it follows that the Fourier transform $G(\omega)$ of the memoryless model noise's $g(\tau)$ is constant. Letting $\Gamma(\omega)$ be the Fourier transform of the signal $\gamma(t)$ itself, taking the Fourier transform of Eq. (14.1) yields[1]

$$G(\omega) = |\Gamma(\omega)|^2 .$$

The right-hand side, $|\Gamma(\omega)|^2$, is called the power spectrum of the signal. Therefore, for an "ideal noise signal $\gamma(t)$ that loses its memory instantaneously," the power spectrum is constant across all frequencies. In the context of "frequency spectra of light and their relation to color," white light is defined as "having equal weight across all frequency components." By analogy, such an idealized noise is referred to as **white noise**.

The integral in Eq. (14.1),

$$g(\tau) = \int^T dt \cdot \gamma(t)\gamma(t+\tau) = \left[\overline{\gamma(t)\gamma(t+\tau)}\right]_T$$

is a time average over the interval T, and with a sufficiently large T, one expects that "all possible values of the signal $\gamma(t)$ will be sampled." On the other hand,

$$\langle \gamma(t)\gamma(t+\tau)\rangle = \int d\gamma \cdot p(\gamma) \cdot \gamma(t)\gamma(t+\tau)$$

is an integral over the distribution $p(\gamma)$ of the possible values of γ, and is called the **ensemble average**. Although both represent averaging processes, the time average and ensemble average are defined differently, and are generally not equal. However, if the time average is taken over a sufficiently long duration, it is intuitively expected that "all possible states will be sampled" through the time dependence of the integrand. As a result, the time average should agree with the ensemble average. This agreement is referred to as **ergodicity**.

The function $p(\gamma)$ represents the "distribution over signal values" of the noise $\gamma(t)$. A model that assumes this distribution to be Gaussian (normal) is called **Gaussian noise**:[2]

$$\gamma(t) \sim p_G\left(\gamma; 0, \sigma^2\right) = \frac{1}{\sqrt{2\pi}\sigma} \exp\left[-\frac{\gamma^2}{2\sigma^2}\right]$$

For the moments of the distribution, we have

[1] This is known as the Wiener–Khinchin theorem [1].

[2] The notation $p_G\left(x; \mu, \sigma^2\right)$ represents a normal distribution function (in variable x) with mean μ and variance σ^2. The expression "$x(t) \sim p(x)$" may be confusing to beginners since $p(x)$ has no explicit t dependence. However, it means that the time series $x(t)$ takes random values according to the frequency distribution $p(x)$.

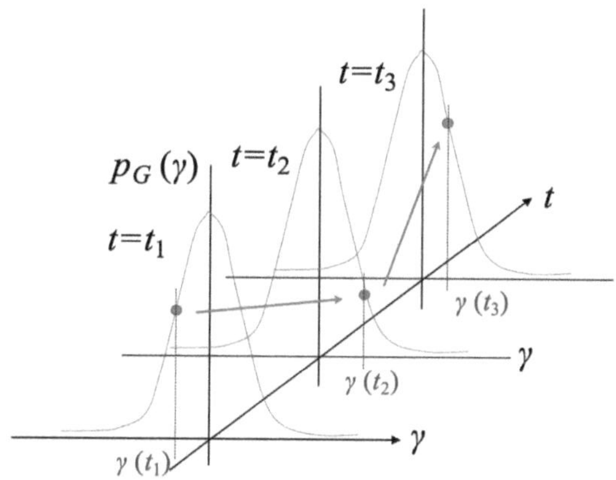

Fig. 14.1 An intuitive image of white Gaussian noise: At each time point t_j, the signal value $\gamma\left(t_j\right)$ is sampled from a Gaussian distribution. As time progresses, the signal evolves as $\gamma\left(t_1\right) \to \gamma\left(t_2\right) \to \cdots$, but in a white Gaussian process, there is no correlation between signal values across this time series

$$\langle\gamma\rangle = \int d\gamma \cdot \gamma \cdot p_G(\gamma) = 0 \quad , \quad \left\langle\gamma^2\right\rangle = \int d\gamma \cdot \gamma^2 \cdot p_G(\gamma) = \sigma^2$$

In the case of a Gaussian distribution, the moment-generating function that characterizes the distribution is completely determined by the first two moments alone [1].

"White noise with a Gaussian distribution" (**white Gaussian noise**) is thus frequently used as an idealized model that allows for analytical simplification (see Fig. 14.1). Including ergodicity, the properties of white Gaussian noise are expressed as

$$\langle\gamma(t)\,\gamma(t+\tau)\rangle = \int d\gamma \cdot p(\gamma) \cdot \gamma(t)\,\gamma(t+\tau) = \sigma^2 \cdot \delta(\tau)$$

or equivalently,

$$\left\langle\gamma(t)\,\gamma\left(t'\right)\right\rangle = \sigma^2 \cdot \delta\left(t-t'\right) \tag{14.2}$$

14.2 Bra–ket Notation

Bra–ket notation is typically introduced in the context of quantum mechanics, and as a result, many beginners misunderstand it as something exclusive to quantum systems. However, it is actually a convenient notational scheme for working in function spaces, and can be applied in both classical and quantum settings wherever basis function expansions are used.[3]

[3] This point is emphasized based on the author's experience teaching in a computer science department. Even students with strong academic abilities but without a background in physics

14.2.1 Introduction to Bra–ket Notation

Bra-ket notation[4] begins by expressing inner products in the form

$$\langle \phi | \psi \rangle = \int dx \cdot \phi^* (x) \cdot \psi (x) \tag{14.3}$$

We also adopt the notational convention of expressing a function $f(x)$ as

$$f(x) = \langle x | f \rangle \tag{14.4}$$

At first glance, this may be read as "the inner product of f and x," which can confuse beginners, what does it mean to take an inner product with x? However, this notation is internally consistent and widely accepted, so for now, let us treat it as a convention and proceed.[5] The above expression is interpreted as: "the abstract vector $|f\rangle$, when projected onto the Euclidean vector space $|x\rangle$, appears as the function $\langle x|f\rangle$, which is $f(x)$."[6] With this notation, it becomes possible to develop a clear and well-structured discussion by drawing analogies to the framework constructed for Euclidean vector spaces [1].

Let us consider the expansion of a function in terms of an orthonormal basis [1]:

$$f(x) = \sum_l f_l \cdot \chi_l (x) \quad , \quad \int dx \cdot \chi_j^* (x) \, \chi_l (x) = \delta_{jl} \, .$$

may shy away from understanding this topic when they see bra–ket notation, wrongly assuming that it belongs to an unfamiliar field.

4 This book introduces bra–ket notation in a rather informal manner, but there are more rigorous formal treatments available [2]. In fact, some of these point out that even authoritative texts have employed the notation in misleading ways. For the time being, however, we prioritize familiarity with the notation at the level used in this book. Later on, it is advisable to study the more formal frameworks, especially the aspects where many people tend to fall into common misunderstandings.

5 The bracket $\langle f|g\rangle$ is often introduced in an elementary context as merely "a notation for inner product." However, more precisely, it arises in the framework of **linear functionals** and **dual spaces**. Specifically, given a vector space element $|g\rangle$, we consider the dual space element $\langle f|$, which maps it to a number $I = \langle f|g\rangle$. This reflects a more mathematically rigorous interpretation grounded in the concept of duality. $\psi(x)$ is interpreted as "a linear functional that assigns a complex number $\psi(x)$ to each element $|x\rangle$ in the vector space." Then, the dual vector $\langle\psi|$ is constructed by pairing this linear functional with $|x\rangle$, which gives meaning to the notation $\psi(x) = \langle x|\psi\rangle$.

6 This interpretation, that a function is an "abstract entity $|f\rangle$," and that its representation is something chosen at human convenience, is a more modern perspective. This underlying idea appears, for example, in analytic continuation within complex analysis [3].

The expansion coefficients $\{f_j\}$ are given by

$$\int dx \cdot \chi_j^* (x) \cdot f (x) = \sum_l f_l \int dx \cdot \chi_j^* (x) \cdot \chi_l (x) = \sum_l f_l \cdot \delta_{jl} = f_j$$

$$\therefore \quad f_j = \int dx \cdot \chi_j^* (x) \cdot f (x) \ .$$

By using the bra–ket notation as in (14.4) and (14.3), the above basis function expansion can be written as

$$\langle x | f \rangle = \sum_l f_l \cdot \langle x | \chi_l \rangle \quad , \quad f_j = \langle \chi_j | f \rangle \tag{14.5}$$

If the basis set is complete, then by substituting the projected components f_j into Eq. (14.5), the original function $f (x) = \langle x | f \rangle$ can be reconstructed. Therefore,

$$\langle x | f \rangle = \sum_l \langle \chi_l | f \rangle \cdot \langle x | \chi_l \rangle \tag{14.6}$$

holds true. By reversing the order of the products on the right-hand side,

$$\langle x | f \rangle = \sum_l \langle x | \chi_l \rangle \langle \chi_l | f \rangle = \langle x | \cdot \sum_l |\chi_l\rangle \langle \chi_l | \cdot | f \rangle$$

this can be regarded as if $\langle x | f \rangle = \langle x | \cdot \hat{1} \cdot | f \rangle$, where the identity operator

$$\hat{1} = \sum_l |\chi_l\rangle \langle \chi_l | \tag{14.7}$$

has been inserted.

Equation (14.7) is a requirement obtained under the assumption of completeness, so it is referred to as the **completeness condition** of the basis $\{|\chi_j\rangle\}$. Once this is remembered, one can insert the completeness condition $\langle x | f \rangle = \langle x | \cdot \hat{1} \cdot | f \rangle$ to formally derive the basis expansion of an arbitrary function, as in Eq. (14.6), making this a convenient notational framework.

14.2.2 Some Expressions Using Bra–ket Notation

Let us write down the expression for the Fourier transform [1]:[7,8]

$$f(x) = \frac{1}{\sqrt{2\pi}} \int dp \cdot e^{ipx} \cdot f(p)$$

On the other hand, in bra–ket notation, $f(x)$ can be expressed as

$$f(x) = \langle x|f\rangle = \int dp \cdot \langle x|p\rangle\langle p|f\rangle = \int dp \cdot \langle x|p\rangle f(p)$$

by inserting the identity relation with respect to the basis $|p\rangle$,

$$\hat{1} = \int |p\rangle\langle p|$$

Comparing this with the standard Fourier expression, we find that

$$\langle x|p\rangle = \frac{1}{\sqrt{2\pi}} \cdot e^{ipx} \tag{14.8}$$

Similarly, applying the same procedure to the expression for the inverse Fourier transform of $f(p)$, we obtain the relation

$$\langle p|x\rangle = \frac{1}{\sqrt{2\pi}} \cdot e^{-ipx} \tag{14.9}$$

as a result.

We express the function using the delta function as

$$f(x) = \int dx_0 \cdot \delta(x - x_0) f(x_0)$$

---

[7] In Fourier transforms, the period 2π of trigonometric functions appears as a normalization factor in the integral. There are different conventions regarding how this 2π is handled: whether to assign $1/2\pi$ to the forward transform, to the inverse transform, or to distribute it evenly as $1/\sqrt{2\pi}$ on both sides. In theoretical contexts, the equal distribution convention is common, so we adopt it in this book as well. Naturally, the results are unchanged regardless of the chosen convention.

[8] The function $f(x)$ is sometimes referred to as the **position representation**, while $f(p)$ is called the **momentum representation**. In quantum mechanics, the wavevector $\mathbf{p}$ corresponds to the momentum vector, which is why this terminology is used. However, the mathematics involved here holds independently of whether quantum mechanics is being discussed or not. Thus, this is not a topic restricted to the domain of quantum theory.

On the other hand, we also have

$$f(x) = \langle x|f\rangle = \langle x|\cdot|f\rangle = \langle x|\int dx_0 \cdot |x_0\rangle\langle x_0||f\rangle$$
$$= \int dx_0 \langle x|x_0\rangle f(x_0)$$

Comparing the two expressions above, we obtain the relation

$$\delta(x - x_0) = \langle x|x_0\rangle \tag{14.10}$$

14.2.3 Operators in the Fourier Representation

Applying the Fourier transform

$$f(x) = \frac{1}{\sqrt{2\pi}}\int dq \cdot f(q) \cdot \exp[iqx]$$

and then operating with the differential operator ∂_x, we obtain

$$\partial_x f(x) = \frac{1}{\sqrt{2\pi}}\int dq \cdot f(q) \cdot \partial_x \left[\exp[iqx]\right]$$
$$= \int dq \cdot iq \cdot f(q) \cdot \frac{\exp[iqx]}{\sqrt{2\pi}}$$

thus, the factor iq appears. Taking Eq. (14.8) into account and rewriting it in bracket notation, we obtain

$$\partial_x \langle x|f\rangle = \int dq \cdot iq \cdot \langle q|f\rangle \cdot \langle x|q\rangle$$
$$= \langle x|\int dq \cdot iq \cdot |q\rangle\langle q|f\rangle$$

For the left-hand side,

$$\partial_x \langle x|f\rangle = \langle x|\,\hat{\partial}\,|f\rangle$$

we can express it using the abstract operator $\hat{\partial}$, which acts on $|f\rangle$ in the abstract space, as

$$\langle x|\,\hat{\partial}\,|f\rangle = \langle x|\int dq \cdot iq \cdot |q\rangle\langle q|f\rangle$$

In particular, by setting $f = p$ and using $\langle q|p\rangle = \delta(p-q)$, we obtain

$$\begin{aligned}\langle x|\,\hat{\partial}\,|p\rangle &= \langle x|\int dq\cdot iq\cdot|q\rangle\langle q|p\rangle\\ &= \langle x|\int dq\cdot iq\cdot|q\rangle\,\delta(p-q)\\ &= \langle x|\,ip\cdot|p\rangle\end{aligned}$$

This can formally be written as

$$\hat{\partial}\,|p\rangle = ip\cdot|p\rangle\ . \tag{14.11}$$

14.3 Pfaffian

In the main text (Sect. 8.2.4), the Pfaffian was described as "a signed sum over terms like $a_{12}a_{34}$, constructed by forming products of index pairs (i, j) with no repeated indices, and summing over all such permutations with appropriate signs." This restriction, "taking products while avoiding index duplication," can be formalized mathematically by introducing the concept of the **wedge product** $\left(e_i \wedge e_j\right)$. In what follows, we explain how this formulation works.

We begin by preparing a linear combination of all index pairs as[9]

$$\Omega = a_{12} + a_{13} + \cdots = \sum_{i<j} a_{ij}\ . \tag{14.12}$$

From this Ω, we can construct powers such as

$$\begin{aligned}\Omega^2 &= (a_{11} + a_{12} + \cdots)\cdot(a_{11} + a_{12} + \cdots)\\ &= a_{11}a_{11} + a_{11}a_{12} + \cdots + a_{12}a_{34} + \cdots + a_{13}a_{24} + \cdots\\ \Omega^3 &= \cdots + a_{12}a_{34}a_{56} + \cdots + a_{13}a_{24}a_{56} + \cdots\end{aligned}$$

These expansions contain terms like $a_{12}a_{34}$ and $a_{12}a_{34}a_{56}$, which are precisely the types of pairwise products that form the building blocks of the Pfaffian. However, in this form, terms such as $a_{11}a_{11}$ or $a_{12}a_{23}$, which contain overlapping indices, are also included in the expansion. We would like to construct the expression so that all such terms are excluded. The question then becomes how to express this requirement formally.

[9] This summation implicitly assumes that $\left\{a_{ij}\right\}$ are elements of an antisymmetric matrix, as reflected in the summation condition $i < j$.

To achieve this, we make use of the wedge product, which satisfies the antisymmetry relation

$$\left(e_i \wedge e_j\right) = -\left(e_j \wedge e_i\right) .$$

Quantities with this antisymmetry property satisfy $(e_i \wedge e_i) = 0$ under index repetition.[10] For higher-order products such as $\left(e_i \wedge e_j \wedge e_k\right)$, this gives

$$(e_1 \wedge e_1 \wedge e_3) = 0$$

which expresses the property that "simultaneous occupancy leads to zero." Moreover, we obtain

$$0 \neq (e_1 \wedge e_2 \wedge e_3) = -(e_1 \wedge e_3 \wedge e_2) = \cdots$$

demonstrating that the sign changes with permutation of indices.

Instead of Eq. (14.12), we now attach wedge products and define

$$\omega = \sum_{i<j} a_{ij} \cdot \left(e_i \wedge e_j\right) .$$

By constructing ω^2, ω^3, and so on, the accompanying wedge products enforce the condition that "simultaneous occupancy leads to zero." As a result, we obtain expressions such as

$$\begin{aligned}
\omega^2 &= a_{12}a_{34} \cdot (e_1 \wedge e_2 \wedge e_3 \wedge e_4) + \cdots + a_{13}a_{24} \cdot (e_1 \wedge e_3 \wedge e_2 \wedge e_4) + \cdots \\
\omega^3 &= a_{12}a_{34}a_{56} \cdot (e_1 \wedge e_2 \wedge e_3 \wedge e_4 \wedge e_5 \wedge e_6) + \cdots \\
&\quad + a_{12}a_{35}a_{46} \cdot (e_1 \wedge e_2 \wedge e_3 \wedge e_5 \wedge e_4 \wedge e_6) + \cdots .
\end{aligned}$$

In these expressions, terms with any repeated indices no longer appear. Each term is a product involving a wedge product of e_j's with all mutually distinct indices. If we reorder the wedge product factors in ascending index order as $(e_1 \wedge e_2 \wedge \cdots \wedge e_{2n})$, a sign appears according to the permutation. As a result, each term takes the form $(-)^P \cdot a_{P_1 P_2} \cdot a_{P_3 P_4} \cdots a_{P_{2n-1} P_{2n}} \cdot (e_1 \wedge e_2 \wedge \cdots \wedge e_{2n})$, and we obtain

$$\omega^n = \sum_P \left[(-)^P \cdot a_{P_1 P_2} \cdot a_{P_3 P_4} \cdots a_{P_{2n-1} P_{2n}}\right] \cdot (e_1 \wedge e_2 \wedge \cdots \wedge e_{2n}) .$$

If the general explanation above still feels abstract, one may find it helpful to work out a concrete example with $n = 2$ (that is, a 4×4 matrix, since the matrix

[10] Since $\left(e_i \wedge e_j\right) + \left(e_j \wedge e_i\right) = 0$, setting $j = i$ leads to $2\,(e_i \wedge e_i) = 0$, hence $(e_i \wedge e_i) = 0$.

order is $2n$). To save space, we abbreviate wedge products as $(e_1 \wedge e_2 \wedge e_3) = e_{123}$, and so on. Then we have

$$\omega = a_{12} \cdot e_{12} + a_{13} \cdot e_{13} + a_{14} \cdot e_{14} + a_{23} \cdot e_{23} + a_{24} \cdot e_{24} + a_{34} \cdot e_{34} \,.$$

From this, we compute

$$\begin{aligned}
\omega^2 &= \begin{pmatrix} a_{12} \cdot e_{12} \\ +a_{13} \cdot e_{13} \\ +a_{14} \cdot e_{14} \\ +a_{23} \cdot e_{23} \\ +a_{24} \cdot e_{24} \\ +a_{34} \cdot e_{34} \end{pmatrix} \times \begin{pmatrix} a_{12} \cdot e_{12} \\ +a_{13} \cdot e_{13} \\ +a_{14} \cdot e_{14} \\ +a_{23} \cdot e_{23} \\ +a_{24} \cdot e_{24} \\ +a_{34} \cdot e_{34} \end{pmatrix} \\
&= \begin{pmatrix} a_{12}a_{34} \cdot e_{12}e_{34} + a_{13}a_{24} \cdot e_{13}e_{24} \\ +a_{14}a_{23} \cdot e_{14}e_{23} + a_{23}a_{14} \cdot e_{23}e_{14} \\ +a_{24}a_{13} \cdot e_{24}e_{13} + a_{34}a_{12} \cdot e_{34}e_{12} \end{pmatrix} \\
&= a_{12}a_{34} \left(e_{1234} + e_{3412}\right) + a_{13}a_{24} \left(e_{1324} + e_{2413}\right) + a_{14}a_{23} \left(e_{1423} + e_{2314}\right) \\
&= 2 \left[a_{12}a_{34} \cdot e_{1234} + a_{13}a_{24} \cdot e_{1324} + a_{14}a_{23} \cdot e_{1423}\right] \\
&= 2 \left[a_{12}a_{34} \cdot e_{1234} + (-)^1 a_{13}a_{24} \cdot e_{1234} + (-)^2 a_{14}a_{23} \cdot e_{1234}\right] \\
&= 2 \left[a_{12}a_{34} - a_{13}a_{24} + a_{14}a_{23}\right] \cdot e_{1234}
\end{aligned}$$

which recovers the expression stated in the main text:

$$\mathrm{Pf}(A) = a_{12}a_{34} - a_{13}a_{24} - a_{14}a_{23} \,.$$

Note that in the transformation to the first line's right-hand side, we use the condition that only terms with no repeated indices are nonzero. In the second line, terms are grouped by rearranging the order of scalar products, for example, using $a_{34}a_{12} = a_{12}a_{34}$. In the third line, we apply cyclic permutation identities such as $e_{ijkl} = e_{jkli}$. In the fourth line, we use index transpositions like $e_{ijkl} = e_{ikjl}$. The second term in the fourth line involves a single transposition, while the third term involves two transpositions, and the corresponding minus signs accumulate accordingly.

In the above case of $n = 2$, we extracted a factor of 2 due to a "swap of two terms," using identities like $a_{34}a_{12} = a_{12}a_{34}$. In the general case, where there are n such swaps, the factor becomes $n!$. In terms of the general formula, this becomes

$$\begin{aligned}
\omega^n &= \sum_P \left[(-)^P \cdot a_{P_1P_2} \cdot a_{P_3P_4} \cdots a_{P_{2n-1}P_{2n}}\right] \cdot (e_1 \wedge e_2 \wedge \cdots \wedge e_{2n}) \\
&= n! \cdot \sum_{P_1 < P_2 < \cdots < P_{2n}} \left[(-)^P \cdot a_{P_1P_2} \cdot a_{P_3P_4} \cdots a_{P_{2n-1}P_{2n}}\right] \cdot (e_1 \wedge e_2 \wedge \cdots \wedge e_{2n})
\end{aligned}$$

The summation enclosed in square brackets is the Pfaffian. Hence, we obtain the canonical definition of the Pfaffian as

$$\frac{\omega^n}{n!} = \text{Pf}(A) \cdot (e_1 \wedge e_2 \wedge \cdots \wedge e_{2n})$$
$$\text{where,} \quad \omega = \sum_{i<j} a_{ij} \cdot \left(e_i \wedge e_j\right) \tag{14.13}$$

as it is found in mathematical literature.

References

1. "Mathematical Methods for Physics and Engineering, Third Edition" (Cambridge University Press, 2006), K.F. Riley, M.P. Hobson, S.J. Bence, ISBN: 978-0521679718
2. "The Principles Of Quantum Mechanics, Fourth Edition" (Oxford University Press, 1988), P.A.M. Dirac, ISBN: 978-0198520115
3. "Visual Complex Analysis" (Clarendon Press, 1999), T. Needham, ISBN: 978-0198534464

Chapter 15
Appendix E: Supplementary Notes on Electronic Structure Theory

In this chapter, we provide an appendix summarizing several technical details related to electronic structure theory. Section 15.1 on the **Cusp theorem** is motivated by the discussion of the cusp condition equation in the main text (Sect. 8.1.2). Section 15.2 on the **Twisted boundary condition** is motivated by the discussion on finite-size errors in Sect. 9.1.2.

15.1 Cusp Theorem

For a wavefunction under a spherically symmetric potential, $\psi(\mathbf{r}) = R_l(r) \cdot Y_{l,m}(\theta, \varphi)$, we assume that the radial part takes the form

$$R_l(r) = r^m \cdot \chi(r) \quad , \quad \chi(0) \neq 0 \, .$$

Then, applying this to Eq. (8.4):

$$\left[\frac{d^2}{dr^2} + \frac{2}{r} \cdot \frac{d}{dr} + \frac{2Z}{r} - \frac{l(l+1)}{r^2} + 2E\right] \cdot R_l(r) = 0 \, .$$

The kinetic energy term becomes

$$\begin{aligned}
T &= \left[\frac{d^2}{dr^2} + \frac{2}{r} \cdot \frac{d}{dr}\right] R_l(r) \\
&= m(m-1) \cdot r^{m-2} \cdot \chi(r) + 2m \cdot r^{m-1} \cdot \chi'(r) + r^m \cdot \chi''(r) \\
&\quad + \frac{2}{r}\left[m \cdot r^{m-1} \cdot \chi(r) + r^m \cdot \chi'(r)\right] \\
&= r^m \cdot \chi''(r) + 2(m+1) \cdot r^{m-1} \cdot \chi'(r) + m(m+1) \cdot r^{m-2} \cdot \chi(r) \, .
\end{aligned}$$

R. Maezono, *Ab initio Quantum Monte Carlo Tutorial*,
https://doi.org/10.1007/978-981-95-4134-8_15

Substituting this into Eq. (8.4), we obtain

$$r^m \cdot \chi''(r) + 2(m+1) \cdot r^{m-1} \cdot \chi'(r) + m(m+1) \cdot r^{m-2} \cdot \chi'(r) + \left[\frac{2Z}{r} - \frac{l(l+1)}{r^2} + 2E\right] \cdot r^m \cdot \chi(r) = 0 .$$

Reorganizing in terms of powers of r, we get

$$r^m \cdot \left[\frac{d^2}{dr^2} + \left(\frac{2(m+1)}{r}\frac{d}{dr} + \frac{2Z}{r}\right) + \left(\frac{m(m+1)}{r^2} - \frac{l(l+1)}{r^2}\right) + 2E\right] \cdot \chi(r) = 0 .$$

First, focusing on the most divergent term, the $1/r^2$ term, we impose the requirement that the divergence from the kinetic energy term, $m(m+1)$, cancels the corresponding term from the potential energy, which depends on l. This leads to the condition $l = m$. Under this condition, the remaining terms become

$$r^l \cdot \chi''(r) + \frac{2(l+1)}{r} r^l \cdot \chi'(r) + \frac{2Z}{r} r^l \cdot \chi(r) + 2E \cdot r^l \cdot \chi(r) = 0$$

$$\therefore \quad \left[\frac{\chi''(r)}{\chi(r)} + \frac{2(l+1)}{r}\frac{\chi'(r)}{\chi(r)} + \frac{2Z}{r} + 2E\right] r^l \cdot \chi(r) = 0$$

To eliminate the divergence arising from the $1/r$ term through cancellation between potential and kinetic energies, we require the following condition, which gives the cusp condition:

$$\left[\frac{\chi'(r)}{\chi(r)}\right]_{r=0} = -\frac{Z}{(l+1)} .$$

15.2 Twisting Boundary Conditions

15.2.1 Mesh Shift

beginners should not confuse regarding the difference between "phase accumulation across a unit cell" and "phase accumulation across a simulation cell." Phase accumulation across a unit cell appears in Bloch's theorem:

$$\psi_k(r+a) = \exp[ika] \cdot \psi_k(r)$$

which defines the **Bloch function labeled by** k [1]. In this context, phase accumulation across the simulation cell (assumed to be L times the unit cell) for a Bloch function labeled by k is

$$\psi_k (r + L \cdot a) = \exp [ik (L \times a)] \cdot \psi_k (r) \ .$$

From here, some beginners tend to make the leap:

$$\psi_k (r + L \cdot a) \stackrel{!}{=} \psi_k (r) \quad \therefore \quad \exp [ik (L \cdot a)] \stackrel{!}{=} 1$$

and thereby "*derive*" the **Born–von Karman periodic boundary condition.**[1] This is **not correct** logic flow. However, what periodicity requires is not that the wavefunction itself "repeats identically," but rather that the **physics described by the wavefunction** repeats identically. That is, the condition should be imposed on the probability density:

$$|\psi_k (r + L \cdot a)|^2 \stackrel{!}{=} |\psi_k (r)|^2 \ .$$

Therefore, what periodicity truly requires is the following condition, which preserves the phase freedom of complex numbers:

$$\psi_k (r + L \cdot a) \stackrel{!}{=} \exp [i \cdot \varphi_T] \, \psi_k (r) \quad \therefore \quad \exp [ik (L \cdot a)] \stackrel{!}{=} \exp [i \cdot \varphi_T]$$
$$\therefore k (L \cdot a) = \varphi_T + 2\pi M \ . \tag{15.1}$$

It is essential to recognize that there is **no intrinsic condition** that must be imposed on the twisting angle φ_T corresponding to the acquired phase $\exp [i \cdot \varphi_T]$. Out of computational convenience, the most straightforward choice is

$$\varphi_T = 0 \quad \cdots \quad \text{(Born-von Karman boundary condition)}$$

This is an **artificially imposed condition**, and simply one of many possible choices for φ_T (others will be discussed below) [2].

When adopting the Born–von Karman condition, Eq. (15.1) leads to

$$k (L \times a) = 0 + 2\pi M$$
$$\therefore \quad k = \frac{2\pi}{a} \times \frac{1}{L} \times M \tag{15.2}$$

This result implies that the k values labeling the Bloch functions correspond to *points on a grid* that discretizes the Brillouin zone (of size $2\pi/a$) into L intervals.

On the other hand, there is nothing wrong with abandoning the Born–von Karman boundary condition and adopting a more general form,

$$\varphi_T \neq 0 \quad \cdots \quad \text{(Twisted boundary condition)}$$

[1] The notation "$\stackrel{!}{=}$" represents the equality sign used in the sense of "...is set as...".

This corresponds to assigning a nonzero φ_T as the phase accumulated upon translation by the simulation cell (**twisted boundary condition** [3, 4]). In this case, instead of Eq. (15.2), we obtain

$$\begin{aligned} k &= \frac{\phi_S + 2\pi M}{a} \times \frac{1}{L} = \frac{\varphi_T}{aL} + \frac{2\pi}{a} \times \frac{1}{L} \times M \\ &= k_S + \frac{2\pi}{a} \times \frac{1}{L} \times M \end{aligned} \tag{15.3}$$

That is, the mesh used to label Bloch functions is given by the Born–von Karman mesh shifted by k_S (**mesh shift**).

15.2.2 Twist Averaging

Electrical conductivity can be understood as the ability of perturbations applied to one region of a wave function to propagate over long distances. This characteristic is quantified by the **phase stiffness** of the wave function [5]. When the phase of the wave function is twisted between two distant points, we consider the stiffness against this twist. In a conductor such as a metal, where the system maintains the influence of the wave function over long distances, a twist at one side propagates to the other side, resulting in a finite phase stiffness. This corresponds to the picture where the **phase coherence** is preserved over long distances. On the other hand, in an insulator, coherence is lost over short distances, so a twist at one end does not affect the far side, and the phase stiffness becomes zero. Thus, a formalism can be established connecting phase stiffness to the transport properties of a system.[2]

When electrical conductivity is understood in this way, metallic systems can be seen as systems where the contribution from the long-wavelength components of the wave function is significant. Therefore, approximations that truncate long-wavelength modes (components with small k) require more careful handling in metallic systems. If the discretized mesh is aligned to the origin of k-space [Born–von Karman boundary condition ($\varphi_T = 0$)], then modes in the interval $k = [0, \Delta k]$, *i.e.*, **long-wavelength modes**, cannot be captured. In metallic systems, where long-wavelength modes play an important role, missing contributions from this interval can become a significant source of error. To address this, a shifted discretized mesh [twisted boundary condition ($\varphi_T \neq 0$)] can be used to partially recover contributions from this region. By randomly selecting the origin shift k_S and averaging over repeated calculations, one can reduce such errors. As shown in Eq. (15.3), the origin shift corresponds to the twist angle φ_T. Hence, this averaging procedure is referred

[2] This idea is related to Kohn's "nearsightedness." A more rigorous formulation was initiated by the work of King-Smith and Vanderbilt and is now known as the "modern theory of polarization" based on Berry phase [6].

to as **twist averaging** [3, 4]. In DFT packages, this use of shifted meshes can also be specified through input files [7].

In the reciprocal space picture, twist averaging can be understood more directly. In metals, the Fermi surface has a more complex shape, and a coarse k-point mesh cannot fully describe such complexity. The idea is that, by using several meshes each with a slightly shifted origin, and averaging over them, one can better capture these intricate features in reciprocal space.

Twist averaging can also be understood from another perspective. A periodic boundary condition fixed at a certain φ_T, such as the Born–von Karman condition, is fundamentally a human-imposed convenience. Nature should not be sensitive to such a choice, yet in practical calculations, the results do depend on φ_T. This dependence arises because an originally infinite system has been approximated by a repeated supercell of length L. We artificially impose a boundary with a fixed phase angle at this length-L point, but in reality, no such boundary exists. If a calculated observable depends on φ_T, it implies that the simulation "feels" this artificial boundary. To cancel out this unnatural dependence, we perform calculations for various values of φ_T, average the results, and thereby remove this bias, obtaining values closer to the true, physical result. This procedure is what is referred to as twist averaging.

References

1. "Fundamentals of Condensed Matter Physics" (Cambridge University Press, 2016), M.L. Cohen, S.G. Louie, ISBN: 978-0521513319
2. "Solid State Physics" (Thomson Learning, 1976), N.W. Ashcroft and N.D. Mermin, ISBN: 978-0030839931
3. "Twist-averaged boundary conditions in continuum quantum Monte Carlo algorithms", C. Lin, F.H. Zong, and D.M. Ceperley, Phys. Rev. E 64, 016702 (2001), https://doi.org/10.1103/PhysRevE.64.016702
4. "Diamond to beta-tin phase transition in Si within quantum Monte Carlo", R. Maezono, N.D. Drummond, A. Ma, and R.J. Needs, Phys. Rev. B 82, 184108 (2010). https://doi.org/10.1103/PhysRevB.82.184108
5. "Electronic Structure: Basic Theory and Practical Methods" (Cambridge University Press, 2004), R.M. Martin, ISBN: 978-0521782852
6. "Berry Phases in Electronic Structure Theory: Electric Polarization, Orbital Magnetization and Topological Insulators" (Cambridge University Press, 2018), David Vanderbilt, ISBN: 978-1107157651
7. https://www.quantum-espresso.org (URL confirmed on 2024.10)

Chapter 16
Appendix F: Notes on Density Functional Theory

This chapter provides a summary of the minimum necessary concepts needed to understand the discussions of density functional theory (DFT) that appear in the main text. Section 16.1 is motivated by the discussions in Sect. 10.1.4 and Sect. 10.2.2. The topics from Sect. 16.2 onward are motivated by the discussion in Sect. 10.1.6.

16.1 Essentials of Density Functional Theory

16.1.1 Minimum Description of Density Functional Theory

When an external electrostatic potential $v_{\text{ext}}(\mathbf{r})$ is applied to a system with charge density $n(\mathbf{r})$, the ground-state energy E_{GS} of the system can be written as

$$\begin{aligned} E_{\text{GS}} &= (\text{Energy intrinsic to the interacting electron system}) \\ &\quad + \int d\mathbf{r} \cdot v_{\text{ext}}(\mathbf{r}) \cdot n(\mathbf{r}) \ . \end{aligned}$$

By writing the term "(Energy intrinsic to the interacting electron system)" as F, and noting that it is determined from the many-body wavefunction $\Psi(\mathbf{r}_1, \cdots, \mathbf{r}_N)$, we can express this as

$$\begin{aligned} &E_{\text{GS}} \left[\Psi_{\text{interacting}}(\mathbf{r}_1, \cdots, \mathbf{r}_N)\right] \\ &\quad = F\left[\Psi_{\text{interacting}}(\mathbf{r}_1, \cdots, \mathbf{r}_N)\right] + \int d\mathbf{r} \cdot v_{\text{ext}}(\mathbf{r}) \cdot n(\mathbf{r}) \, . \end{aligned} \tag{16.1}$$

For later convenience, we attach the subscript "interacting" to the many-body wavefunction to indicate that it pertains to the interacting system. For this theoretical

R. Maezono, *Ab initio Quantum Monte Carlo Tutorial*,
https://doi.org/10.1007/978-981-95-4134-8_16

framework, the following foundation has been established as "density functional theory" [1]:

(1) The term F in Eq. (16.1) can be expressed as a functional of the charge density $n(\mathbf{r})$, that is, $F[n(\mathbf{r})]$,[1]

(2) The dependence of $F[n(\mathbf{r})]$ on $n(\mathbf{r})$ is universal, that is, independent of the specific system,

(3) The exact ground-state energy of the system is obtained by minimizing

$$E_{\mathrm{GS}}[n(\mathbf{r})] = F[n(\mathbf{r})] + \int d\mathbf{r} \cdot v_{\mathrm{ext}}(\mathbf{r}) \cdot n(\mathbf{r})$$

with respect to $n(\mathbf{r})$.

From the above **assertion (1)**, Eq. (16.1) becomes

$$\begin{aligned} E_{\mathrm{GS}}\left[\Psi_{\mathrm{interacting}}(\mathbf{r}_1, \cdots, \mathbf{r}_N)\right] &= F[n(\mathbf{r})] + \int d\mathbf{r} \cdot v_{\mathrm{ext}}(\mathbf{r}) \cdot n(\mathbf{r}) \\ &= E_{\mathrm{GS}}[n(\mathbf{r})] \end{aligned} \tag{16.2}$$

which implies that the ground-state energy is determined solely by the charge density $n(\mathbf{r})$. That is, the ground-state energy can be grasped not in terms of the complex function $\Psi(\mathbf{r}_1, \cdots, \mathbf{r}_N)$ in $3N$-dimensional space, but rather using the real-valued function $n(\mathbf{r})$ in 3D space as the basic quantity. Please note that this is making a "remarkable claim." When studying quantum mechanics, it is emphasized that "the description of quantum phenomena via complex functions, *i.e.*, the existence of phase freedom, is essential." However, for the ground-state energy of an interacting electron system, it is claimed that a formulation in terms of a real-valued function is generally valid. Although this foundation is proven with mathematical rigor using contradiction arguments and other methods, it is a claim that was hard to anticipate before the establishment of this theory in the 1960s.

The above **assertion (2)**, which states that "the dependence of $F[n(\mathbf{r})]$ is universal," is also quite astonishing. Normally, one would expect the nature of electron-electron interactions to be highly diverse—metallic bonding, covalent bonding in insulators, intermolecular forces *etc.*, all differ greatly. When solving the system-

[1] A functional of a function $n(\mathbf{r})$ is written as $F[n]$, omitting the $\mathbf{r}$-dependence, as a common practice. However, this can often confuse beginners unfamiliar with the concept of functionals, blurring the distinction between the function $F(n)$ and the functional $F[n]$. Therefore, in this text, I intentionally write $F[n(\mathbf{r})]$ to serve as a reminder that this is a functional determined by the function $n(\mathbf{r})$. On the other hand, there is a good reason to adopt the standard notation $F[n]$, especially when dealing with functional derivatives. Beginners sometimes mistakenly believe that $n(\mathbf{r}')$ or $n(\mathbf{r}'')$ within the functional are not the target of differentiation because their variables differ from $\mathbf{r}$. Since in functional derivatives such dummy variable differences are irrelevant, standard notation helps avoid confusion. However, this book is intended for readers before that level, so we refrain from using the standard form.

specific many-body Schrodinger equation and obtaining $\Psi_{\text{interacting}}(\mathbf{r}_1, \cdots, \mathbf{r}_N)$, one would anticipate that the functional dependence of $F\left[\Psi_{\text{interacting}}(\mathbf{r}_1, \cdots, \mathbf{r}_N)\right]$ would reflect such diversity in bonding characteristics. However, what the above assertion claims is that although the functional form of $F[\Psi]$ may vary greatly, when viewed in terms of $F[n(\mathbf{r})]$, the diversity pertaining to bonding modes disappears. As long as the bare interactions between particles are the same, it takes on a universal form, regardless of the system. The final **assertion (3)** will later prove to be a particularly useful theorem in practical applications.

It would be convenient if assertions (1) through (3) above had specific names, but due to the complexity of their historical development, they are collectively referred to as **density functional theory** [DFT]. The "Hohenberg–Kohn theorem,"[2] proposed in 1964 by Hohenberg and Kohn, is generally regarded as the foundational basis of DFT. From the perspective of modern practical usage, however, the theory is effectively summarized by the assertions above.

16.1.2 Using a Reference System

In the previous section, we saw that the ground-state energy is determined with the charge density $n(\mathbf{r})$ as the fundamental quantity. This $n(\mathbf{r})$ is obtained by integrating $\Psi_{\text{interacting}}(\mathbf{r}_1, \cdots, \mathbf{r}_N)$, which results from solving the "governing equation with complicated interactions" (Sect. 17.3). The usual understanding is that the "solution of the many-body governing equation" comes first, and the charge density $n(\mathbf{r})$ is obtained secondarily. However, by turning this perspective on its head, we notice that since $E_{\text{GS}}[n(\mathbf{r})]$ is universally determined solely by $n(\mathbf{r})$, if there exists a "simpler problem" that yields exactly the same $n(\mathbf{r})$ as that of the interacting system (the target of consideration), then we can compute $n(\mathbf{r})$ from that simpler problem and still obtain the same $E_{\text{GS}}[n(\mathbf{r})]$.

If so, then we may consider mapping the problem to one involving a virtual non-interacting single-particle system that yields the same density distribution $n(\mathbf{r})$ as that of the interacting system:

$$n(\mathbf{r}) = \sum_j \left|\psi_j(\mathbf{r})\right|^2 \tag{16.3}$$

The idea is to solve the following problem:

$$\left[-\frac{1}{2}\nabla^2 + v_{\text{eff}}(\mathbf{r})\right] \cdot \psi_i(\mathbf{r}) = \varepsilon_i \cdot \psi_i(\mathbf{r}) \tag{16.4}$$

[2] Both are personal names: P. Hohenberg and W. Kohn.

Here, each electron moves independently under an effective potential $v_{\text{eff}}(\mathbf{r})$, and the problem becomes tractable to solve.

That said, one may wonder: "Without solving $\Psi_{\text{interacting}}(\mathbf{r}_1, \cdots, \mathbf{r}_N)$, we cannot know $n(\mathbf{r})$ in advance, so how can we possibly prepare a single-particle problem that yields the same density distribution?" Here, the **assertion (3)** from the previous section becomes powerful. Namely, if one can provide a functional form for $E_{\text{GS}}[n(\mathbf{r})]$ without solving $\Psi_{\text{interacting}}(\mathbf{r}_1 \cdots \mathbf{r}_N)$, then the correct $n(\mathbf{r})$ can be obtained by seeking the minimum of $E_{\text{GS}}[n(\mathbf{r})]$. Now, since we chose to represent $n(\mathbf{r})$ as in Eq. (16.3),[3] we may instead consider minimizing $E_{\text{GS}}\left[\left\{\psi_j(\mathbf{r})\right\}\right]$, and thereby determine the solution. This constitutes the framework of **Kohn–Sham theory**[4] [1].

Although the functional form of $E_{\text{GS}}[n(\mathbf{r})]$ and the effective potential $v_{\text{eff}}(\mathbf{r})$ introduced in Eq. (16.4) remain unspecified, they will be progressively constrained by the guiding principle of Kohn–Sham theory, as we will see in the next subsection. It should also be noted that, in the course of the above discussion, we introduced a set of single-particle orbitals $\left\{\psi_j(\mathbf{r})\right\}$, as in Eq. (16.3), merely as a mathematical convenience for expressing the charge density $n(\mathbf{r})$. Unlike the Hartree–Fock orbitals in the molecular orbital method, these orbitals do not originally carry a strong physical interpretation. Orbitals introduced under such a motivation are referred to as **Kohn–Sham orbitals**.

16.1.3 Kohn–Sham Equation

In Eq. (16.2) of Sect. 16.1.1, the density functional is given by

$$E_{\text{GS}}[n(\mathbf{r})] = F[n(\mathbf{r})] + \int d\mathbf{r} \cdot v_{\text{ext}}(\mathbf{r}) \cdot n(\mathbf{r}) .$$

Here, we express the universal functional $F[n(\mathbf{r})]$ as

$$F[n(\mathbf{r})] = E_{\text{H}} + (\text{other contributions}) ,$$

where the known contribution from classical electrostatic interaction due to the Hartree potential is separated out:

$$E_{\text{H}} = \frac{1}{2}\int d\mathbf{r} \int d\mathbf{r}' \frac{n(\mathbf{r}')\, n(\mathbf{r})}{|\mathbf{r} - \mathbf{r}'|} = E_{\text{H}}[n(\mathbf{r})] .$$

We further separate the "other contributions" as

[3] This choice does not entail any particular loss of generality.

[4] Both are names: W. Kohn and L. Sham.

$$
\text{(other contributions)} = \sum_{n=1}^{N} \int d\mathbf{r} \cdot \psi_j^* (\mathbf{r}) \left[-\frac{1}{2} \nabla^2 \psi_j (\mathbf{r}) \right] + \text{(remaining part of unknown character)} . \quad (16.5)
$$

The quantity separated here,

$$
T_\mathrm{S} = \sum_{j=1}^{N} \int d\mathbf{r} \cdot \psi_j^* (\mathbf{r}) \left[-\frac{1}{2} \nabla^2 \psi_j (\mathbf{r}) \right]
$$

has a formal structure identical to that of the kinetic energy term. However, since the orbital functions $\{\psi_j\}$ are introduced formally as in Eq. (16.3), this T_S is also a formally defined quantity that does not physically represent the kinetic energy. Note that this part may implicitly contain some portion of the electron–electron interaction.

We now assign the name E_XC to the "further uncharacterized part" remaining in Eq. (16.5), and write:

$$
F[n(\mathbf{r})] = T_\mathrm{S} + E_\mathrm{H} + E_\mathrm{XC}[n(\mathbf{r})] .
$$

The subscript "XC" implies that this "uncharacterized part" encompasses the quantum many-body effects of exchange and correlation (eXchange & Correlation). This term is referred to as the **exchange-correlation functional**.

By applying the variational principle so that the density $n(\mathbf{r})$ minimizes the density functional $E_\mathrm{GS}[n(\mathbf{r})]$ in the manner prescribed by Kohn and Sham, the previously undetermined $v_\mathrm{eff}(\mathbf{r})$ in Eq. (16.4) is found to take the following form [1]:[5]

$$
v_\mathrm{eff}(\mathbf{r}) = v_\mathrm{ext}(\mathbf{r}) + \int d\mathbf{r}' \frac{n(\mathbf{r}')}{|\mathbf{r}-\mathbf{r}'|} + \frac{\delta E_\mathrm{XC}[n(\mathbf{r})]}{\delta n(\mathbf{r})} . \quad (16.6)
$$

Substituting this into Eq. (16.4), the one-body problem of the reference system becomes:

$$
\left[-\frac{1}{2}\nabla^2 + v_\mathrm{ext}(\mathbf{r}) + \int d\mathbf{r}' \frac{n(\mathbf{r}')}{|\mathbf{r}-\mathbf{r}'|} + \frac{\delta E_\mathrm{XC}[n(\mathbf{r})]}{\delta n(\mathbf{r})} \right] \cdot \psi_i(\mathbf{r}) = \varepsilon_i \cdot \psi_i(\mathbf{r}) . \quad (16.7)
$$

This equation is known as the **Kohn–Sham equation**.

The procedure for handling the Kohn–Sham equation proceeds as follows: first, once the functional form of the **exchange-correlation potential** is assumed, defined

[5] The effective potential v_eff appears in Eq. (16.4) and is introduced through the identity $T_s = \sum_j \varepsilon_j - \int d^3r \cdot v_\mathrm{eff}(\mathbf{r}) n(\mathbf{r})$ in the minimization procedure.

as the functional derivative of the exchange-correlation functional $E_{XC}[n]$,

$$V_{XC}[n(\mathbf{r})] = \frac{\delta E_{XC}[n(\mathbf{r})]}{\delta n(\mathbf{r})},$$

the form of the Kohn–Sham equation becomes specified. With some proper initial guess for $n(\mathbf{r})$, solving this eigenvalue equation yields the eigenfunctions $\psi_j(\mathbf{r})$, from which the charge density is determined as $n(\mathbf{r}) = \sum_j |\psi_j(\mathbf{r})|^2$. Using this charge density, one reconstructs $V_{XC}[n(\mathbf{r})]$ and repeats the above procedure **self-consistently**. By iterating until convergence, one obtains the solution

$$n(\mathbf{r}) = \sum_j \left|\psi_j^{(\infty)}(\mathbf{r})\right|^2 \qquad E_S = \sum_{i=1}^{N} \varepsilon_i^{(\infty)} \tag{16.8}$$

Note that at this stage, E_S is the energy of the reference system, and is not yet identified as the true ground-state energy E_{GS} of the original problem.

16.1.4 Energy of the System Obtained from Solving the Kohn–Sham Equation

After obtaining the converged self-consistent solution of the Kohn–Sham equation, the energy of the original system is constructed as follows: First, T_S represents the kinetic energy part of the reference system problem (16.4), so by subtracting the potential energy contribution from the reference system energy $E_S = \sum_i \varepsilon_i$, we obtain

$$T_S = E_S - \int d\mathbf{r} \cdot v_{\text{eff}}(\mathbf{r}) \cdot n(\mathbf{r}) \ .$$

By substituting (16.6) into the expression for $v_{\text{eff}}(\mathbf{r})$ in the above equation, we obtain

$$\begin{aligned} T_S &= \sum_i \varepsilon_i - \int d\mathbf{r} \cdot \left[v_{\text{ext}}(\mathbf{r}) + \int d\mathbf{r}' \frac{n(\mathbf{r}')}{|\mathbf{r}-\mathbf{r}'|} + \frac{\delta E_{XC}[n(\mathbf{r})]}{\delta n(\mathbf{r})} \right] \cdot n(\mathbf{r}) \\ &= \sum_i \varepsilon_i - \int d\mathbf{r} \cdot v_{\text{ext}}(\mathbf{r}) \cdot n(\mathbf{r}) - \int d\mathbf{r} \int d\mathbf{r}' \frac{n(\mathbf{r}')\, n(\mathbf{r})}{|\mathbf{r}-\mathbf{r}'|} \\ &\quad - \int d\mathbf{r} \cdot V_{XC}(\mathbf{r}) \cdot n(\mathbf{r}) \ . \end{aligned}$$

Accordingly, the ground-state energy of the system is given by

$$
\begin{aligned}
E_{\mathrm{GS}}[n(\mathbf{r})] &= T_{\mathrm{S}} + \frac{1}{2}\int d\mathbf{r}\int d\mathbf{r}'\,\frac{n(\mathbf{r}')\,n(\mathbf{r})}{|\mathbf{r}-\mathbf{r}'|} + E_{\mathrm{XC}}[n(\mathbf{r})] \\
&\quad + \int d\mathbf{r}\cdot v_{\mathrm{ext}}(\mathbf{r})\cdot n(\mathbf{r}) \\
&= \left(\sum_{i=1}^{N}\varepsilon_i - \int d\mathbf{r}\cdot v_{\mathrm{ext}}(\mathbf{r})\cdot n(\mathbf{r})\right. \\
&\quad \left. - \int d\mathbf{r}\int d\mathbf{r}'\,\frac{n(\mathbf{r}')\,n(\mathbf{r})}{|\mathbf{r}-\mathbf{r}'|} - \int d\mathbf{r}\cdot V_{\mathrm{XC}}(\mathbf{r})\cdot n(\mathbf{r})\right) \\
&\quad + \frac{1}{2}\int d\mathbf{r}\int d\mathbf{r}'\,\frac{n(\mathbf{r}')\,n(\mathbf{r})}{|\mathbf{r}-\mathbf{r}'|} + E_{\mathrm{XC}}[n(\mathbf{r})] \\
&\quad + \int d\mathbf{r}\cdot v_{\mathrm{ext}}(\mathbf{r})\cdot n(\mathbf{r}) \\
&= \sum_{i=1}^{N}\varepsilon_i - \frac{1}{2}\int d\mathbf{r}\int d\mathbf{r}'\,\frac{n(\mathbf{r}')\,n(\mathbf{r})}{|\mathbf{r}-\mathbf{r}'|} + E_{\mathrm{XC}}[n(\mathbf{r})] \\
&\quad - \int d\mathbf{r}\cdot V_{\mathrm{XC}}(\mathbf{r})\cdot n(\mathbf{r})
\end{aligned}
\tag{16.9}
$$

as shown in [1].

Since the discussion has become somewhat lengthy, we briefly summarize the logical flow up to this point:

(A) The ground-state energy can be expressed as a density functional $E_{\mathrm{GS}}[n(\mathbf{r})]$, a fact that has been rigorously established (though its explicit form is not known).
(B) Even without solving for the many-body wavefunction, one can reach the correct answer by performing a variational minimization of the density functional $E_{\mathrm{GS}}[n(\mathbf{r})]$ with respect to $n(\mathbf{r})$.
(C) Suppose that there exists a non-interacting single-particle problem, which gives the same density $n(\mathbf{r})$ as the interacting system:

$$
\left[-\frac{1}{2}\nabla^2 + v_{\mathrm{eff}}(\mathbf{r})\right]\cdot\psi_i(\mathbf{r}) = \varepsilon_i\cdot\psi_i(\mathbf{r})\,, \tag{16.10}
$$

such that $n(\mathbf{r}) = \sum_j \left|\psi_j(\mathbf{r})\right|^2$. By formulating the variational problem in terms of this effective system, one can solve the interacting problem equivalently.

(D) If the density functional is formally decomposed as

$$E_{\mathrm{GS}}[n(\mathbf{r})] = T_{\mathrm{S}} + \int d\mathbf{r} \int d\mathbf{r}' \frac{n(\mathbf{r}')\, n(\mathbf{r})}{|\mathbf{r} - \mathbf{r}'|} + E_{\mathrm{XC}}[n(\mathbf{r})] + \int d\mathbf{r} \cdot v_{\mathrm{ext}}(\mathbf{r}) \cdot n(\mathbf{r}),$$

(where only known contributions are explicitly separated, and the rest absorbed into E_{XC}, preserving generality), then variational minimization leads to

$$\left[-\frac{1}{2}\nabla^2 + v_{\mathrm{ext}}(\mathbf{r}) + \int d\mathbf{r}' \frac{n(\mathbf{r}')}{|\mathbf{r} - \mathbf{r}'|} + \frac{\delta E_{\mathrm{XC}}[n(\mathbf{r})]}{\delta n(\mathbf{r})} \right] \cdot \psi_i(\mathbf{r}) = \varepsilon_i \cdot \psi_i(\mathbf{r}),$$

where v_{eff} has been specified and substituted. By solving this, one obtains the solution to the interacting problem (even though the form of E_{XC} remains unspecified).

(E) Once the eigenvalues $\{\varepsilon_j\}$ from the reference single-particle problem are known, the ground-state energy of the original interacting system can be obtained as

$$E_{\mathrm{GS}}[n(\mathbf{r})] = \sum_{i=1}^{N} \varepsilon_i - \frac{1}{2} \int d\mathbf{r} \int d\mathbf{r}' \frac{n(\mathbf{r}')\, n(\mathbf{r})}{|\mathbf{r} - \mathbf{r}'|} + E_{\mathrm{XC}}[n(\mathbf{r})] - \int d\mathbf{r} \cdot V_{\mathrm{XC}}(\mathbf{r}) \cdot n(\mathbf{r}).$$

It should be noted that in point (C), we stated "assuming such an equivalent problem exists...," but whether this assumption can be rigorously justified requires careful attention [1]. A detailed discussion lies beyond the scope of this text, but it connects to topics symbolized by terms such as "v-representability" and "n-representability."

The framework of DFT begins with the foundational principle described in Sect. 16.1.1 (Hohenberg-Kohn theorem), and the discussion from Sect. 16.1.2 onward concerns how to implement this principle in practice. Although the Kohn–Sham framework is the most widely adopted in practical applications, the implementation methods derived from the principle are not limited to the Kohn–Sham approach, various other formulations also exist [1]. When one first learns the principle of DFT, the most straightforward idea might be to attempt to obtain $E_{\mathrm{GS}}[n(\mathbf{r})]$ explicitly as a functional of $n(\mathbf{r})$, and such approaches have indeed been explored [1].

16.2 Violation of Self-Interaction Cancellation

16.2.1 Cancellation of Self-Interaction

The Hartree–Fock equation given in Eq. (10.1)

$$\left[-\frac{1}{2}\nabla^2 + v_{\text{ext}}(\mathbf{r}) + \sum_{j\neq i}\int d\mathbf{r}'\frac{\left|\psi_j(\mathbf{r}')\right|^2}{|\mathbf{r}-\mathbf{r}'|}\right]\cdot\psi_i(\mathbf{r})$$
$$-\sum_{j\neq i}\left[\int d\mathbf{r}'\psi_j^*(\mathbf{r}')\frac{\delta_{\sigma_i\sigma_j}}{|\mathbf{r}-\mathbf{r}'|}\psi_i(\mathbf{r}')\right]\cdot\psi_j(\mathbf{r})$$
$$=\varepsilon_i\cdot\psi_i(\mathbf{r})$$

contains, on the left-hand side, third and fourth terms expressed as integrals, which represent interaction terms.[6] We write these as

$$V^{(\text{HF})} = \sum_{j\neq i}V_{ij}^{(\text{H})} + \sum_{j\neq i}V_{ij}^{(\text{X})} = V^{(\text{H})} + V^{(\text{X})} \tag{16.11}$$

$$V_{ij}^{(\text{H})} := \int d\mathbf{r}'\frac{\left|\psi_j(\mathbf{r}')\right|^2}{|\mathbf{r}-\mathbf{r}'|}\cdot\psi_i(\mathbf{r}) = \int d\mathbf{r}'\frac{n_j(\mathbf{r}')}{|\mathbf{r}-\mathbf{r}'|}\cdot\psi_i(\mathbf{r}) \tag{16.12}$$

$$V_{ij}^{(\text{X})} := -\int d\mathbf{r}'\psi_j^*(\mathbf{r}')\frac{\delta_{\sigma_i\sigma_j}}{|\mathbf{r}-\mathbf{r}'|}\psi_i(\mathbf{r}')\cdot\psi_j(\mathbf{r})\ . \tag{16.13}$$

The superscripts [HF = Hartree-Fock], [H = Hartree], and [X = eXchange] represent their respective components. The quantity $n_j(\mathbf{r}) = \left|\psi_j(\mathbf{r})\right|^2$ is interpreted as the charge density formed by the j-th orbital.

The summation condition "$i \neq j$" expresses the fact that interactions occur only between distinct particles. However, from the perspective of computational handling, such exclusion conditions are somewhat of an "eyesore." Indeed, the Hartree term

$$V^{(\text{H})} = \sum_{j\neq i}V_{ij}^{(\text{H})}$$

is "so close" to a more convenient form. If only the "$i \neq j$" restriction were lifted, one could write

[6] Correctly speaking, these are interaction terms multiplied by orbital functions.

$$
\begin{aligned}
\tilde{V}^{(\mathrm{H})} := \sum_{i,j} V_{ij}^{(\mathrm{H})} &= \sum_i \int d\mathbf{r}' \frac{1}{|\mathbf{r}-\mathbf{r}'|} \sum_j n_j\left(\mathbf{r}'\right) \cdot \psi_i\left(\mathbf{r}\right) \\
&= \sum_i \int d\mathbf{r}' \frac{n\left(\mathbf{r}'\right)}{|\mathbf{r}-\mathbf{r}'|} \cdot \psi_i\left(\mathbf{r}\right) = \tilde{V}^{(\mathrm{H})}[n]
\end{aligned}
$$

so that it can be expressed using the local density $n\left(\mathbf{r}\right) = \sum_j n_j\left(\mathbf{r}\right)$.

Thus, by using the tilde quantities without the "$i \neq j$" restriction, we write

$$
V^{(\mathrm{H})} = \tilde{V}^{(\mathrm{H})} - V_{ii}^{(\mathrm{H})} \quad , \quad V^{(\mathrm{X})} = \tilde{V}^{(\mathrm{X})} - V_{ii}^{(\mathrm{X})}
$$

and then rewrite $V^{(\mathrm{HF})}$ in Eq. (16.11) as

$$
\begin{aligned}
V^{(\mathrm{HF})} &= \tilde{V}^{(\mathrm{H})}[n] + \tilde{V}^{(\mathrm{X})} - \left(V_{ii}^{(\mathrm{H})} + V_{ii}^{(\mathrm{X})}\right) \\
&= \tilde{V}^{(\mathrm{H})}[n] + \tilde{V}^{(\mathrm{X})} - V^{(\mathrm{SI})}
\end{aligned}
$$

The leftover term with $i = j$ that we have teased out,

$$
V^{(\mathrm{SI})} := V_{ii}^{(\mathrm{H})} + V_{ii}^{(\mathrm{X})} \tag{16.14}
$$

is referred to as the **self-interaction**. From Eqs. (16.12) and (16.13), we find that $V_{ii}^{(\mathrm{H})} = -V_{ii}^{(\mathrm{X})}$. Therefore, in the case of Hartree–Fock theory,

$$
V^{(\mathrm{SI})} = 0 \quad \text{(for Hartree-Fock)} \tag{16.15}
$$

This cancellation is called the **self-interaction cancellation** [2].

16.2.2 *Breakdown of Self-Interaction Cancellation*

Let us once again emphasize that the self-interaction is a quantity that does not physically exist in the first place, but is introduced merely for convenience when constructing $\tilde{V}^{(\mathrm{H,X})}$ with the summation restriction removed. Even though such a quantity has no true physical existence, we allow its formal inclusion under the rationale that "it will cancel out anyway, so there's no problem," this is what is meant by the self-interaction cancellation.

As discussed in the previous section, density functional theory, as its name suggests, is fundamentally based on a "description in terms of n." In this sense, $\tilde{V}^{(\mathrm{H})}[n]$ is a representation that is highly compatible with density functional theory. In contrast, $\tilde{V}^{(\mathrm{X})}$ involves a "nonlocal integral that cannot be simply expressed in terms of $n(\mathbf{r})$,"[7] and only when this is evaluated exactly is the originally valid self-interaction cancellation $V^{(\text{SI})} = 0$ properly ensured.

The nonlocal integral of $\tilde{V}^{(\text{X})}$, however, is computationally cumbersome and costly, and thus, a strategy is employed in which it is approximated using local values of $n(\mathbf{r})$ [**local density approximation**/LDA (local density approximation) [1, 2]]. In this case, the exchange interaction part $V^{(\text{X})}$ deviates from its exact form, and the self-interaction cancellation with $V^{(\text{H})}$ [Eq. (16.15)] is violated. As a result, the self-interaction is no longer cancelled, and a spurious effect that should not originally exist is introduced as a bias. This is the problem known as "**violation of self-interaction cancellation**." A well-known consequence of this bias contamination is the band gap problem, which will be discussed later.

To summarize again:

- **Hartree–Fock method**; Even if self-interaction, which should not originally exist, is included for convenience, it cancels out, so there is no problem.
- **Conventional DFT**; Since self-interaction is not canceled, a self-interaction that should not exist originally is introduced as a bias.

Here, "conventional DFT" refers to DFT that employs exchange-correlation potentials based on the oldest historical approach, namely the local density approximation, which has subsequently been refined.

A point of potential misunderstanding that should be avoided here is that **DFT itself**, which expresses quantities in terms of $n(\mathbf{r})$, is **exact**. It is certainly correct that in the logic described above, "self-interaction cancellation was violated because one tried to express the Fock term using $n(\mathbf{r})$." However, if one skips over the details of this logic and merely grabs the conclusion, a careless interpretation might arise, leading to the **misconception** that "DFT is an *approximation* that expresses things in terms of $n(\mathbf{r})$." To clarify:

- **Density functional theory** (exact); the final $E_G[n]$ can be expressed as a **functional** of $n(\mathbf{r})$, and this is rigorously guaranteed.
- **Local density approximation**; this is an approximation in which the intermediate quantity V_{XC} is expressed as a **function** of $n(\mathbf{r})$.

Please pay attention to the difference between "function" and "functional."

Recalling once more the definitions given in Sect. 10.1: "XC" stands for exchange-correlation, "X" for exchange (effects due to statistics), and "C" for correlation (effects due to wavefunction distortion from interactions), the level of approximation for each scheme can be written as

$$\begin{aligned} V^{(\text{HF})} &= V^{(\text{H})} + V^{(\text{X})}(\text{non-local}) \\ V^{(\text{LDA})} &= \left[V^{(\text{H})} + V^{(\text{X})}(\text{local})\right] + V^{(\text{C})}(\text{local}) \\ V^{(\text{exactKS})} &= V^{(\text{H})} + V^{(\text{XC})} \end{aligned} \tag{16.16}$$

[7] That is, in order to determine the situation at one point, an integral over the entire region is required.

In $V^{(\mathrm{HF})}$, the wavefunction form is fixed to that which is justified in the absence of interaction, so only exchange effects are included, and by definition, correlation effects are excluded [$V^{(\mathrm{C})} = 0$]. In $V^{(\mathrm{LDA})}$, both the exchange term $V^{(\mathrm{X})}$ and the correlation term $V^{(\mathrm{C})}$ are included within the range of approximation.[8] However, due to the localization approximation made for the exchange term in $V^{(\mathrm{LDA})}$, self-interaction cancellation is broken in the portion denoted by $[\cdots]$. On the other hand, in an exact implementation of DFT (if such were possible), V_{XC} should correctly compensate and realize self-interaction cancellation with $V^{(\mathrm{H})}$, since the overall exactness of the theory is guaranteed.[9]

Note that the term "exactKS" rather than "exactDFT" is used here. The reason why the Hartree term appears in practical implementations of DFT is that it is operated based on the Kohn–Sham formulation. There, one isolates the contribution interpreted as the classical interaction $V^{(\mathrm{H})}[n]$ from quantum mechanical interactions and focuses on handling the remaining quantum effects through V_{XC}. This separation offers conceptual clarity. It should also be noted that the exactness of DFT remains intact even after transitioning to the Kohn–Sham framework, without any loss of generality.

16.3 DFT+U Method

16.3.1 Problem of Band Gap Underestimation and Regarding Misconceptions

The underestimation of band gaps in DFT is a well-known issue, and there was a time when it was widely touted as a "fatal flaw that casts doubt on the general validity of DFT."[10] After clarifying the mechanism behind the band gap underestimation problem, several remedies have been developed to address it. Among them, the method most widely known even to non-specialist users is the DFT+U method.[11] The U parameter that appears there is referred to as the "Hubbard's U," and is often

[8] When the author was a graduate student, phrases like "In LDA, electron correlation is neglected" were often heard at conferences, indicating a misunderstanding.

[9] It is known that exchange and correlation effects are coupled in order to fulfill conservation laws at the level of the entire system. Treating V_{X} and V_{C} separately is for convenience; in exact treatments, they are handled as V_{XC}.

[10] Many of these claims were sensationally propagated with "electronic correlation" as a keyword, and it is difficult to say they were based on a sufficient understanding of the mechanism of exchange effects, as described in this section.

[11] Other representative approaches at the level of exchange-correlation functionals include the SIC method (Self-interaction correction) and hybrid functional methods [3]. Beyond the exchange-correlation functional level, the GW method is a notable example [3].

described in textbooks as "related to the U parameter in the Hubbard model used in magnetism theory." The U parameter in the Hubbard model represents the on-site repulsion strength responsible for antiferromagnetism arising from superexchange interactions [4]. In the model of electrons hopping between sites, assuming that "when double occupancy occurs on the same site, the energy increases by U," only antiparallel spin pairs can realize a "virtual intermediate state with an energy increase of U."[12] Antiferromagnetism is then explained as an energy gain through perturbation via this virtual state [4].

The mechanism of antiferromagnetism described by the Hubbard model is extremely clear and leaves a strong impression, and for this very reason, it has ironically become a **source of misunderstanding** regarding DFT+U. In the Hubbard model, $U = 0$ represents a situation with no electron–electron interaction. Many researchers seem to understand the U parameter in DFT+U through an oversimplified explanation such as "U in DFT+U is like that in the Hubbard model" or "DFT originally does not include U, so we add it by hand." In such cases, the image of "$U = 0$ in the Hubbard model (a non-interacting electron system)" lingers, leading to the following chain of misconceptions: "DFT does not include U in the first place" $\rightarrow$ "DFT does not include electron-electron interactions" $\rightarrow$ "DFT does not include electron correlation." This line of reasoning results in an **extremely crude misunderstanding** that "electron correlation, absent in DFT, is forcibly added through the U parameter." As discussed above, DFT is, from its foundational principles, a framework that includes electron–electron interaction. The correlation effects (electron correlation) that arise are certainly imperfectly treated, but they are by no means "ignored." As will be explained below, the U parameter in DFT+U is, if anything, a corrective parameter applied more to exchange effects than to correlation effects.

16.3.2 Mechanism of Band Gap Underestimation

Because the image of "excited levels appear U higher due to on-site occupancy" in the Hubbard model overlaps with the notion that "the band gap is underestimated due to a lack of U," one often sees beginners who interpret the situation as: "in reality, the gap is U wide, but due to insufficient U, the conduction band 'descends,' leading to underestimation of the band gap." However, as we will explain below, the underestimation of the band gap can be understood as a consequence of *self-interaction cancellation deficiency*. In fact, the mechanism works the other way around: it is the valence band that **rises**, causing the band gap to be underestimated. This perspective more accurately reflects the true mechanism behind the phenomenon.

[12] Parallel spin pairs are forbidden to occupy the same site due to the Pauli exclusion principle.

An electron in the valence band is, by nature, supposed to experience interaction from $(N-1)$ other electrons. However, in the formulation that adopts the convenience of including self-interaction, it ends up experiencing interaction from all N electrons. As a result, the Coulomb repulsion is overcounted, causing the energy levels of the valence band to be evaluated unjustly high, and the entire band is consequently pushed up toward the high-energy side. On the other hand, the energy levels that constitute the conduction band should originally be evaluated in the context of an $(N+1)$-electron system, where one extra electron is added. In such a case, the conduction electron is supposed to interact with the other N electrons. In the standard implementation of DFT, however, the conduction band levels are also evaluated as solutions of an N-electron system. In a formulation where self-interaction cancellation is broken, both valence and conduction band levels are treated as experiencing interaction from all N electrons. Thus, the conduction band levels are evaluated as interacting with N electrons, which happens to be quantitatively correct. As a result, there is no upward shift in the conduction band levels. Consequently, the upward-shifted valence band and the unshifted conduction band become closer in energy, leading to an unjustifiably narrowed band gap [5].

16.3.3 Gap Problem Intensified by Localization

The mechanism described in the previous section, underestimation of the band gap due to upward shift of the valence band, occurs in both localized and itinerant systems. However, as described below, the manner in which the N orbital levels are affected differs, and the gap narrowing becomes particularly pronounced in localized systems [3].

In itinerant systems, all orbitals overlap with each other, so the effect of the self-interaction cancellation deficiency is shared among the N orbitals. As a result, each orbital is affected by approximately $1/N$ of a single self-interaction, yielding a situation where many orbitals experience only a slightly raised energy level.

In contrast, in localized systems, there is no mechanism whereby all orbitals share the self-interaction. Instead, a few orbitals near the localized region directly receive the full self-interaction, resulting in significantly raised levels. Since a band is a superposition of these levels, accumulating N tiny shifts in itinerant systems does not result in a large overall shift. However, in localized systems, a few significantly raised levels push up the top of the valence band, narrowing the band gap unjustifiably. Thus, the breakdown of self-interaction cancellation leads to a particularly pronounced narrowing of the gap in localized systems. As discussed earlier, this breakdown arises from the inaccurate evaluation of the exchange term $V^{(X)}$ (see Eq. 16.16). Therefore, it is appropriate to state that the underestimation of the band gap originates more from exchange effects than from correlation effects.

To illustrate the difference between localized and itinerant systems more directly from a mathematical standpoint, the following explanation applies: The extent of self-interaction cancellation deficiency depends on the degree of spatial localization of $\psi(\mathbf{r})$ through the integral expressions in Eqs. (16.12) and (16.13). In the case of localized electrons, the non-integrable function $1/|\mathbf{r}-\mathbf{r}'|$ takes a typical scale of $\sim 1/a$, where a is the lattice constant. Assuming that $\psi(\mathbf{r})$ is spread over each site with a weight of about 1, the integral can be estimated as $\sim 1/a$. On the other hand, in the case of delocalized electronic states, electrons are spread across the entire system with a length scale $L = N \cdot a$. Thus, one must consider a superposition of all possible site-to-site distances that contribute to $1/|\mathbf{r}-\mathbf{r}'|$. In such cases, it is appropriate to treat $\psi(\mathbf{r})$ as a plane wave and to handle $1/r$ using Fourier transformation. In this treatment, the Fourier transform of $1/r$ yields a factor $1/k^2$, and the magnitude of the integral becomes $(1/V)\left(1/k^2\right)$,[13] The wave number k corresponds to the wavelength $\lambda \sim L$ of the electron's extent, so $k = \lambda^{-1}$, and the magnitude of the integral becomes $\left(1/L^3\right)\left(1/L^{-2}\right) \sim 1/L$. Since $L = N \cdot a$, the magnitude of the integration error due to this bias becomes $\sim 1/(N \cdot a)$, yielding an additional factor of $1/N$ compared to the localized case. This factor implies that in itinerant systems, as the particle number $N \to \infty$, the bias, though present, becomes negligibly small.

16.3.4 Remedies for the Underestimation of Band Gaps

As remedies for the underestimation of band gaps, two major strategies can be understood.[14] The first strategy is to implement a mechanism such that, when the contribution from self-interaction ($j = i$) is included, a corrective term is applied to the potential. This corresponds to the DFT+U method, and the U parameter serves as the weight to adjust the magnitude of this correction. This mechanism will be discussed in more detail in the next subsection. A different and more fundamental alternative approach is known as the **exact exchange method** [3]. This method returns to Eq. (16.16), and, just as self-interaction cancellation was realized in $\tilde{V}^{(\mathrm{HF})}$ within Hartree–Fock theory, it seeks to achieve **proper evaluation of the non-local integral** in $\tilde{V}^{(\mathrm{X})}$ so as to recover self-interaction cancellation.

Rather than using "$V^{(\mathrm{XC})}$ expressed by $n(\mathbf{r})$" as in LDA, this strategy instead evaluates the non-local integral in Eq. (16.13) for $V^{(\mathrm{X})}$. Since the exchange part is no longer expressed in terms of $n(\mathbf{r})$ but is explicitly represented in terms of orbital functions, this approach is also referred to as the **orbital-dependent exchange method**. However, the exact exchange method remains computationally expensive due to the cost of integral evaluations and, as such, still largely resides within the realm of fundamental research.

---

[13] $(1/V)$ arises as a factor from the spatial Fourier transformation.

[14] As noted in the footnote of Sect. 16.3.1, there are other strategies beyond these two.

16.3.5 Essential Idea of the DFT+U Method

The problem of underestimated band gaps originates in the inclusion of self-interaction contributions ($j = i$), which should not exist, as discussed in Eqs. (16.11)–(16.14). The idea behind the DFT+U method is to act with a projection operator that flags only the $j = i$ contribution within the integral evaluations that include self-interaction. Then, by adjusting the problematic contributions using a parameter U, the deficiency is corrected.

This projection operator takes a form that increases the potential by U when i and j occupy the same position, and is indeed designed similarly to the "Hubbard U" set when double occupancy occurs in the Hubbard model. However, unlike the setup in the Hubbard model, which consists of "(non-interacting evaluation) + (correction of missing part with Hubbard U)," DFT+U is based on the idea of "removing an *overcounted* contribution from an already interacting evaluation," offering a contrast in the sense that the correction is subtractive rather than additive. As noted in Sect. 16.3.1, the terminology "Hubbard U" can mislead beginners who may confuse the underlying contexts, so caution is necessary. Furthermore, the $j = i$ in the context of DFT+U does not mean that such a configuration arises due to electron hopping (as in lattice models), but rather that a $j = i$ type contribution included in the integral is to be removed.

The U parameter is an adjustable parameter, so in principle, the user must specify a value in the input file. The practical question then becomes how to determine this value. In principle, the goal is to "adjust such that the deviation caused by the self-interaction error is corrected." Among possible deviations, the band gap is the most prominent, so the most straightforward method is to choose U such that the calculated gap agrees with the experimental value. However, it is also known that if U is determined solely to match the band gap, it can result in significant discrepancies in other quantities, such as the lattice constant [6]. Thus, in foundational studies of many-body electronic structure theory, there is ongoing discussion about "which quantity should be matched to determine U" [3].

The question of what value to adopt for the U parameter is a difficult one. At the level of this text, the safest option is to "inherit the values used in previous studies." Recently, tools for determining the U parameter have begun to be included within software packages. Nevertheless, in practical applications, one often encounters the following kinds of difficulties [7]: While it seems natural to assume that "U reflects the extent of self-interaction error, and thus should differ for each material system, so that U in a compound and U in a pure element should differ," one also faces a fundamental dilemma when calculating energy differences such as ΔH: Should different U values be adopted for the compound and the pure element? If so, can any meaningful validity be ascribed to an energy difference derived from such parameter-adjusted values? This remains a complex and open-ended issue, with much room for further discussion.

16.3.6 Summary/to Avoid Misunderstanding of DFT+U

If the issue of self-interaction cancellation failure is described simply as "the problem of self-interaction," it can lead to a number of misunderstandings. For example, people may come to think that there is some kind of "self-interaction that ought to be properly represented," which actually "exists," and that the issue lies in the failure to represent it correctly. This is completely incorrect. Such vague understandings can easily propagate like a game of telephone, giving rise to confusion. It is therefore important to re-emphasize that "self-interaction" is something that does NOT fundamentally exist, but is introduced only for convenience in approximations.

Note that in Eq. (16.16), $V^{(C)}$ is included within $V^{(LDA)}$. The issue of underestimated band gaps arises not because "correlation" is missing, but rather because a simplified form was adopted in which both exchange and correlation are represented in terms of $n(\mathbf{r})$ in a convenient way. This simplification compromises certain essential aspects of the exchange contribution. It is often mistakenly said that "DFT (LDA) has $V^{(C)} = 0$, so DFT+U is a model that adds this missing correlation," or that "the band gap problem arises because 'correlation' is not included, and DFT+U adds correlation in the form of the Hubbard U." However, as explained thus far, the band gap problem is more accurately attributed to an issue with exchange rather than correlation.

References

1. "Density-functional Theory of Atoms And Molecules" (Oxford University Press, 1989), R.G. Parr, W. Yang ISBN: 978-0195092769
2. "Density Functional Theory: A Practical Introduction" (Wiley-Interscience, 2009), D.S. Sholl, J.A. Steckel, ISBN: 978-0470373170.
3. "Strong Coulomb Correlations in Electronic Structure Calculations" (CRC Press, 2000), V.I Anisimov (ed.), ISBN: 978-9056991319
4. "Lecture Notes on Electron Correlation and Magnetism" (World Scientific ,1999) Patrik Fazekas, ISBN: 978-9810224745
5. "Electronic Structure: Basic Theory and Practical Methods" (Cambridge University Press, 2004), R.M. Martin, ISBN: 978-0521782852
6. "Bandgap reduction of photocatalytic TiO_2 nanotube by Cu doping", S.K. Gharaei, M. Abbasnejad, R. Maezono, Sci. Rep. 8, 14192 (2018). https://doi.org/10.1038/s41598-018-32130-w
7. "*Ab initio* thermodynamic properties of certain compounds in Nd-Fe-B system", A.T. Hanindriyo, S. Sridar, K.C. Hari Kumar, K. Hongo, R. Maezono, Comp. Mater. Sci. 180, 109696 (2020). https://doi.org/10.1016/j.commatsci.2020.109696

Chapter 17
Appendix G: Tools Used in Many-Body Perturbation Theory

This chapter serves as an appendix summarizing tools frequently used in many-body perturbation theory, such as second quantization and Green's functions. The material is motivated from the discussion around Sect. 10.2.3 in the main text. In the context of the in-depth discussion on the characteristics of *ab initio* QMC methods in Chap. 10, it is important to understand these methods in contrast to formalisms using field operators. With the goal of understanding Green's functions for field operators in Sect. 17.2.6, this appendix provides a minimal path leading up to that point.

17.1 Overview of the Second Quantization Method

Here we outline the context in which second quantization is introduced, as well as the important consequences derived from it and their placement within the overall logical framework. In order to serve as a conceptual guide when studying the technical details, we provide only references to relevant literature and omit detailed derivations, aiming instead for a concise summary that allows for a bird's-eye view of the entire structure. For those already familiar with the topic, this section may serve to reconfirm the overall logic. For newcomers, it is intended as a guiding map to avoid losing sight of the big picture when delving into detailed study.

17.1.1 Procedure of Canonical Quantization

We begin by reviewing and summarizing the framework of analytical mechanics, which was constructed on the basis of classical mechanics. The governing equations of classical mechanics are derived as the Euler–Lagrange equations, which result

R. Maezono, *Ab initio Quantum Monte Carlo Tutorial*,
https://doi.org/10.1007/978-981-95-4134-8_17

from the variational minimization of the Lagrangian. Once the Lagrangian is given, the generalized momentum is constructed as

$$p = \frac{\partial L(q, \dot{q})}{\partial \dot{q}},$$

and the Hamiltonian H can be introduced via the Legendre transformation,

$$H = p \cdot \dot{q} - L.$$

Once the Hamiltonian is given, the equation of motion for a physical quantity f can be expressed as

$$\frac{df}{dt} = \frac{\partial f}{\partial q}\frac{\partial H}{\partial p} - \frac{\partial H}{\partial q}\frac{\partial f}{\partial p} =: \{f, H\}_{\mathrm{PB}}, \tag{17.1}$$

i.e., in terms of the Poisson bracket with the Hamiltonian. Given the equation of motion written in terms of the Poisson bracket, one can move to the Heisenberg form of quantum mechanics, in which physical quantities are treated as time-dependent operators, by replacing the Poisson bracket with the commutator as

$$\{f, g\}_{\mathrm{PB}} \to \frac{1}{i\hbar}[f, g]. \tag{17.2}$$

Here, the time-dependent operator is defined as[1]

$$\hat{O}_H(t) := \hat{U}^\dagger(t)\,\hat{O}_S\,\hat{U}(t), \quad \hat{U}(t) = \exp\left[-i\frac{\hat{H}}{\hbar}t\right]. \tag{17.3}$$

In a Cartesian coordinate system, the generalized momentum corresponding to the basic variable q coincides with the usual momentum p, and one has

$$\{q, p\}_{\mathrm{PB}} = \frac{\partial q}{\partial q}\frac{\partial p}{\partial p} - \frac{\partial p}{\partial q}\frac{\partial q}{\partial p} = 1,$$

which translates to the quantum mechanical commutation relation

$$\left[\hat{q}, \hat{p}\right] = i\hbar. \tag{17.4}$$

This replacement constitutes the transition to quantum mechanics, and the procedure is referred to as **canonical quantization**.

---

[1] In contrast to quantum mechanics formulated in the Schrodinger picture, when one imposes the time evolution of physical observables in the Heisenberg picture as given by Eq. (17.3), it is possible to derive the equation of motion for the operators. By comparing the resulting expression with the classical form using Poisson brackets, one finds that the replacement given in Eq. (17.2) indeed holds, thereby establishing the correspondence. See also [1].

Once the Lagrangian is specified, the subsequent procedures and mathematical structures are universally applicable, regardless of the governing equations. Thus, for systems not governed by classical mechanics, it is still possible to construct a Lagrangian that yields the equations of motion as Euler–Lagrange equations and to discuss canonical quantization via the Hamiltonian and Poisson brackets. More generally, one can define a Lagrangian density $\mathcal{L}(\varphi(\mathbf{r},t), \dot{\varphi}(\mathbf{r},t))$ for the field variables, and proceed to formulate the Hamiltonian and Poisson brackets. This framework is known as the "classical field theory" [2]. When canonical quantization is applied to such a classical field formulation, one expects to transition into **quantum field theory**.[2]

17.1.2 Canonical Quantization of Schrodinger Field

When the Schrodinger equation

$$i\hbar\frac{\partial}{\partial t}\psi(\mathbf{r},t) = H_1^{(1)}\cdot\psi(\mathbf{r},t) \tag{17.5}$$

is interpreted as a field equation,[3] this equation of motion can be derived from the Euler–Lagrange equation with respect to the fundamental variable ψ for the Lagrangian

$$L = \int d^3x\left[\psi^*(\mathbf{r},t)\,i\hbar\frac{\partial}{\partial t}\psi(\mathbf{r},t) - \psi^*(\mathbf{r},t)\,H_1^{(1)}\cdot\psi(\mathbf{r},t)\right] \tag{17.6}$$

as shown. Starting from this Lagrangian and applying canonical quantization, the field variable ψ and its canonical momentum $i\hbar\psi^*$ are promoted to operators as in Eq. (17.4), leading to the following algebraic relation between operators:

$$\left[\hat{\psi}(\mathbf{r},t), \hat{\psi}^\dagger(\mathbf{r}',t)\right] = \delta^3(\mathbf{r}-\mathbf{r}') \tag{17.7}$$

This commutation relation is required as a quantum-theoretical prescription for the field operators [4, 5].

From the Lagrangian in Eq. (17.6), one can derive the canonical momentum and, via the Legendre transformation, obtain the Hamiltonian of the field.[4] This

[2] In this context, ambiguities in the theory that are not resolved by canonical quantization become a subject of discussion [3].

[3] The superscript (1) on the Hamiltonian denotes that it is a first-quantized Hamiltonian. The subscript "1" indicates it is a one-body Hamiltonian.

[4] This is distinct from $H^{(1)}$ in Eq. (17.5); it refers instead to a newly constructed "field Hamiltonian" derived through the present procedure.

"Hamiltonian of the field" contains the field variable $\psi(\mathbf{r}, t)$, and by replacing it with the operator $\hat{\psi}(\mathbf{r}, t)$, we obtain the "second-quantized field Hamiltonian" $\hat{H}_1^{(2)}$. The time dependence of the operator $\hat{\psi}(\mathbf{r}, t)$ is defined by Eq. (17.3), that is,

$$\hat{\psi}(\mathbf{r}, t) = e^{i\hat{H}t/\hbar} \cdot \hat{\psi}(\mathbf{r}) \cdot e^{-i\hat{H}t/\hbar}.$$

The time evolution equation that governs this time dependence is

$$i\hbar \cdot \frac{\partial}{\partial t}\hat{\psi}(\mathbf{r}, t) = \left[\hat{\psi}(\mathbf{r}, t), \hat{H}_1^{(2)}\right]. \tag{17.8}$$

This is the procedure of "canonical quantization based on Eqs. (17.1) and (17.2)." By evaluating the commutator on the right-hand side, we indeed obtain

$$i\hbar \cdot \frac{\partial}{\partial t}\hat{\psi}(\mathbf{r}, t) = H_1^{(1)} \cdot \hat{\psi}(\mathbf{r}, t). \tag{17.9}$$

This result reproduces a form identical to Eq. (17.5), though their meaning is different. In other words, it shows that the Heisenberg equation in second quantization yields an expression symbolically consistent with the Schrodinger equation in first quantization.[5]

In the above discussion, we refer to the framework of the Schrodinger equation, obtained by quantizing classical theory via the correspondence principle, *i.e.*, conventional quantum mechanics, as **first quantization**, and distinguish it from the quantization treated in this section. The quantization in this section involves regarding the Schrodinger equation as a field equation and applying canonical quantization to it. Since this constitutes an additional quantization step, the resulting framework is called **second quantization**.

17.1.3 Treating of Many-Body Systems

The governing equation of many-body quantum mechanics is expressed in terms of the many-body wavefunction as

$$i\hbar\frac{\partial}{\partial t}\Psi_{\pm}^{(1)}(x_1, \cdots, x_N; t) = H_N^{(1)} \cdot \Psi_{\pm}^{(1)}(x_1, \cdots, x_N; t) \tag{17.10}$$

The many-body Hamiltonian for an N-particle system is given in the form

[5] These derivations are carried out via the eigenfunction expansion given in Eq. (17.14).

$$H_N^{(1)} = \sum_{j=1}^{N} h\left(x_j\right) + \frac{1}{2}\sum_{i,j} V\left(x_i - x_j\right)$$

This equation must be solved under the constraint condition with respect to particle exchange:

$$\Psi_{\pm}^{(1)}\left(\cdots, \mathbf{r}_j, \cdots, \mathbf{r}_i, \cdots\right) = (\pm)\cdot\Psi_{\pm}^{(1)}\left(\cdots, \mathbf{r}_i, \cdots, \mathbf{r}_j, \cdots\right).$$

The burden of "solving while taking this constraint into account" is replaced by the algebraic relations of operators, so that the constraint is automatically satisfied. This corresponds to the formulation of many-particle quantum mechanics using second quantization, as discussed below.

As a general algebraic result, the commutation relation in Eq. (17.7) yields the raising and lowering structure:

$$\hat{\psi}^{\dagger}|0\rangle \sim |1\rangle \quad , \quad \hat{\psi}^{\dagger}|n\rangle \sim |n+1\rangle \quad , \quad \hat{\psi}|n\rangle \sim |n-1\rangle \, , \tag{17.11}$$

as well studied. The algebra in Eq. (17.7) was originally discovered from the insight of canonical quantization. We now consider its generalization to

$$\left[\hat{\psi}(t,\mathbf{x}), \hat{\psi}^{\dagger}(t,\mathbf{y})\right]_{\pm} = \delta^3(\mathbf{x}-\mathbf{y}) \tag{17.12}$$

where

$$[A, B]_{\pm} := AB \pm BA$$

Here, the minus sign denotes the usual commutator, and the plus sign denotes the anti-commutator. Even when this algebra is generalized to include the anti-commutation relation, the ladder structure as in Eq. (17.11) still holds. For the anti-commutation case, one finds that $\hat{\psi}^{\dagger}\hat{\psi}^{\dagger}|n\rangle = 0$, which expresses the Pauli exclusion principle forbidding double occupancy.

Equation (17.11) can be interpreted as indicating that $\hat{\psi}^{\dagger}$ creates a particle, while $\hat{\psi}$ annihilates a particle. Taking this meaning into account, we define a quantity from the second-quantized operators:

$$\Psi_{\pm}^{(1)}(x_1, \cdots, x_N; t) = \frac{1}{\sqrt{N!}}\langle 0|\,\hat{\psi}(x_1,t)\cdots\hat{\psi}(x_N,t)\cdot\hat{c}_1^{\dagger}\cdots\hat{c}_N^{\dagger}\,|0\rangle \, . \tag{17.13}$$

It can be shown that this expression is equivalent to the symmetrized or anti-symmetrized multi-particle wavefunction:

$$\Psi_{\pm}^{(1)}(x_1, \cdots, x_N; t) = \frac{1}{\sqrt{N!}}\sum_{P}(\pm)^P \varphi_1\left(x_{P_1}, t\right)\cdots\varphi_N\left(x_{P_N}, t\right) ,$$

where the field operator is expanded in terms of eigenfunctions as

$$\hat{\psi}(x,t) = \sum_l \hat{c}_l(t) \cdot \varphi_l(x). \tag{17.14}$$

Let $\hat{H}^{(2)}(t)$ denote the Hamiltonian operator in the second-quantized formalism. Then, the time-evolution equation for the field operator in the Heisenberg picture should be given as

$$i\hbar \frac{\partial}{\partial t}\hat{\psi}(t,x) = \left[\hat{\psi}(t,x), \hat{H}^{(2)}(t)\right] \tag{17.15}$$

in accordance with Eq. (17.8). If the field operator $\hat{\psi}(t,x)$ satisfies the commutation relations of Eq. (17.12), and the raising/lowering relations such as those of Eq. (17.11) derived from it, then we are led to the question of what form $\hat{H}^{(2)}(t)$ should take in order to satisfy Eq. (17.15). As a guiding principle, when the many-body wave function $\Psi_{\pm}^{(1)}(x_1, \cdots, x_N; t)$ is constructed as in Eq. (17.13), the operator $\hat{H}^{(2)}(t)$ should be defined in such a way that this wave function satisfies the many-body Schrodinger equation of Eq. (17.10). To state the conclusion first, it is found that the condition is indeed satisfied if

$$\begin{aligned}\hat{H}^{(2)} &= \int d^3x \cdot \hat{\psi}^\dagger(x)\, h(x)\, \hat{\psi}(x) \\ &+ \frac{1}{2}\int d^3x d^3y \cdot \hat{\psi}^\dagger(x)\, \hat{\psi}^\dagger(y)\, V(x-y)\, \hat{\psi}(y)\, \hat{\psi}(x)\end{aligned} \tag{17.16}$$

as demonstrated in [4, 5].

It should be noted here that the second-quantized Hamiltonian $\hat{H}^{(2)}(t)$ does not carry an index N, unlike $H_N^{(1)}$. In the framework of second quantization, there is no explicit dependence on the particle number N. In the first-quantized formulation based on many-body wave functions, an N-particle system is described as a realization probability in a $3N$-dimensional configuration space. However, the physical phenomenon at hand is a single point in three-dimensional space involving N particles, and describing it as a point in $3N$-dimensional space is conceptually obscure. Moreover, a formulation based on fixed N is inherently unsuitable for problems in statistical mechanics, where N may fluctuate, or for quantum phase transition problems, where the phase, rather than the number of particles, is sharply defined. These concerns are among the motivations for the development of field-theoretical approaches formulated in terms of second quantization [6].

17.2 Overview of Green's Functions

The term "Green's function" appears in various contexts in theoretical materials science. One source of confusion for beginners is that Green's functions are often introduced in apparently unrelated settings, with seemingly different definitions despite using the same terminology. For example, they are introduced in the context of linear response (Sect. 17.2.1), as propagators in the main text (Sects. 7.3, 17.2.2), as inverse operators (Sect. 17.2.3), or as complex functions used to express spectral decompositions in seemingly abrupt fashion (Sect. 17.2.5). They are also introduced using field operators in a different chapter (Sect. 17.2.6). In light of this, we present here a summary to provide a coherent overview and clarify the interconnections among these approaches.

17.2.1 Green's Functions Used for Homogeneous Equations

A basic and natural context in which Green's functions are introduced is in linear response theory, which deals with the response of linear, time-invariant systems. If the external input or signal is expressed, as shown below, as a superposition of delta functions, then the output response, due to the linearity and time invariance of the system, can also be expressed as a superposition of the responses to delta function inputs.

For an **inhomogeneous equation** expressed by a linear operator $\hat{L}_x$,

$$\hat{L}_x \cdot \psi(x) = f(x),$$

if the inhomogeneous term $f(x)$ is expressed as

$$f(x) = \int dx' \cdot f\left(x'\right) \cdot \delta\left(x - x'\right),$$

then the inhomogeneous equation can be rewritten as

$$\hat{L}_x \cdot \psi(x) = \int dx' \cdot f\left(x'\right) \cdot \delta\left(x - x'\right). \tag{17.17}$$

The variable in the differential equation is x, so the "function" on the right-hand side is $\delta\left(x - x'\right)$. We interpret the right-hand side as a superposition of delta functions, with $f\left(x'\right)$ as amplitude. When the inhomogeneous term is a single $\delta\left(x - x'\right)$, we write:

$$\hat{L}_x \cdot G\left(x - x'\right) = \delta\left(x - x'\right),$$

where $G\left(x-x'\right)$ corresponds to the response to a single delta function. If we regard $f\left(x'\right)$ on the right-hand side of Eq. (17.17) as the "coefficient of a linear combination," then due to the linearity and shift-invariance of $\hat{L}_x$, a linear combination constructed under the same rule as the right-hand side of Eq. (17.17)

$$\psi\left(x\right)=\int dx'\cdot f\left(x'\right)\cdot G\left(x-x'\right) \tag{17.18}$$

should serve as the solution $\psi\left(x\right)$ to the inhomogeneous equation. In this way, the concept of the **Green function** $G\left(x-x'\right)$ is introduced in an elementary manner.

What often confuses beginners is the situation where this framework of the Green function is applied to a

$$\hat{L}_x\cdot\psi\left(x\right)=0$$

i.e., a **homogeneous equation**. Even in such a case, textbooks abruptly state that "we consider the corresponding Green function."

$$\hat{L}_x\cdot G\left(x-x'\right)=\delta\left(x-x'\right). \tag{17.19}$$

However, if there is no inhomogeneous term, then the logic used to derive Eq. (17.18) cannot be developed, and this becomes the source of confusion.

This confusion can in fact be dispelled by a small whisper about "how to interpret Eq. (17.19)": That is, to reconsider Eq. (17.17) as

$$\begin{aligned}\hat{L}_x\cdot\psi(x)&=f(x')\cdot\delta(x-x')\\&=\begin{cases}0 & (x\neq x')\\ f(x') & (x=x')\end{cases}\end{aligned}$$

in terms of its superposition. Although it is not easily recognized when written with the delta function, we can reinterpret it such that, at almost all points except $x=x'$, it becomes a homogeneous equation. The nonzero constant value $f(x')$ given at $x=x'$ corresponds to something that can be regarded as an "initial displacement." In other words, Eq. (17.17) can be viewed as "the free response (response governed by the homogeneous equation) when an initial displacement $f(x')$ is given at $x=x'$."[6]

Let us recall the elementary treatment of free oscillations in a spring-mass system and confirm what corresponds to the Green function in that context. First, one realizes that free oscillation cannot be expressed as a Green function solution to the homogeneous equation obtained by simply setting the external force to zero.

[6] Indeed, Green functions for spring-mass systems or LCR circuits describe dynamics known as free responses [7]. The Green function for the diffusion equation also expresses "the subsequent diffusion dynamics of a particle distribution that was initially localised at $x=x'$ as a delta function" [7].

There appears to be no oversight in the derivation of Eq. (17.17) using the Green function, so this outcome seems correct. Then, the homogeneous equation with vanishing external force corresponds to setting $f\left(x'\right) = 0$, and the solution derived from Eq. (17.18) would then be $\psi(x) = 0$—the trivial solution. This clearly does not yield the nontrivial free-oscillation solution to the homogeneous equation. The standard procedure of solving a homogeneous equation for nonzero solutions, and determining integration constants from initial or boundary conditions, should be understood not as "simply setting the external force to zero," but rather as "assigning a delta function on the right-hand side."

Based on the above, let us reconsider the meaning of expressing the inhomogeneous term in an inhomogeneous equation as

$$f(x) = \int dx' \cdot f\left(x'\right) \cdot \delta\left(x - x'\right).$$

This represents the forcing function $f(x)$ as a superposition of various initial conditions, each weighted by $f\left(x'\right)$. In a proverbial sense, the inhomogeneous equation is being recast as a superposition of homogeneous equations (free responses), each with a different initial condition.

17.2.2 Green Functions and Propagators

Reflecting on the meaning of Green functions discussed in the previous section helps to soften the initial discomfort often felt by beginners, such as: "Why does the unknown function $\psi(x)$ in the original equation depend on a single variable, while the green function $G\left(x - x'\right)$ appears to depend on two variables?" From the discussion in the previous subsection, the Green function $G\left(x - x'\right)$ can be understood as representing the weight of the contribution at x from each boundary condition imposed at $x = x'$. In this light, we may interpret the solution

$$\psi(x) = \int dx' \cdot f\left(x'\right) \cdot G\left(x - x'\right)$$

as expressing the effect propagated from x' to x with weight $f\left(x'\right)$. The green function $G\left(x - x'\right)$ thus describes propagation from x' to x, and is referred to as a **propagator kernel**.

17.2.3 Green Functions and Inverse Operators

For an inhomogeneous equation in spacetime

$$\hat{L}\left(\partial_t, \nabla_x\right) \cdot \psi(x) = f(x),$$

the corresponding Green function satisfies

$$\hat{L}(\partial_t, \nabla_x) \cdot G(x - x'; t - t') = \delta(x - x')\,\delta(t - t').$$

Applying a Fourier transform, this becomes

$$\hat{L}(-i\omega, i\mathbf{k}) \cdot G(\mathbf{k}; \omega) = 1,$$

which implies that

$$G(\mathbf{k}; \omega) = \hat{L}^{-1}(-i\omega, i\mathbf{k}).$$

In other words, the Green function corresponds to the inverse operator of $\hat{L}$.

17.2.4 Green Functions and Self-Adjoint Operators

An elementary introduction to Green functions was given in Sect. 17.2.1, based on their role in linear, time-invariant systems. However, even if the system is not time-invariant (*i.e.*, not shift-invariant), as long as $\hat{L}$ is a **Sturm-Liouville** type **self-adjoint operator** [7], the solution can still be expressed using a Green function. This will be shown as follows.

For the Green function problem

$$\hat{L} \cdot \psi(x) = f(x) \quad , \quad \hat{L} \cdot G(x - x') = \delta(x - x') \tag{17.20}$$

if the operator $\hat{L}$ is self-adjoint, then

$$\left\langle \hat{L} \cdot \psi \middle| G(x - x') \right\rangle = \left\langle \psi \middle| \hat{L} \cdot G(x - x') \right\rangle$$

holds true [7]. By substituting Eq. (17.20) into both sides at the positions where $\hat{L}$ acts,

$$\langle f | G(x - x') \rangle = \langle \psi | \delta(x - x') \rangle$$

is obtained. Writing this explicitly using integrals,

$$\int dx' \cdot f(x')\, G(x - x') = \int dx' \cdot \psi(x')\, \delta(x - x')$$

$$\therefore \quad \psi(x) = \int dx' \cdot f(x')\, G(x - x')$$

This yields the same form as Eq. (17.18), which was derived in the elementary introduction of Green functions based on linear response theory.

17.2.5 Spectral Representation of the Green Function

For the eigenvalue problem

$$\hat{L} \cdot \psi(x) = -\lambda \cdot \psi(x) \tag{17.21}$$

consider the case where the eigenvalues and eigenfunctions are given by $\{\lambda_j, u_j(x)\}_j$. The Green function corresponding to Eq. (17.21) is defined in a λ-dependent form as

$$\left[\hat{L} + \lambda \cdot \hat{I}\right] \cdot G\left(x, x'; \lambda\right) = \delta\left(x - x'\right).$$

Then it can be shown that [7]

$$G\left(x, x'; \lambda\right) = \sum_{n=0}^{\infty} \frac{u_n(x)\, u_n^*\left(x'\right)}{\lambda - \lambda_n}.$$

This expression is derived under the assumption that $\hat{L}$ is a self-adjoint operator.

Using the bra–ket notation,

$$\begin{aligned} G\left(x, x'; \lambda\right) &= \sum_{n=0}^{\infty} \frac{\langle x|u_n\rangle \left\langle u_n|x'\right\rangle}{\lambda - \lambda_n} \\ &= \langle x| \sum_{n=0}^{\infty} \frac{|u_n\rangle \langle u_n|}{\lambda - \lambda_n} \left|x'\right\rangle \\ &= \langle x| \hat{G}(\lambda) \left|x'\right\rangle, \end{aligned}$$

we introduce the Green operator as

$$\hat{G}(\lambda) = \sum_{n=0}^{\infty} \frac{|u_n\rangle \langle u_n|}{\lambda - \lambda_n}.$$

Extending λ to a complex number z, we define

$$\hat{G}(z) = \sum_{n=0}^{\infty} \frac{|u_n\rangle \langle u_n|}{z - \lambda_n},$$

which leads to a representation that can be evaluated via residue integration. For the real eigenvalues $\{\lambda_j\}_j$, the poles in the above residue integral lie along the real axis. In this case, the standard prescription in complex analysis is to displace the poles slightly off the real axis, writing

$$\hat{G}(z) = \sum_{n=0}^{\infty} \frac{|u_n\rangle \langle u_n|}{z - \lambda_n \pm i\delta},$$

where δ is a positive infinitesimal.

17.2.6 Green's Function for the Field Operator

The time dependence of the field operator $\hat{\psi}(\mathbf{r}, t)$ is defined in the Heisenberg picture as

$$\hat{\psi}(\mathbf{r}, t) = e^{i\hat{H}^{(2)}t/\hbar} \cdot \hat{\psi}(\mathbf{r}) \cdot e^{-i\hat{H}^{(2)}t/\hbar}. \tag{17.22}$$

In the absence of interactions, its dynamics reduces to

$$i\hbar \cdot \frac{\partial}{\partial t}\hat{\psi}(\mathbf{r}, t) = H_0^{(1)} \cdot \hat{\psi}(\mathbf{r}, t) \tag{17.23}$$

as shown in Eq. (17.9).[7] The Green function corresponding to this dynamics is given by

$$\left[i\hbar\frac{\partial}{\partial t} - H_0^{(1)}\right] G\left(\mathbf{r}, t; \mathbf{r}', t'\right) = \hbar \cdot \delta\left(t - t'\right) \delta\left(\mathbf{r} - \mathbf{r}'\right), \tag{17.24}$$

and it can be shown that

$$G\left(\mathbf{r}, t; \mathbf{r}', t'\right) := -i\left\langle T_t\left[\hat{\psi}(\mathbf{r}, t)\, \hat{\psi}^\dagger\left(\mathbf{r}', t'\right)\right]\right\rangle \tag{17.25}$$

satisfies Eq. (17.24) [5]. Here, the expectation value expressed by the notation $\left\langle \hat{A}(t)\right\rangle$ is defined as the thermal average in statistical mechanics:

$$\left\langle \hat{A}\right\rangle = \frac{1}{\Xi}\mathrm{Tr}\left[\exp\left[-\beta\hat{H}^{(2)}\right] \cdot \hat{A}\right], \quad , \quad \Xi = \mathrm{Tr}\left[\exp\left[-\beta\hat{H}^{(2)}\right]\right]. \tag{17.26}$$

[7] The relationship between the first-quantized Hamiltonian $H^{(1)}$ and the second-quantized Hamiltonian $H^{(2)}$ is as described in Sect. 17.1.3.

The operator T_t is called the time-ordering operator, which arranges the operators so that those with later time arguments appear to the left.

In this way, when there is no interaction, the Green function $G\left(\mathbf{r}, t; \mathbf{r}', t'\right)$ defined in Eq. (17.25) acts as the Green function for the time evolution of the field operator. As described in Sect. 17.2.2, a Green function carries the meaning of a propagator. Looking at Eq. (17.25), it corresponds to an interpretation such as "creating a particle at time t' and annihilating it at time t" and thus takes the form that can be read as a propagator for particle creation and annihilation.

The Green function in Eq. (17.24) follows, as a general property of Sturm–Liouville-type self-adjoint operators, that its poles coincide with the eigenvalues of the operator $\hat{H}^{(1)}$ (Sect. 17.2.5). That is, by locating the poles ε_k of the Green function in Eq. (17.25),

$$G(z) \sim \sum_k \frac{1}{z - \varepsilon_k} \tag{17.27}$$

the single-particle energy levels can be obtained.[8]

17.2.7 Thermal Green Function

In the left-hand side of Eq. (10.21),

$$G\left(\mathbf{r}, \tau; \mathbf{r}', \tau'\right) \sim \left\langle T_\tau\left[\hat{\psi}^\dagger(\mathbf{r}, \tau)\, \hat{\psi}\left(\mathbf{r}', \tau'\right)\right]\right\rangle,$$

the symbol " $\sim$" is used instead of " =." This is because the Green function we have been discussing using the Roman letter "G" was originally introduced in Eq. (17.25) as

$$G\left(\mathbf{r}, t; \mathbf{r}', t'\right) := -i\left\langle T_t\left[\hat{\psi}(\mathbf{r}, t)\, \hat{\psi}^\dagger\left(\mathbf{r}', t'\right)\right]\right\rangle,$$

a function of *real time* t. Around Eq. (10.16), we *temporarily* replaced it with a τ-dependent form by defining $\tau = it/\hbar$ as a provisional tool:

$$\mathcal{G}\left(\mathbf{r}, \tau; \mathbf{r}', \tau'\right) = -\left\langle T_\tau\left[\hat{\psi}^\dagger(\mathbf{r}, \tau)\, \hat{\psi}\left(\mathbf{r}', \tau'\right)\right]\right\rangle,$$

which is a Green function with *imaginary time* τ as the argument. This is referred to as the **thermal Green function**.

[8] The Green function takes the form $G\left(x, x'\right) = G\left(x - x'\right)$, where t and $\mathbf{r}$ are represented by x. By performing a Fourier transform with respect to $y = x - x'$, we obtain $G(ik)$, and ik is then treated as a complex variable z.

The thermal Green function is introduced in the context of evaluating the partition function Ξ appearing in Eq. (17.26) as

$$\begin{aligned}\Xi &= \mathrm{Tr}\left[e^{-\beta H}\right] = \mathrm{Tr}\left[e^{-\beta H_0}e^{\beta H_0}e^{-\beta H}\right] \\ &= \Xi_0 \cdot \mathrm{Tr}\left[\frac{e^{-\beta H_0}}{\Xi_0}e^{\beta H_0}e^{-\beta H}\right] =: \Xi_0 \cdot \left\langle e^{\beta H_0}e^{-\beta H}\right\rangle_0 \\ &=: \Xi_0 \cdot \langle U(\beta)\rangle_0 \end{aligned} \tag{17.28}$$

and evaluated by perturbative expansion (Sect. 10.3.1) using the operator defined in Eq. (10.19),

$$\hat{U}(\tau) = \sum_{n=0}^{\infty}\frac{(-)^n}{n!}\int_0^{\tau}\cdots\int_0^{\tau} d\tau_1\cdots d\tau_n \cdot T_\tau\left[\hat{H}_I(\tau_1)\cdots\hat{H}_I(\tau_n)\right].$$

By substituting Eq. (17.16) into $\hat{H}_I(\tau_k)$,

$$\hat{H}_I = \frac{1}{2}\int d^3x d^3y \cdot \hat{\psi}^\dagger(\mathbf{r})\,\hat{\psi}^\dagger(\mathbf{r}')\,V(\mathbf{r}-\mathbf{r}')\,\hat{\psi}(\mathbf{r}')\,\hat{\psi}(\mathbf{r})$$

and applying the **Bloch–De Dominicis theorem**, the N-body expectation value is decomposed into products of two-body contractions, in which the two-body terms $\mathcal{G}(\mathbf{r}, \tau; \mathbf{r}', \tau')$ appear.

The thermal Green function is shown to be periodic in τ [4, 5, 8]. Accordingly, the τ-dependence of $\mathcal{G}$ is replaced by a dependence on $i\omega_n$ through Fourier series expansion.[9] This $\mathcal{G}(i\omega_n)$ can be expressed in the form of a spectral representation involving eigenvalues of the Hamiltonian, as in Eq. (17.27) (**Lehmann representation**). Likewise, the Fourier transform $G(\omega)$ of the real-time Green function can also be represented via the Lehmann expansion. By comparing the two, it is concluded that one can first obtain the *thermal Green function* $\mathcal{G}$ and then, via a variable substitution (analytic continuation) $i\omega_n \to \hbar\omega$, derive the real-time Green function G [4, 5, 8]. It is known that evaluating the thermal Green function is mathematically more tractable than evaluating the real-time Green function.[10] Thus, the common procedure is to evaluate diagrams using the thermal Green function and then perform an analytic continuation to obtain the real-time Green function (*i.e.*, the propagator).

---

[9] $i\omega_n$ is referred to as the **Matsubara frequency**.

[10] The evaluation of the thermal Green function using the Bloch–De Dominicis theorem (finite temperature), which makes use of thermal averages, is easier than the perturbative expansion of the real-time Green function using **Wick's theorem** (operator identities).

17.3 Charge Density and Pair Distribution Function

$|\Psi(\mathbf{r}_1, \cdots, \mathbf{r}_N)|^2$ is interpreted as the probability of finding N electrons at $(\mathbf{r}_1, \cdots, \mathbf{r}_N)$ in a single snapshot. Then, the following integral

$$
\begin{aligned}
n(\mathbf{r}) &= \int |\Psi(\mathbf{r}_1, \cdots, \mathbf{r}_N)|^2 \cdot \delta(\mathbf{r} - \mathbf{r}_1) \cdot d\mathbf{r}_1 \cdots d\mathbf{r}_N \\
&= \int |\Psi(\mathbf{r}, \cdots, \mathbf{r}_N)|^2 \cdot d\mathbf{r}_2 \cdots d\mathbf{r}_N
\end{aligned}
$$

focuses on $\mathbf{r}_1$, summing over all possibilities for $\mathbf{r}_{j\neq 1}$. The quantity $n(\mathbf{r})$ is therefore interpreted as a scaled probability of finding an electron at $\mathbf{r}_1 = \mathbf{r}$. By considering a similar treatment for $\mathbf{r}_{j\neq 1}$ and recalling the indistinguishability that "which $\mathbf{r}_j$ it is" cannot be discriminated,

$$
\rho(\mathbf{r}) = \frac{\int |\Psi(\mathbf{r}_1, \cdots, \mathbf{r}_N)|^2 \cdot \sum_{j=1}^{N} \delta(\mathbf{r} - \mathbf{r}_j) \cdot d\mathbf{r}_1 \cdots d\mathbf{r}_N}{\int |\Psi(\mathbf{r}_1, \cdots, \mathbf{r}_N)|^2 d\mathbf{r}_1 \cdots d\mathbf{r}_N} = \left\langle \sum_{j=1}^{N} \delta(\mathbf{r} - \mathbf{r}_j) \right\rangle
$$

is found to define the charge density.[11]

Next, consider the integral

$$
\begin{aligned}
g(\mathbf{r}, \mathbf{r}') &\propto N(N-1) \\
&\times \frac{\int |\Psi(\mathbf{r}_1, \cdots, \mathbf{r}_N)|^2 \delta(\mathbf{r} - \mathbf{r}_1) \delta(\mathbf{r}' - \mathbf{r}_2) \cdot d\mathbf{r}_1 \cdots d\mathbf{r}_N}{\int |\Psi(\mathbf{r}_1, \cdots, \mathbf{r}_N)|^2 d\mathbf{r}_1 \cdots d\mathbf{r}_N}.
\end{aligned}
$$

The delta functions in the integrand serve to fix $\mathbf{r}_1 = \mathbf{r}$ and $\mathbf{r}_2 = \mathbf{r}'$, while summing over all possibilities for the remaining $\mathbf{r}_j$. Among all N coordinates $\mathbf{r}_j$, there are $N(N-1)$ possible ways to choose a pair assigned to $\mathbf{r}$ and $\mathbf{r}'$. However, due to the antisymmetry of Eq. (4.3), any such pair gives the same value of $|\Psi^2|$. Hence, it suffices to evaluate the integral for the choice $\mathbf{r}_1 = \mathbf{r}$, $\mathbf{r}_2 = \mathbf{r}'$ and multiply the result by $N(N-1)$, which explains the presence of this prefactor.

$$
g(\mathbf{r}, \mathbf{r}') \propto \left\langle \sum_{j\neq j'} \delta(\mathbf{r} - \mathbf{r}_j) \delta(\mathbf{r}' - \mathbf{r}_{j'}) \right\rangle
$$

thus represents the probability of finding another particle at $\mathbf{r}'$, given that one particle is placed at $\mathbf{r}$, and is understood as an extension of the charge density. This

[11] The denominator is the normalization factor for the many-body wave function.

summation can be rearranged as[12]

$$\begin{aligned}\left\langle\sum_{j\neq j'}\delta\left(\mathbf{r}-\mathbf{r}_j\right)\delta\left(\mathbf{r}'-\mathbf{r}_{j'}\right)\right\rangle &= \left\langle\sum_{j}\delta\left(\mathbf{r}-\mathbf{r}_j\right)\left\{\sum_{j'}\delta\left(\mathbf{r}'-\mathbf{r}_{j'}\right)-\delta\left(\mathbf{r}'-\mathbf{r}_j\right)\right\}\right\rangle \\ &= \left\langle\sum_{j}\delta\left(\mathbf{r}-\mathbf{r}_j\right)\sum_{j'}\delta\left(\mathbf{r}'-\mathbf{r}_{j'}\right)\right\rangle \\ &\quad -\delta\left(\mathbf{r}'-\mathbf{r}\right)\left\langle\sum_{j}\delta\left(\mathbf{r}-\mathbf{r}_j\right)\right\rangle\end{aligned}$$

so that

$$g\left(\mathbf{r},\mathbf{r}'\right)\propto\left\langle\hat{\rho}\left(\mathbf{r}\right)\hat{\rho}\left(\mathbf{r}'\right)\right\rangle-\delta\left(\mathbf{r}'-\mathbf{r}\right)\left\langle\hat{\rho}\left(\mathbf{r}\right)\right\rangle \tag{17.29}$$

and the right-hand side is seen to be a quantity proportional to the product of charge densities. Accordingly, we define the **pair distribution function** as the normalized expression

$$\begin{aligned}g\left(\mathbf{r},\mathbf{r}'\right) &= \frac{1}{\rho\left(\mathbf{r}\right)\rho\left(\mathbf{r}'\right)}\left\langle\sum_{j}\delta\left(\mathbf{r}-\mathbf{r}_j\right)\sum_{j'}\delta\left(\mathbf{r}'-\mathbf{r}_{j'}\right)\right\rangle \\ &= \frac{1}{\rho\left(\mathbf{r}\right)\rho\left(\mathbf{r}'\right)}\left[\left\langle\hat{\rho}\left(\mathbf{r}\right)\hat{\rho}\left(\mathbf{r}'\right)\right\rangle-\delta\left(\mathbf{r}'-\mathbf{r}\right)\left\langle\hat{\rho}\left(\mathbf{r}\right)\right\rangle\right].\end{aligned}$$

This pair distribution function is, again, the probability, normalized to 1, of finding another particle at $\mathbf{r}'$ given that one particle is located at $\mathbf{r}$.

The pair distribution function can be written in terms of the operator expression inside the expectation value as

$$\hat{g}\left(\mathbf{r},\mathbf{r}'\right)=\hat{\rho}\left(\mathbf{r}\right)\hat{\rho}\left(\mathbf{r}'\right)-\delta\left(\mathbf{r}'-\mathbf{r}\right)\hat{\rho}\left(\mathbf{r}\right).$$

In second quantization, the density operator is given by $\hat{\rho}\left(\mathbf{r}\right)=\hat{\psi}^{\dagger}\left(\mathbf{r}\right)\hat{\psi}\left(\mathbf{r}\right)$, so we have

$$\hat{g}\left(\mathbf{r},\mathbf{r}'\right)\propto\hat{\psi}^{\dagger}\left(\mathbf{r}\right)\hat{\psi}\left(\mathbf{r}\right)\hat{\psi}^{\dagger}\left(\mathbf{r}'\right)\hat{\psi}\left(\mathbf{r}'\right)-\delta\left(\mathbf{r}'-\mathbf{r}\right)\hat{\psi}^{\dagger}\left(\mathbf{r}\right)\hat{\psi}\left(\mathbf{r}\right).$$

[12] The factor $\delta\left(\mathbf{r}'-\mathbf{r}_j\right)$ appearing in the second term is transformed into $\delta\left(\mathbf{r}'-\mathbf{r}\right)$ in the second line, since the action of $\delta\left(\mathbf{r}-\mathbf{r}_j\right)$, which multiplies it, substitutes $\mathbf{r}_j=\mathbf{r}$.

Using the anticommutation relations of second quantization, this can be rewritten as

$$\hat{g}\left(\mathbf{r},\mathbf{r}'\right) \propto \hat{\psi}^{\dagger}\left(\mathbf{r}\right)\hat{\psi}^{\dagger}\left(\mathbf{r}'\right)\hat{\psi}\left(\mathbf{r}'\right)\hat{\psi}\left(\mathbf{r}\right) ,$$

which gives the second-quantized operator form of the pair distribution function.

References

1. “Schrödinger’s Mechanics” (World Scientific, 1989), D.B. Cook, ISBN: 978-9971507602
2. “The Quantum Theory of Fields” (Cambridge University Press, 2005), S. Weinberg, ISBN: 978-0521670531
3. “Classical Mechanics” (Pearson, 2013) H. Goldstein, C.P. Poole, J.L. Safko, ISBN: 978-1292026558
4. “Quantum Theory of Many-Particle Systems” (Dover Publications, 2003), A.L. Fetter, J.D. Walecka, ISBN: 978-0486428277
5. “Bussei-butsuri no tame no Bano Riron, Green Kansu (Second Edition)” (in Japanese, Saiensu-Sha Co.,Ltd, 2024) M. Ogata, ISBN: 978-4781916132
6. “The Story of Spin” (University of Chicago Press, 1998), S. Tomonaga, ISBN: 978-0226807942
7. “Mathematical Methods for Physics and Engineering, Third Edition” (Cambridge University Press, 2006), K.F. Riley, M.P. Hobson, S.J. Bence, ISBN: 978-0521679718
8. “Quantum Many-particle Systems” (Perseus, 1998), J.W. Negele, H. Orland, ISBN: 978-0738200521

Chapter 18
Appendix H: Overview of Many-Body Perturbation Theory

In this chapter, we compile supplementary notes regarding the method of Feynman diagrams, which appear throughout the main text beginning from Sect. 10.2.3. Feynman diagram methods appear throughout this book on many-electron theory, as well as in various fields of condensed matter theory. However, they are introduced in different contexts: for instance, in the treatment of electron–electron interactions as in this book, or in contexts such as electron–phonon interactions and response function evaluations in linear response theory. This diversity can make it difficult for beginners to grasp the method in a unified manner. Therefore, this appendix summarizes the essential points with an emphasis on illustrating their interrelations, keeping detailed explanations to an outline or selected excerpts.

18.1 Self-Consistent Treatment of Diagrams

18.1.1 Geometric Series Representation of Diagrammatic Sums

In the perturbative expansion using Feynman diagrams, the procedure for evaluating the propagator G that incorporates interaction effects reduces to a summation over diagrams that are connected, as illustrated in Fig. 10.5. Since the wavy interaction lines correspond to the interaction V_{ij}, a diagram containing a single such line represents a first-order perturbation, while a diagram with two such lines corresponds to a second-order contribution. Among these diagrammatic sums, there exist diagrams, highlighted in the figure with a box, that can be separated into multiple parts by cutting through a propagator line (*i.e.*, the solid arrow lines, not the wavy lines), and those that cannot. Diagrams that cannot be separated into parts by cutting a single propagator line are referred to as **proper diagrams** (see Fig. 10.5).

R. Maezono, *Ab initio Quantum Monte Carlo Tutorial*,
https://doi.org/10.1007/978-981-95-4134-8_18

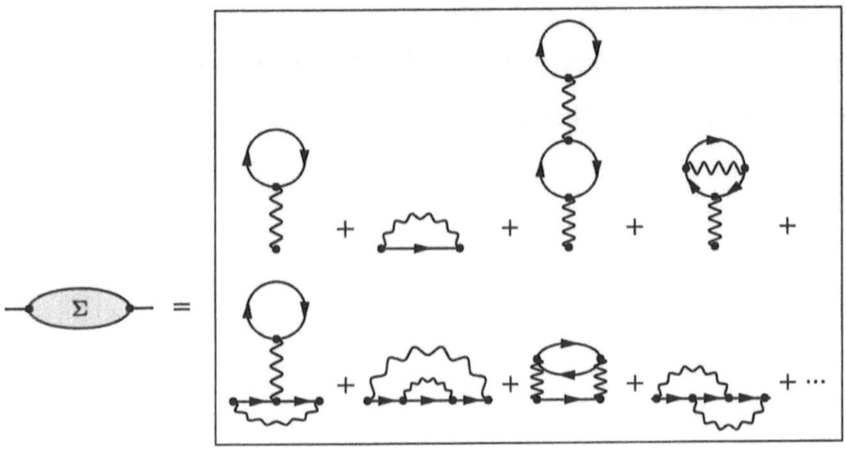

Fig. 18.1 Suppose that we are able to sum over all the proper diagrams. We denote this partial sum by the symbol Σ

As shown in Fig. 18.1, suppose that we are able to sum over all the proper diagrams. Let us denote this partial sum by the symbol Σ. Then, the perturbative expansion in Fig. 10.5 can be treated in a highly sophisticated manner in the form of a geometric series: Among all the diagrams that should be included in the perturbative expansion, the subset of non-proper diagrams consists of those that can be cut by a single propagator line. Therefore, by summing all possible variations where the sum of proper diagrams Σ is connected by propagator lines, we effectively include both proper and non-proper diagrams in total. This idea is illustrated in Fig. 18.2. The "dressed propagator" on the left-hand side represents the Green function that incorporates all diagrams (each term in the perturbative expansion), taking into account the interaction effects. Accordingly, this "dressed propagator," which is a superposition involving Σ, can be expressed in a remarkably recursive form using itself (G):

$$G = G_0 + G_0 \cdot \Sigma \cdot G$$

This diagrammatic relation for the propagator is known as the **Dyson equation**. This relation can be formally solved as

$$(1 - G_0 \cdot \Sigma)\, G = G_0 \quad , \quad \therefore \quad G = \frac{G_0}{1 - G_0 \cdot \Sigma} = \frac{1}{G_0^{-1} - \Sigma} \ .$$

From Eq. (17.27), the non-interacting Green function is

$$G_0^{-1}(p) = z - \varepsilon_p \ .$$

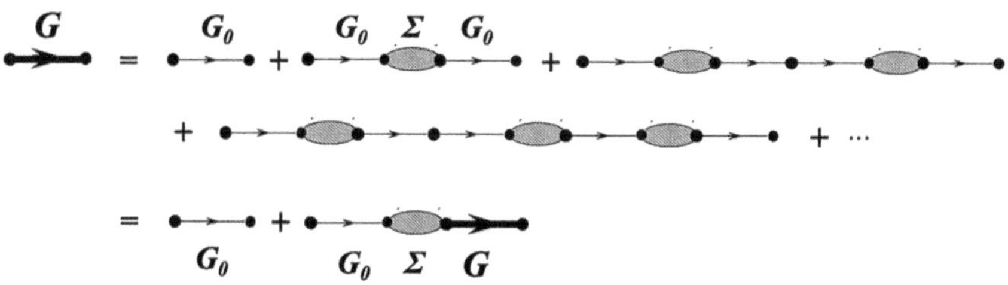

Fig. 18.2 By using the self-energy Σ shown in Fig. 18.1, the contributions to be summed into the dressed propagator (the Green function) G, incorporating interactions, can be recursively reorganized in terms of G itself

Therefore, the Green function incorporating interactions perturbatively is evaluated as[1]

$$G(p) = \frac{1}{z - \varepsilon_p - \Sigma_p}. \tag{18.1}$$

This result indicates that the quasiparticle spectrum is shifted by Σ_p, which is why Σ_p is referred to as the self-energy. The self-energy Σ_p is generally a complex number, with its real part giving the **energy shift**, and its imaginary part giving the life-time (**damping**).

18.1.2 Skeleton Diagrams

Many-body perturbation theory is, at its core, an expansion from lower to higher orders and thus can naively be regarded as a procedure in which the expansion is truncated at some finite order. There was a time when a theorist's skill was measured by "how cleverly one could devise a truncation scheme." However, as we have seen in the evaluation of the Dyson equation illustrated in Fig. 18.2, certain types of diagrams can be incorporated recursively and self-consistently up to infinite order without any truncation, using a geometric series approach. In modern many-body perturbation theory, the focus has shifted toward such formulations, where diagrammatic sums are handled cleverly through self-consistent schemes like the Dyson equation, aiming not for finite-order truncations, but rather for convergent solutions that incorporate contributions to all orders [1, 2].

As discussed in Sect. 10.3.2, the diagrammatic expansion of the Green function can be derived by "cutting" a propagator of each diagram in the diagrammatic

[1] The subscript p refers to the momentum index in $G(\mathbf{p})$, which is the Fourier transform of $G(\mathbf{r}, \mathbf{r}')$. Since the proper diagram series Σ is composed of diagrams connected by $G(\mathbf{p})$, it shares the same momentum index.

Fig. 18.3 By taking a simple proper diagram (as shown on the left-hand side), and replacing its propagators with the "dressed propagators" from Fig. 18.2 (*i.e.*, the effective Green functions incorporating interactions), one can imagine that reverting these "dressed propagators" back to the original bare propagators would generate a variety of more complex diagrams, as illustrated on the right-hand side

expansion of the partition function. Therefore, one typically examines the perturbative expansion of the partition function to determine **which classes of diagrams to include** and **which physical processes each class corresponds to**. Moreover, the perturbative terms in the partition function are represented by "closed, connected diagrams" without external lines. Let us now consider the contributions from the ring-type diagrams shown in Fig. 18.3.

In Fig. 18.2, we introduced the concept of a "dressed propagator" incorporating interactions. Consider now the simple ring diagram constructed from thick lines shown on the left-hand side of Fig. 18.3. If we replace these thick lines with the collection of diagrams built from the original bare propagators, it becomes apparent that a variety of more complex diagrammatic structures emerge, as depicted on the right-hand side of Fig. 18.3. In this way, the summation over many diagrams drawn using the bare propagator G_0 can be effectively encompassed by a smaller number of more fundamental diagrams constructed with the dressed propagator G. These more fundamental diagrams have a "skeletal" character and are referred to as **skeleton diagrams**.

Based on the above, the procedure for the many-body perturbation theory of quasiparticles can be reorganized as follows: Among the diagrammatic expansions for the partition function Ξ, only the skeleton diagrams, such as those in Fig. 18.3, are enumerated. The propagators within these are replaced from the bare G_0 to the interaction-renormalized effective G, that is, they are "dressed." By collecting only the closed skeleton diagrams drawn with dressed propagators, and then substituting each occurrence of the dressed propagator G with the full diagrammatic expansion built from bare propagators G_0, one effectively includes all possible diagrams contributing to the partition function Ξ. By "cutting" a propagator G_0 in these diagrams, one recovers the complete set of diagrams for the self-energy Σ as shown in Fig. 18.1.[2]

[2] One might wonder how to convert a diagram drawn with dressed propagators back into one with bare propagators and then "cut" it appropriately. In practice, however, the procedure of "cutting" corresponds to the mathematical operation of "taking the functional derivative with respect to G_0" [1, 2]. There is no need to explicitly redraw the diagrams using bare propagators.

18.1.3 Vertex Functions and Effective Interactions

As discussed above, the diagrammatic expansion of the Green function can be derived by "cutting" the diagrams that appear in the perturbative expansion of the partition function. As mentioned at the end of the previous section, by summing over the skeleton diagrams in the perturbative expansion of the partition function, one obtains the self-energy Σ, which allows us to understand the effect of particle-particle interactions as changes in the spectrum according to Eq. (18.1). this way, one typically discusses how far to include diagrams in the perturbative expansion of the partition function and which diagrams correspond to which physical processes.

As stated at the end of the previous subsection, the sum of skeleton diagrams and the self-energy Σ are related diagrammatically as illustrated in Fig. 18.4a. The first two diagrams shown in Fig. 18.4b represent the lowest-order (first-order) skeleton diagrams with respect to the interaction V_{ij}. When these two terms are incorporated into the self-energy Σ, it can be shown that they correspond, respectively, to the Hartree term and the Fock term [2–4].

To go beyond the Hartree–Fock approximation and incorporate electron correlation, it is necessary to include the higher-order skeleton diagram contributions, represented by the "$+\cdots$" in the first row of Fig. 18.4b, to evaluate the self-energy.

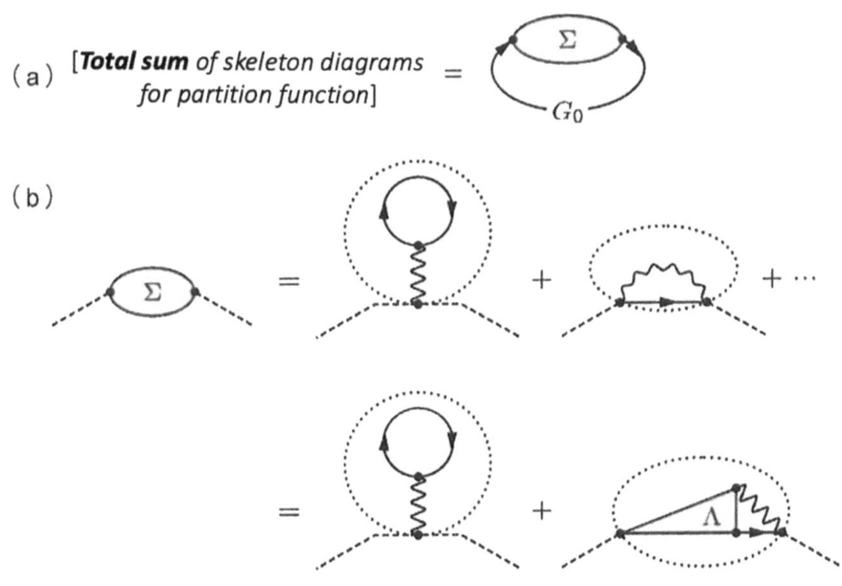

Fig. 18.4 The summation over skeleton diagrams and the self-energy Σ are diagrammatically related as illustrated in panel (**a**). The diagrammatic set contributing to the self-energy Σ, composed solely of skeleton diagrams constructed from bold (interacting) propagators, can be collectively represented by the symbol Λ shown in the second row of panel (**b**). Only the portion enclosed by the dashed line contributes to Σ, while the dotted lines indicate auxiliary connections that represent where propagators attach to the diagram

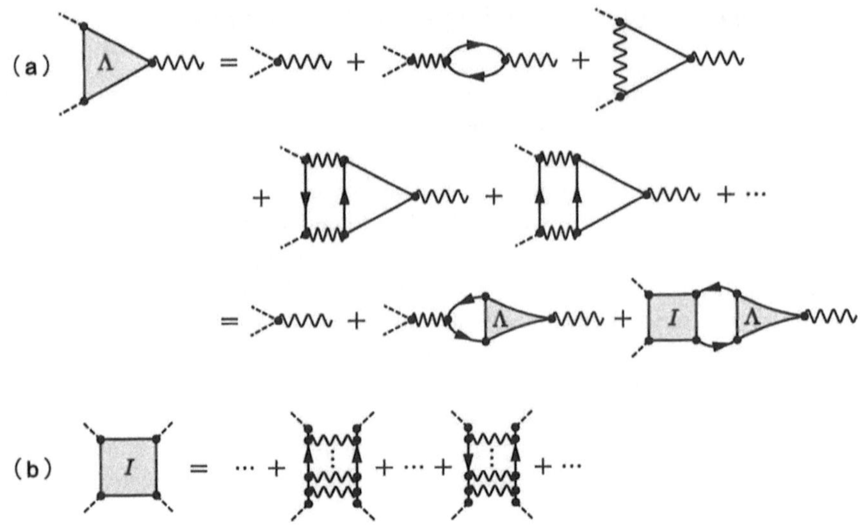

Fig. 18.5 When the various contributions contained within the vertex part Λ introduced in Fig. 18.4 are written out explicitly (the first row of panel (**a**)), they can be recast into a recursive equation as shown in the second row of panel (**a**), using the representative symbol I, referred to as the irreducible effective interaction. The diagrammatic components included in this effective interaction I are illustrated in panel (**b**) (the dashed lines indicate the connection points for propagators and serve as auxiliary guides)

Recalling that only skeleton diagrams composed of bold (interacting) propagators are to be included in this summation, the additional higher-order diagrams can be summarized using a vertex function Λ, as shown in the second row of Fig. 18.4b. This function plays a role analogous to the representative diagrammatic sum used in the treatment of the self-energy Σ in Fig. 18.1. The diagrammatic relationship is expressed as:

$$\Sigma = V \cdot G + \Lambda \cdot V \cdot G \, .$$

We refer to Λ as the **vertex function**, which represents the collection of contributions such as those shown in the first row of Fig. 18.5a. This set of diagrams is recursively expressed as a self-referential summation, shown in the second row of the same figure. The corresponding relation is written as

$$\Lambda = \delta + VGG\Lambda + IGG\Lambda \, .$$

The part denoted by I is referred to as the **irreducible effective interaction**, which includes contributions such as the ladder-type diagrams shown in Fig. 18.5b, for example. The diagrammatic equations governing the vertex function Λ and the effective interaction I are called the **Bethe-Salpeter** equations.

As discussed above, in many-body perturbation theory, one constructs effective propagators and vertex functions by summing diagrammatic contributions up to an appropriate level. Between the resulting effective propagator and vertex function, certain consistency relations must be satisfied, such as the Takahashi–Ward identity.[3] In conventional cutoff-based approximations, one must carefully choose which diagrams to include so that this identity is satisfied. On the other hand, in the self-consistent approaches described at the beginning of this section, it is possible to formulate a scheme in which the identity is automatically fulfilled.[4]

18.2 Diagrams and Electromagnetic Response

18.2.1 Dielectric Polarization and Diagrams

We consider the response of an electronic system when subjected to an external electromagnetic field. Within the regime of infinitesimal responses where a linear approximation is valid, the corresponding response coefficients (susceptibilities) can be treated within the framework of quantum statistical mechanics. This approach is known as linear response theory and is commonly referred to under names such as the **Kubo formula**. When the physical quantities corresponding to the stimulus and response are denoted by $\hat{A}$ and $\hat{B}$ respectively, the response coefficient is given by

$$K_{BA}\left(t-t'\right)=\frac{i}{\hbar}\left\langle\left[B_H(t),A_H(t')\right]\right\rangle\theta\left(t-t'\right),$$

as is known [5, 6]. Here, $B_H(t)$ and $A_H(t)$ represent the time dependence in the Heisenberg representation, as defined in Eq. (17.3), and $K_{BA}\left(t-t'\right)$ is referred to as the **two-point Green function**. The meaning of the statistical average $\langle\cdots\rangle$ in the two-point Green function is the same as that defined in Eq. (17.26). The corresponding thermal Green function for the two-point Green function is given by

$$L_{BA}\left(\tau\right):=\left\langle T\hat{B}\left(\tau\right)\hat{A}\right\rangle,$$

where the τ-dependence of $B\left(\tau\right)$ follows the same form as in Eq. (10.16). transform of the thermal Green function L_{BA} and applying the analytic continuation $i\omega\to\omega+i\delta$, one obtains the two-time Green function. This procedure follows the same reasoning described in Sect. 17.2.7.

$$\begin{aligned}\hat{B}\left(\mathbf{r},\tau\right)&=e^{\tau\hat{H}}\hat{B}\left(\mathbf{r}\right)e^{-\tau\hat{H}}\\&=e^{\tau\hat{H}}\cdot e^{-\tau\hat{H}_0}e^{\tau\hat{H}_0}\cdot\hat{B}\left(\mathbf{r}\right)\cdot e^{-\tau\hat{H}_0}e^{\tau\hat{H}_0}\cdot e^{-\tau\hat{H}}\end{aligned}$$

[3] An identity that enforces gauge invariance, which must be preserved by the theory.

[4] Known as the Baym–Kadanoff conserving approximation [1, 2].

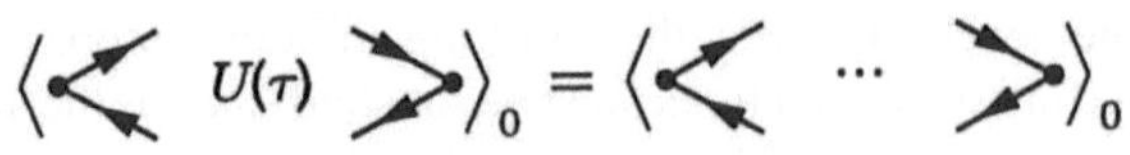

Fig. 18.6 The response coefficient K_{00}, which pertains to the electrostatic response, takes the form of a closed diagram composed of a propagator flowing into and out from the two endpoints $\mathbf{r}$ and $\mathbf{r}'$, respectively, with various diagrammatic contributions arising from the perturbation expansion connected between them

$$= \hat{U}^{\dagger}(\tau) \cdot \hat{B}(\mathbf{r}, \tau) \cdot U(\tau),$$

where the operator $\hat{U}(\tau) = e^{\tau \hat{H}_0} e^{-\tau \hat{H}}$ is introduced, following the same treatment treatment discussed around Eq. (10.17) in the context perturbation expansion.

The response of a material to an electric field is fundamentally characterized by the response coefficient K_{00}, which quantifies the degree to which an induced charge ρ_{ind} appears in response to a test charge ρ_{test} placed in the material:[5]

$$K_{00} \sim \langle [\rho(r), \rho(r')] \rangle .$$

The operator for the density $\rho(r)$ is given by $\hat{\rho}(\mathbf{r}) = \hat{\psi}^{\dagger}(\mathbf{r}) \hat{\psi}(\mathbf{r})$, so that

$$K_{00} \sim \left\langle \hat{\psi}^{\dagger}(\mathbf{r}, \tau) \hat{\psi}(\mathbf{r}, \tau) \cdot \hat{\psi}^{\dagger}(\mathbf{r}', \tau') \hat{\psi}(\mathbf{r}', \tau') \right\rangle$$

is the quantity to be evaluated. Applying the perturbative expansion scheme for the τ dependence using $\hat{U}(\tau) = e^{\tau \hat{H}_0} e^{-\tau \hat{H}}$, one finds that the resulting computation can be represented by a summation over diagrams, as illustrated in Fig. 18.6. That is, it results in structures in which propagators flow into and out from the two endpoints $\mathbf{r}$ and $\mathbf{r}'$, respectively, and various diagrammatic contributions emerging from the perturbative expansion are inserted between them.

In Sect. 18.1.2, we discussed that when a "cut is made on the propagator" in the diagrams contributing to the partition function, it results in the "Green function with interaction effects included." On the other hand, let us now consider making a cut on the "interaction line" instead of the propagator. In this case, the diagrammatic sum takes the form shown in Fig. 18.7, which we denote by Π_{tot}. This Π_{tot} exactly corresponds to the diagrammatic sum required in Fig. 18.6, and we find that

$$K_{00} \sim \Pi_{\text{tot}} .$$

[5] In general framework, the discussion is formulated in terms of a four-vector current $j_{\mu} \sim (\rho, \mathbf{j})$, which combines the charge density and current. The response coefficients then form the tensor $K_{\mu\nu} \sim \langle [j_{\mu}(r), j_{\nu}(r')] \rangle$. Among these, the 00 component corresponds to the charge–charge correlation function, and hence, K_{00} describes the charge density response.

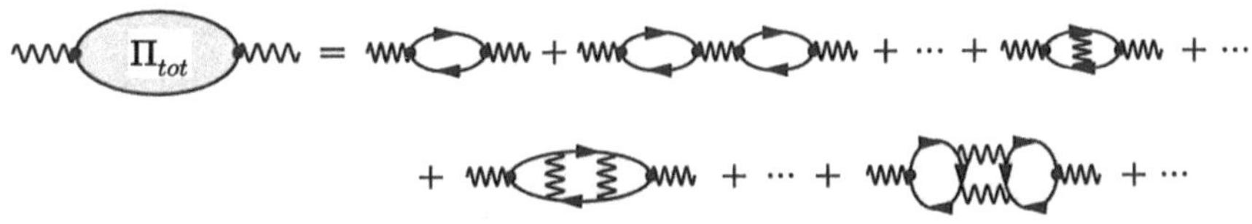

Fig. 18.7 Diagrams corresponding to polarization. Instead of cutting the propagator in the diagrams contributing to the partition function consider making a cut along the interaction line. The resulting diagrammatic sum Π_{tot} corresponds precisely to the sum of diagrams needed in Fig. 18.6, and it is found to correspond to the response coefficient associated with electrostatic response

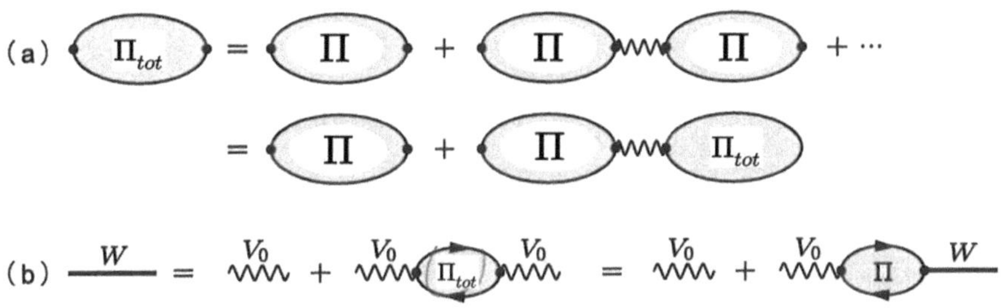

Fig. 18.8 Proper part in polarization diagrams. The diagrammatic sum Π_{tot} in Fig. 18.7 can be regarded as a repetition of the proper part Π, and thus can be rearranged into a recursive relation analogous to the Dyson equation, leading to the solution in Eq. (18.2) [Panel (1)]. Looking again at Fig. 18.7, it becomes evident that the effective interaction W, incorporating the screened interaction, can be obtained as shown in Panel (2). This diagrammatic sum also satisfies a recursive relation, which can be solved as in Eq. (18.6)

The diagrammatic sum Π_{tot} in Fig. 18.7 can be treated in the same manner as before, by considering it as a repetition of the proper part Π , as illustrated in Fig. 18.8a. This leads to a recursive relation similar to the Dyson equation. The relation can be written as

$$\Pi_{\text{tot}} = \Pi + \Pi \cdot V_0 \cdot \Pi_{\text{tot}} \quad , \quad \therefore \Pi_{\text{tot}} = \frac{\Pi}{1 - V_0 \cdot \Pi} \tag{18.2}$$

and solved accordingly.

Let us recall the basics of **macroscopic electromagnetism**. In that context, the electric field $\mathbf{E}$, the electric displacement field $\mathbf{D}$, the dielectric constant ε, the test charge ρ_{test}, and the induced charge ρ_{ind} were related by the equations

$$\nabla \cdot \mathbf{E} = 4\pi \cdot (\rho_{\text{test}} + \rho_{\text{ind}}) \quad , \quad \nabla \cdot \mathbf{D} = 4\pi \cdot \rho_{\text{test}} \quad , \quad \mathbf{D} = \varepsilon \cdot \mathbf{E}$$

as described in [7]. The induced charge ρ_{ind} defines the electric polarization $\mathbf{P}$ via

$$\rho_{\text{ind}} = -\nabla \cdot \mathbf{P}$$

also noted in [7]. Thus, we have

$$
\begin{aligned}
\nabla \cdot \mathbf{E} &= 4\pi \cdot (\rho_{\text{test}} + \rho_{\text{ind}}) = 4\pi \cdot (\rho_{\text{test}} - \nabla \cdot \mathbf{P}) \\
\nabla \cdot (\mathbf{E} + 4\pi \mathbf{P}) &= 4\pi \cdot \rho_{\text{test}} = \nabla \cdot \mathbf{D} \\
\therefore \quad \mathbf{D} &= \mathbf{E} + 4\pi \mathbf{P} \\
\therefore \quad \varepsilon &= \frac{D}{E} = 1 + \frac{4\pi \mathbf{P}}{\mathbf{E}} \,. \qquad (18.3)
\end{aligned}
$$

On the other hand, using the terms test charge ρ_{test} and induced charge ρ_{ind}, we can write

$$
\varepsilon = \frac{D}{E} = \frac{\rho_{\text{test}}}{\rho_{\text{ind}} + \rho_{\text{test}}} = \frac{1}{(\rho_{\text{ind}}/\rho_{\text{test}}) + 1} \,.
$$

The induced and test charges are connected via the response function K_{00} as

$$
\begin{aligned}
\rho_{\text{ind}} &= K_{00} \cdot V_0 \cdot \rho_{\text{test}} \\
\therefore \quad \frac{\rho_{\text{ind}}}{\rho_{\text{test}}} &= V_0 \cdot K_{00} \sim V_0 \cdot \Pi_{\text{tot}} \,.
\end{aligned}
$$

Taking into account Eq. (18.2), we obtain

$$
\varepsilon = \frac{1}{V_0 \cdot \Pi_{\text{tot}} + 1} = \frac{1}{\dfrac{V_0 \cdot \Pi}{1 - V_0 \cdot \Pi} + 1} = 1 - V_0 \cdot \Pi \qquad (18.4)
$$

By comparing this equation with Eq. (18.3), we obtain

$$
1 + \frac{4\pi \mathbf{P}}{\mathbf{E}} = 1 - V_0 \cdot \Pi \qquad (18.5)
$$

This means that the sum of proper diagrams Π included in Π_{tot} (Fig. 18.7), corresponds to the macroscopic polarization $\mathbf{P}$ in classical electrodynamics. From this observation, in many-body perturbation theory, diagram sets such as those in Fig. 18.7 are referred to as **polarization**.

By revisiting Fig. 18.7, one can observe that the effective interaction W, in which the bare interaction is renormalized, can be obtained as shown in Fig. 18.8b. The diagrammatic sum corresponding to this W can also be expressed recursively as

$$
\begin{aligned}
W(k, \omega_n) &= V_0(k) + V_0(k) \cdot \Pi(k, \omega_n) \cdot W(k, \omega_n) \quad , \\
\therefore \quad W(k, \omega_n) &= \frac{V_0(k)}{1 - V_0(k) \cdot \Pi(k, \omega_n)} \,. \qquad (18.6)
\end{aligned}
$$

18.2.2 Screening and Ring Diagrams

The first two diagrams on the right-hand side of Fig. 18.7 are single- and double-ring diagrams, respectively. Let us consider the sum $\Pi^{(\text{ring})}$ obtained by summing only these ring-type diagrams, single, double, triple, and so on, to infinite order. By substituting $\Pi = \Pi^{(\text{ring})}$ into Eq. (18.6) and into Eq. (18.6) and taking the static limit $|\omega| \ll v_F |\mathbf{q}|$,[6] we find that the spatial dependence yields

$$W(r) = \frac{1}{r}\exp[-\kappa r]$$

which describes a **screened interaction** [8]. This implies that the bare interaction $v(\mathbf{r}) = e^2/|\mathbf{r}|$ is effectively shortened in range by the exponential factor $\exp[-\kappa r]$ due to many-body effects.

The condition $\Pi = \Pi^{(\text{ring})}$ corresponds to choosing $\Pi^{(\text{proper})}$ as the first single-ring diagram on the right-hand side of Fig. 18.7. By substituting this into Eq. (18.4) and setting the condition $\varepsilon = 0$, *i.e.*, the condition under which spontaneous polarization arises even without applying an electric field $\mathbf{E}$, we can obtain the condition for the emergence of **plasma oscillation**, where a long-wavelength charge density modulation occurs at very high frequency [8]. Plasma oscillation is, in elementary terms, obtained by applying the approximation that neglects the second term in the equation of motion for the wavevector component of charge density

$$\rho_{\mathbf{q}} = \sum_j \exp\left(i \cdot \mathbf{q} \cdot \mathbf{r}_j\right)$$

as given by

$$\begin{aligned}\frac{d^2\rho_{\mathbf{q}}}{dt^2} &= -\sum_j \left(\mathbf{q}\cdot\mathbf{v}_j\right)^2 \exp\left[-i\mathbf{q}\cdot\mathbf{r}_j\right] \\ &-\sum_{j,k}\sum_{\mathbf{q}'} \frac{4\pi e^2}{mq'^2 V}\left(\mathbf{q}\cdot\mathbf{q}'\right)\exp\left[i\mathbf{q}'\cdot\left(\mathbf{r}_j-\mathbf{r}_k\right)-i\mathbf{q}\cdot\mathbf{r}_j\right] \end{aligned} \tag{18.7}$$

as described in [9]. The rationale behind this approximation lies in the perspective that, when focusing on a specific phase $\mathbf{q}$, the numerous contributions from other phases $\mathbf{q}'$ will interfere with random phases and cancel out with opposite signs. Therefore, this approximation is referred to as the **Random Phase Approximation** (RPA).

[6] This refers to the static limit of the Fourier transform $G(\mathbf{q}, \omega)$ of the propagator $G\left(\mathbf{r}-\mathbf{r}', \tau-\tau'\right)$ with respect to time and space. Note that this limit ($|\omega| \ll v_F|\mathbf{q}|$) means "when focusing on a much slower time variation scale compared with the fast frequency that characterizes the electronic energy scale."

Since the result derived from RPA is reproduced, the approximation that incorporates the contributions of ring diagrams as an infinite sum is called the **RPA**. In the context of perturbative expansion, each term of the ring diagram contributes a divergence; however, because the signs alternate with each order, it is expected that the infinite series will converge conditionally, thus leading to the RPA result.[7]

References

1. "Self-Consistent Approximations in Many-Body Systems", G. Baym, Phys. Rev. 127, 1391 (1962), https://doi.org/10.1103/PhysRev.127.1391
2. "Tatai Mondai" (in Japanese, Asakura Publishing, 1999) Y. Takada, ISBN: 978-4254136791
3. "Tatai Mondai" (in Japanese, Baifukan Co., Ltd, 1975) F. Takano, ISBN: 978-4563024185
4. "Quantum Monte Carlo Approaches for Correlated Systems" (Cambridge University Press, 2017), F. Becca, S. Sorella, ISBN: 978-1107129931
5. "Many-Particle Physics, Third Edition", (Springer, 2000/10/31), G.D. Mahan, ISBN: 978-0306463389
6. "Bussei-butsuri no tame no Bano Riron, Green Kansu (Second Edition)" (in Japanese, Saiensu-Sha Co.,Ltd, 2024) M. Ogata, ISBN: 978-4781916132
7. "Classical Electrodynamics" (Wiley, 1998), J.D. Jackson, ISBN: 978-0471309321
8. "Quantum Theory of Many-Particle Systems" (Dover Publications, 2003), A.L. Fetter, J.D. Walecka, ISBN: 978-0486428277
9. "Elementary Excitations In Solids" (CRC Press, 2019), D. Pines, ISBN: 978-0367091705
10. "Quantum Many-particle Systems" (Perseus, 1998), J.W. Negele, H. Orland, ISBN: 978-0738200521

[7] This is known as the theory of Gell–Mann–Brueckner or Bohm–Pines, where the statement is made that the long-range nature of the bare $1/r$ interaction is pushed to high-energy collective excitations (*i.e.*, plasma oscillations) due to many-body effects, and individual low-energy excitations become short-ranged. When the interaction is described in terms of a force-mediating boson, the bare $1/r$ interaction corresponds to a massless boson. The fact that the interaction becomes short-ranged implies that the mediating boson acquires mass. This is tied to the presence of plasma oscillations (as massive modes), and more generally, it is understood as the Higgs mechanism [10].

Index

R. Maezono, *Ab initio Quantum Monte Carlo Tutorial*,
https://doi.org/10.1007/978-981-95-4134-8

L

M

N

O

MIX
Papier aus verantwortungsvollen Quellen
Paper from responsible sources
FSC® C105338

If you have any concerns about our products,
you can contact us on
ProductSafety@springernature.com

In case Publisher is established outside the EU,
the EU authorized representative is:
Springer Nature Customer Service Center GmbH
Europaplatz 3, 69115 Heidelberg, Germany

Printed by Libri Plureos GmbH
in Hamburg, Germany